CELL BIOLOGY RESEARCH PROGRESS

NEURONAL CELL APOPTOSIS

Cell Biology Research Progress

Additional books in this series can be found on Nova's website
under the Series tab.

Additional E-books in this series can be found on Nova's website
under the E-books tab.

CELL BIOLOGY RESEARCH PROGRESS

NEURONAL CELL APOPTOSIS

CARL J. SCHMID
AND
JASON L. WOLFE
EDITORS

Nova Science Publishers, Inc.
New York

For permission to use material from this book please contact us:
Telephone 631-231-7269; Fax 631-231-8175
Web Site: http://www.novapublishers.com

NOTICE TO THE READER

Additional color graphics may be available in the e-book version of this book.

Library of Congress Cataloging-in-Publication Data

Neuronal cell apoptosis / editors, Carl J. Schmid and Jason L. Wolfe.
p. cm.
Includes index.
ISBN 978-1-61324-633-7 (hardcover)
1. Neurons. 2. Apoptosis. I. Schmid, Carl J. II. Wolfe, Jason L.
QP356.25N474 2011
612.8'1046--dc23

2011014549

Published by Nova Science Publishers, Inc. ✝ New York

Contents

Preface

Neuronal apoptosis is a normal physiological process that is an essential and integral part of neurogenesis. The massive overproduction of neurons that occur during the development of the brain is followed by a programmed cell death of roughly one-half of the originally formed neurons. In this book, the authors present topical research in the study of neuronal cell apoptosis, including comparative microarray analysis of multiple neurodegenerative models; death of neuronal and glial cells induced by photodynamic treatment; cell death during cerebellar development and the fundamental mechanisms and pathophysiological implications of neuronal apoptosis.

Chapter I - Neurons have unique complex morphological and functional features. They have a characteristic cell body with plethora of surface ion channels, and greatly elongated processes (axons and dendrites), that consequentially transport signals for intracellular trafficking creating amazing web of connectivity. Could these features be considered as a double sword?. Although, neurons are considered among the most long-lived cell types, but they are encoded by various genetic programs involved in initiation, signaling, and execution of cell death during development and adult life. Thus, two main types of neuronal apoptosis can be distinguished, based on their timing and cellular targets; an early physiological apoptosis of dividing progenitors or immature neurons that is referred as naturally occurring neuronal death (NOND), and a late cell apoptois that strikes mainly postmitotic neurons and involved in the inappropriate elimination of cells under neuropathological conditions. The term apoptois of cell death was derived originally from the Greek words apo- "away" and ptosis- "fall". Apoptosis was defined as a distinct mode of programmed cell death (PCD) on the basis of a series of characteristic ultrastructural features according to a sequence of events including nuclear and cytoplasmic condensation, cell fragmentation and finally phagocytosis. In this type of death, the cells are responsible of their own demise, a reason for which apoptosis is commonly referred to as a “cell suicide”. During vertebrate brain development, about 50% of the neurons do not survive into adulthood, most of which die by apoptosis, resulting in the loss of the cell and all its synapses and processes. Why so many neurons are generated in excess and eliminated shortly thereafter remains to be understood. This normal neurodegeneration is thought to be important for matching the size of neuronal groups to the size of their targets (other groups of neurons to which they are connected by axons and nerve terminal synapses). Although, the brain is only 2–3% of total body mass, but it consumes 20% of body oxygen. Neurons are particularly susceptible to oxidative damage due to high levels of polyunsaturated fatty acids in their membranes and relatively low activity of endogenous antioxidant enzymes. Moreover, because neurons have limited glycolytic capacity, so they are highly dependent on aerobic oxidative phosphorylation by mitochondria for their energetic needs. The postmitotic differentiated neurons are completely devoid of replicative capability. Indeed they are believed to enter an "extended G0 phase" which, unlike other cell types, is irreversible. Contrary to this dogma, a growing body of evidence indicates that neurons have the capacity to re-enter the cell cycle. However, while neurons retain the ability to re-activate the cell cycle, engagement of the cell cycle machinery rarely leads to neuronal proliferation but typically induces apoptosis. Several different stimuli can initiate the apoptotic death of neurons. However, the finding that common morphological and biochemical alterations are observed independently upon the event that triggers apoptosis, suggests that most apoptotic pathways converge on a restricted number of common effector pathways. Basically two major pathways can be differentiated by the relative timing of caspase activation and mitochondrial release of cytochrome c. In the first, which is exemplified by activation of death receptors, an effector caspase is activated prior to mitochondrial alterations. In the other, cytochrome c is released from the mitochondrial intermembrane space prior to caspase activation. Regarding such molecular consequences, apoptosis is regulated by wide array of molecules within cells, somc arc antiapoptotic proteins (such as Bcl-2, Bcl-xL) and Proapoptotic proteins (such as Bax, Bcl-xS, Bad, Bid, Bik, caspases, p53). The molecular mechanisms for apoptosis involve the participation of at least three groups of proteins that are made from three different gene families. One set of proteins is the caspase family of cysteine-containing, aspartate-specific proteases (14 members have been identified to date). A second group includes the death promoting and death-suppressing proteins in the Bcl-2 family. A third group of proteins is designated as the inhibitor of apoptosis protein (IAP) family. Is the future is promising for exploring the causes of neurodegenerative disorders and identifying treatments? Unfortunately, until now there are no effective treatments are available to prevent or radically cure any of them. Thus more complete understanding of the characteristics and mechanisms of neuronal apoptosis within the CNS could provide such hope.

Chapter II - Apoptosis is the earliest identified mode of cell death characterized by its closely coordinated signaling mechanisms that put the cell to its demise without disruption to its microenvironment. It is the most reported cause of neuronal death seen in the pathogenesis of numerous neurodegenerative disorders. The current focus of this review revolves around the significance of the major biological processes (e.g. oxidative stress, endoplasmic reticulum/lysosomal disruption, mitochondrial respiratory dysfunction and ubiquitin-proteasome system inhibition) at work in the induction of neuronal apoptosis during neurodegeneration in various central nervous system diseases. To further complement and substantiate the existing findings, comparative microarray analysis was conducted on five neurodegenerative global gene profiles from the authors' laboratory to provide comprehensive temporal elucidation of the implication of these neuropathological mechanisms during neuronal apoptosis at the transcriptional level. These transcriptomic profiles were derived from the application of widely employed pharmacological agents on *in vitro* rodent models to mimic neuropathological hallmarks. They comprise of lactacystin (proteasomal inhibitor), Glu (excitotoxicity inducer), nitric oxide (source of reactive nitrergic and oxygen species), hypochlorous acid (source of reactive chlorinergic and oxygen species) and rotenone (mitochondrial complex I inhibitor) commonly used to model chronic neurodegenerative diseases such as Alzheimer's disease, amyotrophic lateral sclerosis,

Parkinson's disease and Down syndrome as well as acute neurological disorders such as stroke and traumatic brain injury. Mediation of neuronal injury and apoptosis by any one agent is not mutually exclusive to any particular neurodegenerative condition, but at most times demonstrates simultaneous involvement across the neurodegenerative disease spectrum. This is an indirect indication of a significant commonality of signaling cascades in the pathogenesis of most neurodegenerative diseases, and that apoptosis plays a major role in the trigger of these downstream mechanisms. Subsequently, this provides the opportunity for identification of novel biological targets whose manipulation would provide beneficial broad-spectrum neurodegenerative disease therapeutic intervention.

Chapter III - Photodynamic therapy (PDT), an inducer of oxidative stress, is used for treatment of cancer including brain tumors. To study PDT effect on neurons and glial cells, authors used a simple model object - an isolated crayfish mechanoreceptor consisting of only two neurons enveloped by glial cells. PDT abolished neuronal activity and caused necrosis of neurons and necrosis and apoptosis of glial cells. Percent of necrotic neurons and glial cells progressively increased after PDT. At the ultrastructural level, the swelling of mitochondria, dictyosomes and endoplasmic reticulum was observed in neurons and glial cells after PDT. At 1 hour postirradiation these alterations were expanded throughout the cells. Neurons lost the "tigroid" morphology that was a result of segregation of the cytoplasm by Nissl bodies enriched with ER, mitochondria, ribosomes and polysomes, which are involved in protein synthesis and transport along neurites. The structures involved in glia-to-neuron (neuronal submembrane cisterns, glial protrusions, double-wall vesicles), intra-glial (tubular lattices) and intra-neuronal transport (Nissl bodies, Golgi apparatus, microtubular bundles) were also impaired at the earlier stages of the stretch receptor damage. Local inactivation of a neuron body by a focused laser beam enhanced glial apoptosis but not necrosis that was induced by following photosensitization. Therefore, the stretch receptor neuron prevented glial apoptosis. In order to find molecular signals involved in the neuroglial interactions, authors tested neurotrophic factors NGF, BDNF and GDNF. NGF and GDNF protected glial cells from PDT-induced necrosis and apoptosis. Inhibition of protein kinase JNK enhanced PDT-induced apoptosis of glial cells and prevented the anti-apoptotic effect of NGF. Hence, JNK was involved in the NGF-mediated protection of glial cells from PDT-induced apoptosis. Unlike, another MAP kinase, p38, was involved in PDT-induced glial necrosis. Calmodulin, calmodulin-dependent kinase II, phospholipase C, protein kinase C and adenylate cyclase also participated in PDT-induced necrosis of glial cells. Phospholipase C and glycogen synthase kinase also participated in PDT-induced apoptosis of glial cells, whereas adenylate cyclase, JNK, protein kinases A and C played the anti-apoptotic role in glial cells. The involvement of these signaling pathways in responses of neurons and glial cells to PDT suggests that their modulation can modify PDT efficacy. The scheme of participation of diverse inter- and intracellular signaling processes in responses of neurons and glial cells to PDT-induced oxidative stress has been created.

Chapter IV - Development of the nervous system not only involves the generation of new neurons and glia, but also the death of cells that are no longer required and/or are produced in excess. Because the process of cell death follows a complex framework of regulated events during normal development, it is usually referred to as programmed cell death. Programmed cell death is a gene-regulated process that plays important roles in several normal and pathological conditions. It takes several different forms, but, most commonly, eukaryotic cells die following a conserved series of molecular and cellular modifications usually referred to as apoptosis, a form of "cell suicide" most often found during embryonic development, but also in normal cell and tissue turnover during adulthood and senescence. Another form of cell death is autophagy, a catabolic process involving the degradation of cellular components through the lysosomal machinery. There is no conclusive evidence as regarding the involvement of autophagy in apoptosis, and a causative relationship between autophagy and cell death has not yet been established in full. Nevertheless, observations of cells with autophagic features undergoing programmed cell death have led to define an autophagic type of death (also known as cytoplasmic cell death or type II cell death), with morphological features somehow different from those of apoptosis. Proliferative and/or apoptotic events have been extensively characterized not only during embryonic development but also in several areas of the post-natal and adult brain such as the cerebellar cortex. The post-natal cerebellar cortex undergoes significant structural remodeling mainly because of the generation, differentiation, migration, and apoptosis of the granule cells. Therefore, it has been widely investigated as a general model for studying neuronal cell apoptosis. The existence of two waves of apoptotic cell death has been demonstrated in cerebellum and other areas of the central nervous system, affecting different types of neurons in normal development and in certain pathological conditions. The first wave regards neuronal progenitors and pre-migratory neuroblasts, the second post-migratory neuroblasts and mature neurons. In parallel, autophagic phenomena have also been observed to mainly affect the granule cells and the Purkinje neurons. These may be of particular relevance in the granule cells, for the regulation of the cellular levels of B cell lymphoma-2, one of the most important anti-apoptotic proteins in mammals.

Chapter V - Brain cells fundamentally differ, structurally, biochemically and functionally from other cell types of the body. In the brain, neuronal apoptosis is a normal physiological process that is an essential and integral part of neurogenesis. The massive overproduction of neurons that occur during the development of the brain is followed by a programmed cell death of roughly one-half of the originally formed neurons. This precisely controlled neuronal demise is highly conserved in diverse organisms. However, apoptosis when becomes aberrant has been implicated in various neurological disorders as diverse as Parkinson's disease, Alzheimer's disease, Huntington's disease, and amyotrophic lateral sclerosis, spinal muscular atrophy, diabetic encephalopathy, and ischemic and traumatic brain injuries. At the fundamental level, the main focus of the recent research has been in the area of understanding the functional mechanisms of the death receptor mediated 'extrinsic' and mitochondria mediated 'intrinsic' apoptotic pathways, and the roles of caspases and protein members of the Bcl-2 family. Neuronal apoptosis is classically viewed as an active and highly regulated form of neuronal death, whereas neuronal necrosis is considered as an accidental form of cell death. Also, the recent concept of a programmed necrosis has considerable scope in neuronal cell death while autophagy, a caspase independent apoptosis related mechanism in which the cytoplasm is destroyed by lysosomes has been implicated in cell death as well as in cell protection. The main purpose of this chapter is to provide the basic details and update the current understanding of neuronal apoptosis and its implications in neurodegenerative, ischemic and traumatic conditions. Thus, this chapter focuses on two major areas (a) the fundamental mechanisms of neuronal apoptosis and (b) the pathophysiological implications of neuronal apoptosis.

Chapter VI - The cells that make up the nervous system die at two distinct times during the lifetime of an organism. These periods of neuron loss occur during embryonic development and with advancing age, occurring at an accelerated rate in neurodegenerative disorders. The major risk factor for many neurodegenerative diseases is ageing. During development, neuronal apoptosis is essential for the correct shaping and hard-wiring of the nervous system. However, in a number of neurological conditions, apoptosis causes a devastating, and currently irreversible, loss of neurons. This review considers the extracellular and intracellular signals that determine which neurons live, and which neurons die, during development. It then asks the question of whether the authors' understanding of apoptosis during neurodevelopment can lead to novel treatments to protect the nervous system from neuron loss due to pathology.

Chapter VII - The term apoptosis, from the Greek word for "falling off" of leaves from a tree, is used to describe a process in which a cell actively participate in its own destructive processes. Apoptosis, or programmed cell death, is a physiological process observed in many organs and cells, which may also occur in pathological conditions such as diabetic neuropathy. The apoptotic program is characterized by certain morphological features, that include changes in the plasma membrane such as loss of membrane asymmetry and attachment, condensation of the cytoplasm and nucleus, and internucleosomal cleavage of DNA. In the final stages, the dying cells becomes fragmented into "apoptotic bodies" which are rapidly eliminated by phagocytic cells without eliciting significant inflammatory damage to surrounding cells.

Chapter VIII - Neuronal apoptosis is an underlying cause of many neurodegenerative diseases such as amyotrophic lateral sclerosis and Parkinson's disease, as well as, in acute neuronal injury such as stroke. The two main pathways that induce apoptosis in neurons are the intrinsic (mitochondrial) and extrinsic death receptor cascades. These pathways are not mutually exclusive and the intrinsic apoptotic cascade can function as an important amplification loop to the extrinsic apoptotic pathway. Evidence also suggests that endoplasmic reticulum (ER) stress is a significant contributor to neuronal apoptosis. Central to each of these pathways is the B-cell lymphoma-2 (Bcl-2) family of proteins which function as crucial regulators of programmed cell death. The Bcl-2 family is comprised of three distinct subgroups: the pro-survival subgroup (e.g., Bcl-2 and Bcl-xL); the pro-apoptotic, multi-domain subgroup (Bax and Bak); and the Bcl-2 homology-3 domain (BH3)-only pro-apoptotic subgroup (e.g., Bim, Bad, and Puma). Regulation of apoptosis is largely mediated by finely tuned interactions between pro-survival and pro-apoptotic Bcl-2 family proteins. An essential role of Bcl-2 family proteins in neuronal survival has been underscored by evidence demonstrating that either loss of pro-survival Bcl-2 function or gain of Bax/Bak function are causally linked to neuronal apoptosis and neurodegeneration. More recently, a prominent role has been established for BH3-only proteins in neuronal apoptosis associated with chronic neurodegenerative diseases or acute neuronal injury. At least thirteen BH3-only proteins have been described in mammals and evidence suggests that these proteins function as sensors of intracellular and extracellular cell death signals that tip the delicate balance between pro-survival and pro-apoptotic Bcl-2 proteins towards apoptosis. In this review, authors discuss the roles of BH3-only proteins in the intrinsic, extrinsic, and ER stress apoptotic cascades, as well as hierarchical BH3-only pathways and regulation of BH3-only proteins. Authors also examine specific evidence of a role for BH3-only proteins in *in vitro* models of neuronal apoptosis, animal models of neurodegeneration or acute neuronal injury, and in human neuropathology. Given their crucial roles in neuronal apoptosis, BH3-only proteins represent a novel class of therapeutic targets for the treatment of neurodegenerative diseases and acute neuronal traumas such as stroke.

In: Neuronal Cell Apoptosis
Editors: C. J. Schmid and J. L. Wolfe
ISBN: 978-1-61324-633-7

Chapter I

Neuronal Cell Apoptosis

Amany A. Abdin
Department of Pharmacology, Faculty of Medicine,
Tanta University, Egypt

Abstract

Neurons have unique complex morphological and functional features. They have a characteristic cell body with plethora of surface ion channels, and greatly elongated processes (axons and dendrites), that consequentially transport signals for intracellular trafficking creating amazing web of connectivity. Could these features be considered as a double sword?. Although, neurons are considered among the most long-lived cell types, but they are encoded by various genetic programs involved in initiation, signaling, and execution of cell death during development and adult life. Thus, two main types of neuronal apoptosis can be distinguished, based on their timing and cellular targets; an early physiological apoptosis of dividing progenitors or immature neurons that is referred as naturally occurring neuronal death (NOND), and a late cell apoptois that strikes mainly postmitotic neurons and involved in the inappropriate elimination of cells under neuropathological conditions. The term apoptois of cell death was derived originally from the Greek words apo- "away" and ptosis- "fall". Apoptosis was defined as a distinct mode of programmed cell death (PCD) on the basis of a series of characteristic ultrastructural features according to a sequence of events including nuclear and cytoplasmic condensation, cell fragmentation and finally phagocytosis. In this type of death, the cells are responsible of their own demise, a reason for which apoptosis is commonly referred to as a “cell suicide”.

During vertebrate brain development, about 50% of the neurons do not survive into adulthood, most of which die by apoptosis, resulting in the loss of the cell and all its synapses and processes. Why so many neurons are generated in excess and eliminated shortly thereafter remains to be understood. This normal neurodegeneration is thought to be important for matching the size of neuronal groups to the size of their targets (other groups of neurons to which they are connected by axons and nerve terminal synapses). Although, the brain is only 2–3% of total body mass, but it consumes 20% of body oxygen. Neurons are particularly susceptible to oxidative damage due to high levels of polyunsaturated fatty acids in their membranes and relatively low activity of endogenous antioxidant enzymes. Moreover, because neurons have limited glycolytic capacity, so they are highly dependent on aerobic oxidative phosphorylation by mitochondria for their energetic needs. The postmitotic differentiated neurons are completely devoid of replicative capability. Indeed they are believed to enter an "extended G0 phase" which, unlike other cell types, is irreversible. Contrary to this dogma, a growing body of evidence indicates that neurons have the capacity to re-enter the cell cycle. However, while neurons retain the ability to re-activate the cell cycle, engagement of the cell cycle machinery rarely leads to neuronal proliferation but typically induces apoptosis. Several different stimuli can initiate the apoptotic death of neurons. However, the finding that common morphological and biochemical alterations are observed independently upon the event that triggers apoptosis, suggests that most apoptotic pathways converge on a restricted number of common effector pathways. Basically two major pathways can be differentiated by the relative timing of caspase activation and mitochondrial release of cytochrome c. In the first, which is exemplified by activation of death receptors, an effector caspase is activated prior to mitochondrial alterations. In the other, cytochrome c is released from the mitochondrial intermembrane space prior to caspase activation. Regarding such molecular consequences, apoptosis is regulated by wide array of molecules within cells, some are antiapoptotic proteins (such as Bcl-2, Bcl-xL) and Proapoptotic proteins (such as Bax, Bcl-xS, Bad, Bid, Bik, caspases, p53). The molecular mechanisms for apoptosis involve the participation of at least three groups of proteins that are made from three different gene families. One set of proteins is the caspase family of cysteine-containing, aspartate-specific proteases (14 members have been identified to date). A second group includes the death promoting and death-suppressing proteins in the Bcl-2 family. A third group of proteins is designated as the inhibitor of apoptosis protein (IAP) family.

Is the future is promising for exploring the causes of neurodegenerative disorders and identifying treatments? Unfortunately, until now there are no effective treatments are available to prevent or radically cure any of them. Thus more complete understanding of the characteristics and mechanisms of neuronal apoptosis within the CNS could provide such hope.

When Kerr et al. at the early 1960's reported observation of a distinct form of cell death; they initially used the term "shrinkage necrosis" to describe it. Subsequently they developed the term "apoptosis" that derived from the Greek apo- "away" and ptosis- "fall", pointing to the role of this type of cell death in normal tissue turnover resembling picture of falling of the leaves. Apoptosis in a scientific term was originally defined as a distinct mode of "programmed cell death (PCD)" on the basis of a series of characteristic stereotyped, morphologically well defined phases including; chromatin and cytoplasmic condensation, DNA fragmentation, plasma membrane blebbing without rupture of the plasma membrane followed by phagocytosis as a sequence of events depend on activation of an intrinsic program of gene expression (Kerr et al., 1972). In this type of death, cells are responsible of their own demise where the apoptotic machinery are constitutively expressed or generated by the cell itself, a reason for which apoptosis is commonly referred to as a "cell suicide" (Lossi and Merighi, 2003). As early as 1951, the first note that neuronal cell daeth played a major role in vertebrate embryogenesis has been reported (Glucksmann, 1951). The adult brain contains about 10^{11}-10^{12} neurons, which are supported and protected by at least twice as many neuroglial cells. There are several types of glial cells, oligodendrocytes, microglia and astrocytes (Emerit et al., 2004). Neurons have unique complex morphological and functional features. They have a characteristic cell body with plethora of surface ion channels, and greatly elongated processes (axons and dendrites), that consequentially transport signals for intracellular trafficking creating amazing web of connectivity. Could these features be considered as a double sword? The decisions of neurons for life or death are dependent upon wide range of either physiological or pathological factors that could be exogenous or endogenous (Becker and Bonni, 2004). Although, neurons are considered among the most long-lived cell types, but they are encoded by various genetic programs involved in initiation, signaling, and execution of cell death decide cells' fate during development and adult life (Nicotera, 2004). Normally, the mature neurons that enter the posmitotic state, unlike other types of cells as hepatocytes, do not have the ability for regeneration because they cannot enter back into the cell cycle and undergo cell division (Martin, 2003). Their general inability to divide explains why aging and neurodegenerative disorders are functional loss related process where neurons die, without chance to be replaced. While the brain is only 2–3% of total body mass, but it consumes 20% of body oxygen and depends on glucose for its energy production as it needs about 4×10^{21} molecules every minute (Emerit et al., 2004), and in addition neurons have limited glycolytic capacity, so this provides an explanation why neurons are highly dependent on aerobic oxidative phosphorylation by mitochondria for their energetic needs (Moreira et al., 2010). Moreover, neurons are particularly sensitive and susceptible to oxidative damage due to high levels of polyunsaturated fatty acids in their membranes and relatively low activity of endogenous antioxidant enzymes (Miller et al., 2009). In this context, neurons like any other cells can die in two different ways; namely apoptosis and necrosis based on their own distinct morphological, biochemical and molecular behavior (Martin, 2003). Unlike the uncontrolled necrotic process, apoptosis is always viewed as an insular organized event, concerning single cells rather than groups of adjacent cells. Once the apoptotic cascade is completed, no trace is left as the shrunken remainders are phagocytosed and their constituents are recycled, and further no other cells are compromised (Waldmeier, 2003). In this context, microglial cells are the residential macrophages of the brain that is responsible for clearing the apoptotic neurons. They have the ability to recognize, uptake through scavenger receptor, and degrade them intracellularly froming the recognized apoptotic bodies that are degraded by lysosomal proteases (Stolzing and Grune, 2004). In general, lower concentrations of toxins particularly if delivered over a more protracted time course; induce apoptosis while higher levels induce necrosis (Bonfoco et al., 1995). Thus, the switch of apoptotic processes to necrosis has been argued if the damage is too extensive and/or the concerned cell is energetically unable to properly complete apoptosis (Waldmeier, 2003). Nowadays, multiple variants of PCD have been described; one of them is that cultured neurons sometimes do not show many of the typical signs of apoptosis referred to as "paratosis" (Sperandio et al., 2000). Another group reported evidence that apoptosis, once initiated, can be aborted so the term "abortosis", resulting in survival of the affected cell (Raina et al., 2001), albeit with unknown long-term fate and functionality. Finally, when the biochemical processes involved in cellular apoptosis occur in synaptic terminals of neurons, this could result in only localized degeneration of synapses known as "synaptosis", without obligatory and/or immediate death of the corresponding neuron (Mattson and Duan, 1999). Apoptosis must be verified from "autophagy", the cellular process involves self regulated degradation of intracellular proteins and organelles by their own lysosomal pathways, so the term "autophagocytosis", unlike apoptosis where process is referred to as "heterophagocytosis" as it depends on secondary lysosome activation of tissue macrophages (Larsen and Sulzer, 2002). During autophagy macromolecules are converted into energy substrates and amino acids are recycled, for purpose of removing and recycling damaged organelles and as a means of providing nutrients during starvation to maintain cellular homeostasis.. However, autophagy and apoptosis are not exclusive processes because they can share signaling elements and can occur simultaneously in a particular tissue or cell (Lemasters et al., 2002).

I. Types of Neuronal Cell Apoptosis

Based on their timing and cellular targets, two main types of neuronal apoptosis can be distinguished; an early physiological cell death of dividing progenitors or immature neurons and a late cell death that strikes mainly postmitotic neurons under pathological conditions, independently of synaptogenesis (Buss et al., 2006). Although, the underlying molecular and cellular mechanisms of both types seem to differ, in particular the extrinsic cues and receptors involved, but eventually all of them converge to induce a final common cellular machinery characteristic for classical apoptotic cascades (Yuan et al. 2003; Bredesen et al. 2006).

I.1. Early Apoptosis of Progenitor Neurons

As in most other parts of the developing body (Baehrecke 2002), apoptosis has been recognized to occur normally during natural development of neurons. So, this type is usually referred to as the naturally occurring neuronal death (NOND). It is generally assumed that about half of the neurons produced during neurogenesis die before completion of the central nervous system maturation, and this process affects nearly all classes of neurons (Lossi and Merighi, 2003). Animals are born with excess numbers of neurons. The naturally occurring neuronal death (NOND) is viewed to be an essential process by which these supernumerary neurons are eliminated at specific times. In this way, patterning of the brain is achieved, where size of the neuronal groups match size of their targets (other groups of neurons to which they are connected by axons and nerve terminal synapses) as well as to their own synaptic inputs from other regions (Martin, 2003). Most strikingly, the overgrowth is mainly confined to the forebrain, including sometimes an increase in the thickness and surface of the cerebral cortex. These data support the idea that even modest modifications in the size of the progenitor pool during its exponential growth can directly affect the final size and shape of the mammalian forebrain, which can occur by influencing the proliferation, but also the survival of neural progenitors (Vanderhaeghen and Cheng, 2010). This Early apoptosis of progenitor neurons appears to be closely linked to cell cycle regulation (%5D"Lossi and %5 D"Merighi, 2003). This developmental neuronal death (particularly of spinal and sympathetic ganglion neurons and motor neurons) is thought to be controlled by the supply of sustaining neurotrophic factors that are synthesized by the target axon associated glial cells and neighboring neurons (Raff et al., 1994; Miller and Kaplan, 2001). The neurotrophic theory has clearly established that neurons compete for limited amounts of target-derived trophic factors that have a protective role against apoptosis (%5BAuthor%5D"Lossi and %5BAuthor%5D"Merighi, 2003). It is believed that only those neurons survive that receive the proper trophic input from their targets (Becker and Bonni, 2004). Neurotrophins are a family of neurotrophic factors that includes nerve growth factor (NGF), brain-derived neurotrophic factor (BDNF), and neurotrophins (NT)-3 and -4. (Barde, 1994; Huang and Reichardt, 2001). Neurotrophins are synthesized as precursors that can be cleaved either intracellularly by furin and other proconvertases, or extracellularly by enzymes present in the environment such as plasmin, which is activated by cleavage of plasminogen by tissue plasminogen activator (tPA), as well as specific matrix metalloproteinases (MMPs). The mechanisms that regulate cleavage are not yet fully understood, yet are clearly of great importance (Friedman, 2010). However, even under normal physiological conditions, some of these factors may be secreted in their uncleaved form as proneurotrophins. Although, the cleaved "mature" form of neurotrophins serve to mediate neuronal survival, differentiation, and synaptic function through binding with high affinity to Trk family of receptor tyrosine kinases, but the uncleaved pro-neurotrophins bind preferentially to another set of p75 neurotrophin receptor ($p75^{NTR}$) that was proved to have a clearly defined role in signaling apoptosis (Lee et al., 2001). Over the years, these factors have been shown to have many other functions that are determined by the cellular context and suggest a greater complexity in their function than was previously appreciated. The role of Trk has been well defined to include activation of the PI_3 kinase-Akt, ras-MAP kinase, and PLCγ signaling pathways (Kaplan and Miller 2000; Patapoutian and Reichardt 2001). Whereas, $p75^{NTR}$ binds in a complex with the coreceptor sortilin, but their signaling pathways are still not clearly defined. This recent findings give rise to the concept that the balance of secreted mature and proneurotrophins has a critical impact on neuronal survival or death. Such opposing functions in the brain as survival or apoptosis, depending on tuned balance between which form of the protein is secreted and which receptor and signaling pathway is activated (Friedman, 2010). the relative levels of the Trk and p75NTR receptors may alter the balance of signaling and determine whether proNGF induces survival or apoptosis (Masoudi et al., 2009).

I.2. Late Apoptosis of Postmitotic Neurons

Later on life, apoptosis on one side is viewed as a way to eliminate damaged and dysfunctional neurons, in which case it would represent a "defense mechanism". On the other side, the damage inflicted by various environmental and/or genetic factors might lead to aberrations in the apoptotic machinery in certain neurons and thus cause "death by mistake" (Waldmeier, 2003). Death of postmitotic neurons is a widely recognized phenomenon that plays a crucial role in sculpting and maintaining the architecture of the mature CNS. Moreover, a growing body of evidence is accumulating to indicate that apoptosis is also responsible for the loss of neurons associated with physiological aging (Kaufmann et al., 2001). It is worth mentioning that the adult brain also requires a constant trophic input for appropriate function. In the adult CNS a distinct set of pathogenic stimuli rather than the lack of trophic support, predominate as inducers of apoptosis. One of the most widely studied extrinsic insults triggering neuronal cell death in the mature nervous system is the neurotransmitter glutamate (Zipfel et al., 2000). Neurotrophic factors and spontaneous neuronal activity in newly generated postmitotic neurons might serve to limit the extent of neuronal apoptosis and perhaps correct cell cycle exit abnormalities in neurons depending on their nature. As neurons form connections, the regulation of the cell cycle machinery may be handed over to target-derived neurotrophic factors and trans-synaptic neuronal activity that regulate neuronal survival and death during the period of neuronal connectivity

(%5BAuthor%5D"Becker and %5BAuthor%5D"Bonni, 2004). Although the main source of trophic factors for mature neurons is considered to arise locally from glial cells and synaptic partners, recent evidence suggests that hormonal-like influences from distant sources may also be important. These include not only relatively well-characterized steroid hormones that cross the brain barriers, but also blood-borne protein growth factors able to cross the barriers and exert unexpected, albeit specific, trophic actions (Torres-Aleman, 2000). Among these the insulin-like growth factor-I (IGF-I), fibroblast growth factor-2 (FGF-2) or the neurotrophins (NTs), exert a tonic trophic input on brain cells and likely are important players in the regulation of cell death of postmitotic neurons (Torres-Aleman et al., 1998). It is in fact possible that cells are instructed to die by switching on the apoptotic machinery, or cells intrinsically committed to death are rescued by switching off their suicide program. The molecular machinery of late apoptotic death of postmitotic neurons more commonly follows the mitochondrial pathway of intracellular signal transduction, but the death receptor pathway may also be involved. The signal transduction pathways acting upstream formation of the apoptosome are however still not completely characterized (Bredesen and Rabizadeh, 1997). Differentiated neurons are postmitotic cells completely devoid of replicative capability. Most mammalian CNS neurons enter the postmitotic state during embryonic life. In doing so some dividing cells exit from the cell cycle and enter a phase of mitotic quiescence commonly referred to as the G0 phase. Indeed postmitotic neurons are believed to enter an "extended G0 phase" which, unlike other cell types, is irreversible. However, investigation of the mechanisms of neuronal apoptosis has led to the unexpected discovery that, in many instances, the quiescent and dormant cell cycle machinery is resuscitated, giving rise to the concept of "cell cycle re-entry" as an additional key mechanism of neuronal apoptosis (Lossi and Merighi, 2003).

II. Cell Cycle Re-entry in Neuronal Apoptosis

It had been thought for a long time that once neurons differentiate; they desist permanently from cell cycle processes and that cell cycle proteins are downregulated in postmitotic neurons (Hayes et al., 1991). Contrary to this dogma, increasing evidence indicates that neurons have the capacity to re-enter the cell cycle. However, while neurons retain the ability to re-activate the cell cycle, engagement of the cell cycle machinery rarely leads to neuronal division and proliferation but typically the signaling pathways are "rewired" to induce apoptosis (Feddersen et al., 1992; Wartiovaara et al., 2002). Re-activation of cell cycle has been emerged as a key component of neuronal apoptosis in development and disease. Recently, a direct interaction between the cell cycle machinery and the cell-intrinsic apoptotic program was demonstrated in postmitotic neurons (Konishi et al., 2002). Consistent with this idea are the observation that brain tumors of neuronal origin are of quite rare incidence indicating resistance of neurons to proliferate in response to oncogenic transformation (Heintz, 1993). Re-entry into the cell cycle is not specific to particular neurodegenerative diseases or distinct neuronal populations, but rather characterizes a common pathway involved in neuronal apoptosis. Re-activation of components of the cell cycle is often observed in dying neurons and may contribute to neuronal apoptosis in the developing brain and to neuronal degeneration in the adult brain (Husseman et al., 2000; Liu and Greene, 2001). In the developing mammalian nervous system, much of neuronal death in some populations of neurons, including the cerebellar granule neurons takes place before neurons make connections and is subject to regulation by neurotrophic factors and neuronal activity largely in a target-independent manner (Wood et al., 1993). It is conceivable that under these circumstances naturally occurring cell death may result from random genetic abnormalities during cell cycle exit that are inherited in the newly generated but irreversibly postmitotic neurons. As the nervous system matures, synaptic communication might effectively silence the cell cycle machinery in neurons. However, disruption of synaptic functions in the adult brain, upon exposure to ischemic and neurodegenerative pathogenic stimuli might de-repress the dormant cell cycle machinery and thus induce apoptosis in mature neurons via mechanisms that are shared with those underlying neuronal apoptosis in the developing brain (Becker and Bonni, 2004). Interestingly, apoptosis and the cell cycle share some common regulatory participants including retinoblastoma (Rb) protein, E2 Transcription Factor (E2F), and p53, and exhibit morphological similarities of cell rounding, and chromatin condensation (King and Cidlowski, 1995). This suggests that both processes might be regulated to some extent by similar molecular mechanisms. Moreover, cell cycle re-activation in developmental and disease-related neuronal apoptosis bear many similarities including the induction of components of the G1 and G2 phases of the cell cycle and the lack of chromosomal segregation and cytokinesis. Basically, the cell cycle is typically divided into four phases: G1 (first gap), S (DNA synthesis), G2 (second gap), and M (mitosis). Progression through these phases is regulated by the sequential expression, activation, and inhibition of two balancing systems namely; cyclin-dependent kinase (CDK) complexes with their activating cyclins subunits and cyclin-dependent kinase inhibitors (CDKIs) (Lees, 1995). Cells that exit the cell cycle remain in a quiescent G0 state. In proliferating cells, progress through the cell cycle is tightly controlled by checkpoints. These checkpoints function as molecular switches that ensure critical events in one phase of the cell cycle are completed before the next phase can be entered, thereby coordinating cell growth with cell proliferation. Cells arrested at any of the checkpoints will either return to G0 or re-differentiate (if arrested at G1 checkpoint), or the cells will die by apoptosis (Nagy, 2000; Liu and Greene, 2001). The latter is a mechanism to ensure that non-repairable DNA modifications are not passed on to the progeny of a damaged cell. In mammals, seven CDKs and eight cyclins have been described to date (Grana and Reddy, 1995; Lees, 1995; Morgan, 1997). It has been proposed that the appearance of these cycle-related proteins represents an abortive attempt of the post-mitotic cells to return to the cell cycle (Hass,1994)Two families of proteins inhibit the activity of cyclin-CDK complexes. Members of the Kip/Cip family of proteins (p21, p27 and p57) regulate the activity of all the G1 cyclin-CDK complexes and, to a lesser extent, cyclin B/CDC2, while members of the INK4 family (p16, p15, p19 and p18) specifically inhibit CDK4 and CDK6. As is the case for the CDKs and cyclins, the activity of CDKIs is tightly regulated by transcription, translation, and ubiquitin-mediated proteolysis (Sherr and Roberts, 1999). During development, downregulation of CDKs and cyclins, as well as induction of CDKIs lead to cell cycle exit and differentiation. In particular, p19 and p27 are required to maintain the quiescent state of neurons in the cortex and hippocampus (Zindy et al., 1999). The activation of the mitotic kinase CDC2, also termed CDK1, promotes neuronal cell death in part by phosphorylating the BH3-only protein Bad. However, CDC2 is likely to induce neuronal apoptosis by phosphorylating additional targets. In future studies, it will be important to identify additional substrates of CDC2 and other cell cycle proteins that mediate neuronal apoptosis and perhaps neuronal degeneration. Remarkably, CDC2 is abnormally expressed in degenerating neurons in NFT diseases including AD, where CDC2 has been proposed to phosphorylate tau and hence directly contribute to the pathogenesis of AD neurodegeneration. An important question that remains to be addressed is how cell cycle re-entry is coordinated with other pro-apoptotic signaling pathways in promoting neuronal cell death. Aside from re-activated cell cycle proteins, the protein kinases GSK-3, JNK, and p38 MAPK have been suggested to promote neuronal apoptosis and neuronal degeneration (Hetman et al., 2000; Harper and LoGrasso, 2001; Hongisto et al., 2003). It will be important to determine whether the activity of CDC2 and other cell cycle proteins is regulated by these protein kinases or whether cell cycle proteins act upstream of the these apoptotic kinase pathways in neurons. An interesting hypothesis that has guided studies of neuronal cell death in the adult brain is that increased oxidative stress leads to DNA damage that in turn triggers cell-cycle re-entry and apoptosis (Becker and Bonni, 2004).

III. Cellular and Molecular Machinery of Apoptosis

As other types of cells, neuronal apoptosis is regulated by specific molecules within cells (table 1). A variety of genes and their protein products have been uncovered, which can influence or determine the progression of apoptosis and serve as markers of the process (Tatton and Olanow), Structurally, the molecular mechanisms for apoptosis involve the participation of at least three distinct groups of proteins that are made from three different gene families. One group of proteins is the caspase family of cysteine-containing, aspartate-specific proteases (15 members have been identified to date). Some are "initiators" regulating the upstream phase (including caspase-2, caspase-8, caspase-9, caspase-10) and the others are "effectors" for the downstream terminal phase (including caspase-3, caspase-6, caspase-7). A second group includes the death promoting and death-suppressing proteins in the Bcl-2 family (e.g., Bcl-2, Bcl-X_L, Bcl-X_S, Bax, Bak, Bad, Boo, Bid).

Table 1. Apoptosis regulatory proteins

Bcl-2 family	Antiapoptotic proteins: Bcl-2, Bcl-xL, Boo Proapoptotic proteins: Bax, Bcl-xS, Bad, Bid, Bik
Caspase family	Initiators: caspase-2, caspase-8, caspase-9, caspase-10 Executioners: caspase-3, caspase-6, caspase-7 Cytokine processors: caspase-1, caspase-4, caspase-5, caspase-11, caspase-12, caspase-14
Tumor suppressor family	p53, p63, p73
IAP family	NAIP, IAP1, IAP2, XIAP

A third group of proteins is designated as the inhibitor of apoptosis protein (IAP) family. Classifying all of these machinery molecules based on their functional role give rise to two sets; one includes antiapoptotic proteins (such as Bcl-2, Bcl-X_L, Boo) and the other includes proapoptotic proteins (such as Bax, Bcl-X_S, Bad, Bid, Bak, caspases, p53) (Martin, 2003).

III.1. Bcl-2 Family

The Bcl-2 family of proteins (b-cell lymphoma/leukemia gene 2) is a critical regulator of cell death (Adams and Cory, 1998; Gross et al., 1999). Apoptosis regulation by the bcl-2 family is a very complex and exciting process owing to the diverse activities of the members of this family. Membership into the family of Bcl-2-related proteins is defined by homology domains within their amino acid sequences (Martin, 2003). All members of the Bcl-2 family of proteins contain at least one of the four conserved Bcl homology domains BH1–4. The Bcl-2 protein family is subdivided into the anti-apoptotic "multidomain" proteins containing BH1–4 (Bcl-2, Bcl-X_L, Boo), the pro-apoptotic "multidomain" proteins possessing homology in the domains BH1–3 (Bad, Bax, Bcl-X_S, Bak), and the "BH3-only" pro-apoptotic protein (Bid). These domains function in the interactions (i.e., binding) between members (Becker and Bonni, 2004). The members of the Bcl-2 family that of the most abundant expression in mammalian nervous tissue; are Bcl-2, Bcl-X_L and Bax (Tatton and Olanow, 1999). Bcl-2 family members exist as monomers (single proteins) that form homodimers (two of the same proteins bound together), heterodimers (two different proteins bound together), and higher order multimers (more than two interacting proteins). For example, Bax forms homodimers or forms heterodimers with either Bcl-2 or Bcl-X_L. When Bax is present in excess, it antagonizes the antiapoptotic activity of Bcl-2. The formation of Bax homodimers promotes apoptosis, whereas Bax heterodimerization with either Bcl-2 or Bcl-X_L blocks apoptosis. Thus, the complex steady-state array of protein–protein interactions among members of the Bcl-2 family and the ratio of the pro- and anti-apoptotic proteins functions in dictating the fate of a given cell to live or die by apoptosis. (Martin, 2003; Becker and Bonni, 2004). Bcl-2 is widely expressed in the nervous system and is localized to the outer mitochondrial membrane, endoplasmic reticulum, and nuclear membrane (Krajewski et al., 1993). Subcellular localization studies have shown that the anti-apoptotic members of the Bcl-2 family (Bcl-2, Bcl-X_L, Boo) reside on the mitochondrial outer membrane, while the pro-apoptotic family members (Bad, Bax, Bid, Bcl-X_S, Bak) may be either cytosolic or present on the cytoplasmic surface of the outer mitochondrial membrane (Zimmermann et al., 2001). Studies in primates have shown that Bcl-2 is primarily expressed during development in immature neurons and that its expression levels decline dramatically after birth (Bernier and Parent, 1998). In contrast, Bcl-X_L is expressed in a stable manner in the adult mammalian CNS (Gonzalez-Garcia *et al.,* 1995). Concurrently, the increased Bcl-2 levels may reflect the induction of a survival program in neurons that have sustained damage that is not sufficient to cause apoptosis, while increased Bax levels may mean that the neuron has progressed to the early stages of apoptosis (Tatton and Olanow, 1999). Thus, it is proposed that Bcl-X_L might be an even more attractive candidate for antipoptotic therapy (Hartmann et al., 2002). The exact molecular mechanisms by which Bcl-2 proteins control apoptosis are still not entirely clear. However it was proved to produce wide array of effects including balance and shift in cellular redox potential in neurons to a more reduced state by suppression of production of reactive oxygen species (ROS) (Kane et al., 1993; Hockenberry et al., 1993) and elevation of reduced glutathione (GSH) levels and its redistribution to the nucleus with correction of reduced glutathione/oxidized glutathione and $NADH/NAD^+$ ratios (Ellerby et al., 1996; Voehringer et al., 1998). Moreover, Bcl-2 has beneficial effects on mitochondrial dysfunction that linked to apoptotic pathways. These effects include; enhancement of mitochondrial membrane potential and improvement of ATP/ADP ratios (Hennet et al., 1993), inhibition of mitochondrial calcium release with stimulation of mitochondrial calcium uptake potential of neural cells (Baffy et al., 1993; Murphy et al., 1996), upstream suppression of mitochondrial transition pore and the subsequent release of cytochrome *c* and therefore preventing caspase activation (Kluck et al., 1997; Yang et al., 1997). In addition, Bcl-2 inhibits apoptotic Bax channel-forming activity (Antonsson et al., 1997). The related protein Bcl-X_L also blocks cytochrome *c* release by either direct binding to it (Kharbanda et al., 1997) or via improvement of mitochondrial dysfunction by inhibition of mitochondrial swelling and inhibition of loss of mitochondrial membrane potential by regulating proton flux (Shimizu et al., 1998).

III.2. p53

The tumor suppressor and nuclear transcription factor p53 is a tetramer phosphoprotein that can regulate several major cellular functions including gene transcription, DNA synthesis, DNA repair, cell cycle regulation, senescence, and cell death (Culmsee and Mattson, 2005). It is considered one of the most important cellular sensors as it plays a critical role as a transducer of DNA damage into G1 cell cycle arrest and/or apoptosis. Constitutively, cells contain fairly low levels of p53, since the inactive form of the protein is highly unstable short-lived with half life is approximately 10–30 min) (Soussi, 2000). Several stimuli including DNA damage are required for activation of p53 with further stabilization and accumulation. For activation, it requires orchestrated post-translational modifications, including either phosphorylation predominantly at N-terminal residues, acetylation of residues in the C-terminal region, or poly(ADP) ribosylation (Malanga et al., 1998; Appella and Anderson, 2001; Brooks and Gu,2003). Such modifications either increase the stability of p53 or directly enhance its DNA-binding affinity (Culmsee and Mattson, 2005). Although a large number of kinases is capable to phosphorylate p53, response to DNA damage is most likely mediated by DNA-dependent protein kinase (DNA-PK) which is a substrate of certain caspases, the product of the ataxia-teleangectasia gene (ATM) (Lee and McKinnon, 2000). At least 20 phosphorylation sites and 4 acetylation sites have been identified that are important for p53 accumulation and binding of p53 to promoter regions of specific target genes (Appella and Anderson, 2001). The subcellular localization of p53 is determinant for its role. Normally, its localization changes during neuronal maturation; from the nucleus in immature cells, to the cytoplasm in fully differentiated neurons (Eizenberg et al., 1996). In response to appropriate stimuli, p53 translocates to the nucleus and plays a regulatory role in directing primary neurons toward differentiation or apoptosis (Lossi and Merighi, 2003). While p53 has a clear role in neuronal apoptosis of the mature nervous system following injury as evidenced in many conditions such as ischemia and neurodegenerative diseases. (Crumrine et al., 1994; Shahbazian et al., 2001), but its role in apoptosis of progenitor neurons is not detected. The explanation why p53 is required for neuronal cell death in these systems but not for neuronal apoptosis during development comes from observation that under these pathogenic conditions, oxidative stress is a prominent feature for DNA damage (Becker and Bonni,

2004). Oxidative DNA damage consists of highly specific chemical lesions such as hydroxyl radical-modified bases (8-hydroxyl-20-deoxyguanosine), apurinic/apyrimidinic abasic sites (AP sites), and strand breaks. In addition, oxidative damage of proteins involved in cell cycle regulation or DNA repair may contribute to accumulating DNA damage and finally to activation of p53 (Culmsee and Mattson, 2005). Several lines of evidence converge to indicate that neurons could undergo apoptosis by p53-dependent or –independent pathways, according to the stimulus responsible for genotoxic stress. However, it is clear that p53 has a significant role in apoptosis that follows DNA damage in vivo (Wood and Youle, 1995) and is also a fundamental component of the $p75^{NTR}$ apoptotic signal cascade (Aloyz et al., 1998). One of the actions of this pluripotent transcription factor is to cause upregulation of pro- and downregulation of antiapoptotic members of the Bcl-2 family; another is induction of mitochondrial enzymes locally generating reactive oxygen species (ROS) (Waldmeier, 2003). The biological functions of p53 are largely mediated through transactivation of specific target genes that are selectively operated depending on the type and severity of the stress stimulus and on the specific amino acid residues of p53 that are phosphorylated or acetylated (Hofseth et al., 2004). For example, p53 can cause cell cycle arrest by inducing the expression of p21 and can mediate apoptosis by inducing the expression of Bax and the BH3-only protein PUMA that cause mitochondrial membrane permeabilization (Jeffers et al., 2003). Cell culture studies have established strong correlation between p53 expression and excitotoxic neuronal death induced by glutamate via N-methyl-D-aspartate (NMDA) receptor (Cregan et al., 1999). Recent data demonstrated that p53 inhibition did not alter the sustained increase in intracellular calcium levels in glutamate-exposed neurons (Xiang et al., 1996); instead, it interfered with early steps of the cell death program triggered by calcium. Although p53 was originally thought to be unique, it is now known to be part of a larger gene family including the recently discovered p63 and p73 (Courtois et al., 2004). Both p63 and p73 share key functional domains with p53, but their genes encode several isoforms that lack the N-terminal transactivation domain and can function as dominant negatives and even interfere with p53 function by forming mixed complexes with p53.overexpressed (Levrero et al., 2000; Lohrum and Vousden, 2000). While p53 has a pro-apoptotic function in neurons, the p53-related protein p73 has anti-apoptotic activity in selected neuronal populations. Mice deficient for p73 exhibit enhanced developmental apoptosis of sympathetic neurons in the PNS (Pozniak et al., 2000), as well as of cortical neurons in the CNS. In addition, the postnatal development of the olfactory bulb is perturbed in *p73−/−* mice, suggesting a role for p73 in the maintenance of some adult neuronal populations (Pozniak et al., 2002).

III.3. Caspases

Caspases are considered the pivotal executioners of apoptotic cascade. They were formerly referred to as interleukin converting enzyme (ICE)-like proteases, but nowadays they are termed caspases, where the "c" letter denoting cysteine protease and the "aspase" referring to their tendency to cleave the substrates after the aspartate residue (Cohen, 1997) As many as 15 different caspases have been identified in mammalian nervous tissue (Chowdhury et al., 2008). Based on the prodomain structural and functional characteristics of them, caspases are classified into three sets. The first one referred as apoptotic "initiators" with a long prodomain (including caspase-2, -8, -9, -10), whose activation characteristically sets in motion a caspase cascade. The second set of caspases is the apoptotic "executioners or effectors" those contain short prodomains (including caspase-3, -6, -7) where they eventually degrade vital cellular proteins. The third set is termed inflammatory caspases with large prodomains (including caspase-1, -4, -5, -12, -13, -14) comes from their role in cytokine maturation and inflammatory responses (Budihardjo et al., 1999; Waldmeier and Tatton, 2004; Pop and Salvesen, 2009), where they are implicated in proteolytic processing and bioactivation of the proforms of certain interleukins (ILs) (Reed, 1999). Typical "non-apoptotic" members such as caspase-1, -4, and -5 are proposed to induce "pyroptosis," a form of death associated with massive activation of inflammatory cells (Labbé and Saleh, 2008). The only truly remaining non-apoptotic human candidate may be caspase-14, a mediator in keratinocyte differentiation (Denecker et al., 2007). Initially, caspases are synthesized as zymogens and then cleaved concomitant with their activation, through distinct activation pathways can be categorized as recruitment-activation that is mediated by homophilic CARD-domain interaction, autocatalytic activation, or as trans-activation by a previously activated caspase (a so-called "caspase cascade"). The initiators and executioners possess different mechanisms of activation; those are dimerization for the initiators and interdomain cleavage for the executioners (Chowdhury et al., 2008). Caspase activation is tightly controlled by many proteins. One of them is the FLICE inhibitory proteins (FLIPs) that mimic the prodomains of caspases but lacking the enzyme active sites, so in binding to the cytoplasmic protein complex (disc) of death receptors, they modulate/inhibit the recruitment/activation of caspases. In addition, cells also encode an inhibitor protein, XIAP that belongs to IAPs (inhibitor of apoptosis protein) family that in direct way bind to and inhibit caspases (Hardwick, 2002). In this context, the IAPs family is X-chromosome-linked proteins including IAP, IAP1, IAP2, and NAIP (neuronal apoptosis inhibitory protein). The primary mechanism of IAPs is the prevention of proteolytic processing of specific caspases, where procaspase-9 is the major target of IAPs suppressor activity. However, IAPs do not prevent caspase-8-induced proteolytic activation of the downstream procaspase-3. IAPs can also block apoptosis by reciprocal interactions with the nuclear transcription factor NFκ-B. Once activated, caspases act on nuclear proteins, cytoskeletal proteins, or cytosolic proteins (Martin, 2003). Over 250 proteins have been found to be cleaved by caspases during the execution of apoptosis (Nicholson, 2004). The caspase specificity is based on the recognition of a stretch of four amino acid residues, the last one of them corresponding to the aspartic residue after which the cleavage occurs (Bulat N, Widmann, 2009). Aside from their targets from the structural proteins (such as actin and lamins) and DNA repair enzymes), caspases also cleave regulatory proteins (such as inhibitor of caspase-activated DNase, focal adhesion kinase and mitogen-activated protein (MAP) kinases) (Honig and Rosenberg, 2000), to inhibit or enhance their activities or to amplify the apoptotic signal through inactivation of apoptotic inhibitors (such as IAPs and the anti-apoptotic Bcl-2 proteins) (Olson and Kornbluth, 2001). In addition to promotion of apoptotic cell death, caspases are reported to participate in another distinct signaling pathway for activation of proinflammatory cytokines (Salvesen and Dixit, 1997). In view of their beneficial roles; caspases are proved to be required for normal neuronal development; where the deletion of Apaf-1, caspase-9, or caspase-3 results in excessive numbers of neurons, particularly in the brain cortex and other brain regions as well (Hardwick, 2002).

III.4. AIF (Apoptosis-Inducing Factor)

Apoptosis-inducing factor (AIF) is a conserved NADH oxidase flavoprotein of 67 kD that normally located in the mitochondrial intermembrane space where it displays life supporting mitochondrial activities and essential functions for optimal oxidative phosphorylation (Joza et al., 2001; Miramar et al., 2001). Recent data show that AIF is released from mitochondria by a distinct mechanism and in a slower rate than that of other apoptogenic proteins mainly cytochrome *c*, although they start simultaneously. This slower-rate kinetics (hour-range for AIF and minute-range for other apoptogenic proteins) is contributed to the requirement of a putative secondary event (Munoz-Pinedo et al., 2006) such as proteolytic cleavage by activated cysteine proteases different from caspases, namely cathepsins and calpains (Yuste et al., 2005; Otera et al., 2005) and the mitochondrial uncoupling proteins (UCP) (Sullivan et al., 2004). The possibility that the initial activation of caspases could indirectly trigger cathepsin/calpain activation by enhancing lysosomal and mitochondrial permeabilization (Foghsgaard et al., 2001), is accepted on the basis of the recent report that p53-induced caspase-2 activation may be involved in mitochondrial AIF release (Seth et al., 2005). Overexpression of the antiapoptotic Bcl-2 was found to suppress release of AIF from mitochondria, but has no effect on its apoptogenic activity (Susin et al., 1999; Cregan et al., 2004). Upon apoptosis signaling, AIF translocates to the nucleus, directly binds to DNA with subsequent provocation of chromatin condensation and large scale DNA fragmentation (~50 kbp), in a process that apparently is caspase-independent (Daugas et al., 2000). Human AIF DNA binding is not dependent on a specific nucleotide sequence (Ye et al., 2002), which strongly suggests that other types of direct physical interactions (e.g., electrostatic) might be involved in AIF-mediated DNA and chromatin compaction (Vahsen et al., 2006). Interestingly, in embryonic morphogenesis, genetic inactivation of AIF appears to abolish early neuronal death of the proliferating precursor cells and the young postmitotic neuroblasts (Joza et al., 2001). However, owing to its primary putative oxidoreductase and peroxide scavenging activities, AIF paradoxically could play an antioxidant and anti-apoptotic role by regulating the production of reactive oxygen species (ROS) (Krantic et al., 2007).

These opposing activities of AIF could be explained on a structural base; where the integrity of AIF's C-terminal domain but not that of the oxidoreductase-related (NADH) and FAD-binding N-terminal domains, is required to trigger PCD, following its nuclear translocation. Supporting such concept, AIF was found to differ structurally from all known NADH oxidases, indicating that its mechanism of action is extended behind this oxidoreductase capacity. AIF is enzymatically active as a monomer whereas other NADH oxidases act as di- or tetramers (Miramar et al., 2001; Loeffler et al., 2001). It is apparent that there are many different AIFs have been identified to date and that different ones may be operative in different models of apoptosis.

Another mitochondrial AIF is referred as an interleukin converting enzyme-like (ICE) protease, since it can be inhibited by the same inhibitor agent specific to the ICE family. BCL-2 has also been shown to prevent the release of this AIF, presumably by maintaining closure of the permeability transition pore (PTP) of mitochondria and preserving the mitochondrial membrane potential (MMP, $\Delta\Psi_M$) (Susin et al., 1996).

III.5. Glutamate Excitotoxicity

Neurons communicate by neurotransmission at synapses. Glutamate is the most dominant excitatory amino acids and the primary neurotransmitters in about half of all the synapses in the brain. Physiologically, it is responsible for many neurological functions including cognition, memory, movement and sensation. Neuronal glutamate is released from the pre-synaptic terminal into the synaptic cleft by regulated exocytosis of synaptic vesicles. Concentrations of glutamate at the synaptic cleft have been estimated to be approximately 1 mM, whereas its interstitial concentration is only about 1 mM. Re-uptake of extracellular glutamate is done at the pre-synaptic terminals and adjacent glial cells (Martin, 2003). Glutamate exerts its activity by binding to two distinct glutamate receptors (GluRs). The metabotropic non-NMDA receptors, further divided into the a-amino-3-hydroxy-5-methyl-4-isoxazole propionate (AMPA) and kainate (KA) receptors, mediate entry of Na^+ and K^+ into the postsynaptic neuron, resulting in cytotoxic edema, while activation of the ionotropic NMDA (N-methyl D-aspartate) receptors allows entry of Ca^{2+} into the post-synaptic neuron with subsequent necrotic cell death or apoptosis. (Lipton and Rosenberg, 1994). The intracellular Ca^{2+} concentration is very tightly regulated and even small variations result in the neuron's commitment to apoptosis (Orrenius et al., 2003). Excitotoxic injury refers to trans-synaptic injury due to release of excitatory amines, which occurs in the peripheral and central nervous system (Moritani, et al., 2002). While excitotoxicity had long been thought to result exclusively in necrosis, accumulating evidence suggests that apoptosis is a prominent contributor to excitotoxicity induced neuronal loss (Martin et al., 1998; Zipfel et al., 2000). Under pathological conditions either acute such as epilepsy and stroke or neurodegenerative disorders including Alzheimer's disease (AD), Amyotrophic lateral sclerosis (ALS), Huntington's disease, and Parkinson's disease (PD), the increased levels of glutamate and exitotoxicity were found to be a key event in induction of apoptotic cascade. The postnatal period of brain development is particularly vulnerable to glutamate excitotoxic injury. Excitatory amino acid receptors are abundant in the cortex during the 1st two years of life. The high rate of synaptogenesis results in an overexpression of NMDA receptor that found to dominate in the immature brain when synaptic transmission is weak and extremely plastic. Experimental studies show that too much or too little NMDA receptors can be threatening to developing neurons. During maturation, AMPA receptor and kainic receptor predominate (Lipton and Nakanishi, 1999). Both neuronal culture and animal model data are discordant with regard to whether excitotoxic neuronal death is apoptosis or necrosis. Activation of neuronal GluRs kills neurons by pathways that may involve alterations in cytosolic free Ca^{2+} homeostasis and downstream activation of wide range of enzymes including Ca^{2+}-sensitive proteases, protein kinases, endonucleases, lipases, phospholipases as well as activation of endonucleases (DNA-cleaving enzymes) (Martin, 2003). Of the protein kinases and proteases, several members have been shown to be involved in glutamate-induced neuronal apoptosis cascade, including PKC, MAPK, CaMK, PTK, ERK, Caspase and Calpain. Recently, death-associated protein kinase (DAPK) expression was found to be increased in glutamate-treated cultures exhibiting a prevailing form of neuronal apoptosis in a delayed time course. In the developing central nervous system, DAPK is widely expressed in proliferative and postmitotic regions in brain; and the expression is gradually decreased during postnatal life but still remains high in a number of areas (Yu et al., 2006).

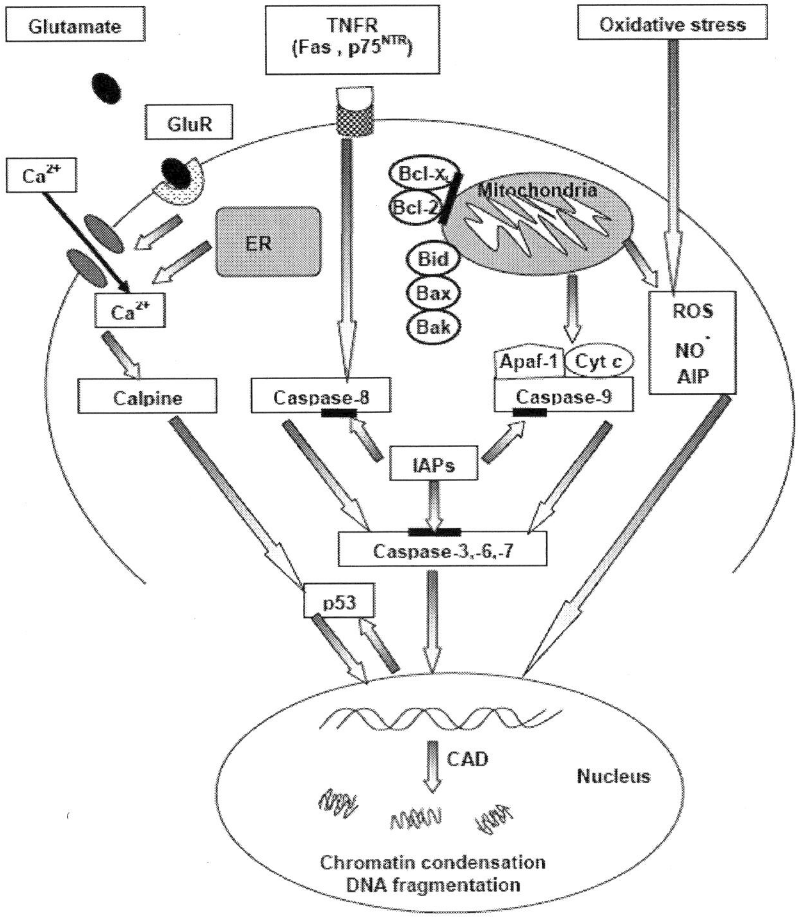

[AIP (apoptosis inducing protein); Apaf-1 (apoptotic protease activating factor-1); Bcl-2, Bcl-XL (anti-apoptotic Bcl-2 family);; Bax, Bak, Bid (pro-apoptotic Bcl-2 family); CAD (Caspase-Activated Deoxyribonuclease); caspase-8, caspase-9 (initiator upstream caspases); caspase-3, caspase-6, caspase-7 (effector downstream caspases); Cyt *c* (cytochrome *c*); ER (endoplasmic reticulum|); GluR (glutamate receptor); IAPs (inhibitor of apoptosis protein family); NO· (nitric oxide); ROS (reactive oxygen species); TNFR (tumor necrosis factor receptor)].

Figure 1. Simplified diagram of the main apoptotic machinery pathways. The extrinsic apoptotic pathway that is initiated by activation of the cell surface death receptors prior to mitochondrial alterations. The intrinsic apoptotic pathway that is mitochondria-dependent prior to caspase activation. The endoplasmic reticulum-mediated pathway.

From these remarks, it could be concluded that the excitotoxic death of neurons occur as hybrid form of cell death midway between apoptosis and necrosis with coexisting characteristics that lie along a structural continuum. This is referred as "apoptosis-necrosis continuum" and is reviewed to be influenced by the high diversity in the expression, localization, and function of GluR subtypes and second messenger systems in the CNS, all of them are determined by the state of neuronal maturity. This comes from observation that excitotoxic degeneration of adult neurons does not occur with the typical apoptotic features compared to that closely resemble those seen during naturally occurring cell death in the developing nervous system (Martin, 2003).

IV. Pathways of Apoptotic Cascade

Understanding neuronal death pathways and their cross-talk not only informs the detailed pathobiology but also suggests novel therapeutic strategies (Nagley et al., 2010) (figure 1). Apoptotic death of neurons could be triggered by several different physiological or pathological stimuli; those could be external (e.g. toxic molecules impinging on the cell) or internal (e.g. inappropriate intracellular protein aggregation). (Martin, 1998; Nagley et al., 2010). However, the finding that common morphological and biochemical alterations are observed independently upon the event that triggers apoptosis, suggests that most apoptotic pathways converge on a restricted number of common effector pathways. Basically, two major pathways can be differentiated by their relative timing of caspase activation and mitochondrial release of cytochrome *c*. The first is the extrinsic apoptotic pathway that initiated by activation of cell surface death receptors such as Fas and TNF receptors (TNFR) prior to mitochondrial alterations; leading to caspase-8 activation, which in turn cleaves and activates downstream caspases such as caspase-3, -6, and -7. The second is the intrinsic apoptotic pathway that is mitochondria dependent; where cytochrome *c* is released from the mitochondrial intermembrane space prior to caspase activation, promoting activation of the initiator caspase-9 which then activates the effector caspase-3 (Martin, 2003; Lossi and Merighi, 2003). The extrinsic and intrinsic pathways are not entirely distinct regarding their signaling and targeting cascads. While in the death receptor pathway apoptosis is triggered by a relatively small number of structurally-related ligands; mitochondrial apoptosis in neurons can be triggered by a variety of structurally-unrelated agents (Sastry and Rao, 2000). Nowadays, there is a belief that a connection between the extrinsic and the intrinsic pathways could exist. Such link was found to be dependent on Bid cleavage (Taylor et al., 2008). Bid that is inactive in its full length form; bearing a recognition site for caspase-8 of the extrinsic pathway, and when processed by this caspase, becomes active and lead to cytochrome *c* release with further activation of caspase-9 of the intrinsic pathway (Bulata and Brain, 2009). Recently, there is growing evidence for other pathway referred to as endoplasmic reticulum (ER)-mediated pathway (Chowdhury et al., 2008).

IV.1. Death Receptor Mediated-Pathway

Death receptors are cell surface receptors that trigger apoptosis. There are several types of death receptors in different tissues, but two members of the tumor necrosis factor receptor (TNFR) family were recently demonstrated to be involved in neuronal death namely; Fas (CD95/Apo-1) and p75 neurotrophin receptor ($p75^{NTR}$). (Raoul et al., 2000). Upon stimulation of death receptors; an intracellular death-inducing signaling complex called the "death domain" (DD) is formed through an amino acid stretch within the carboxy terminus of the receptor. This domain is responsible for coupling the death receptor to either a cascade of caspases, leading to induction of apoptosis, or to the activation of kinase-signaling pathways, resulting in gene expression through NF-κB or activator protein 1 (AP-1) (Zimmermann and Green, 2001).

Fas activation has been proved to be a key step in trophic factor deprivation induced-apoptosis (Raoul et al., 1999). Upon exposure to cell death-triggering stimuli, lymphocytes express at the surface of their membrane the Fas ligand (Fas-L) which binds Fas through an autocrine or paracrine mode. This domain is responsible for coupling the death receptor to either a cascade of caspases induction of apoptosis, or to the activation of kinase-signaling pathways, resulting in gene expression through NF-κB or activator protein 1 (AP-1) (Zimmermann and Green, 2001). It was reported that stimulation of TNFR on cultured dopamine neurons induces nuclear translocation of NFU-B in association with triggering of apoptosis(Hunot et al., 1997). Upon formation of the Fas/Fas-L complex, trimerization of Fas receptor takes place with subsequent recruitment of the Fas-associated death domain (FADD) adaptor protein. Caspases-8 is then dimerized by FADD, promoting caspase-8 activation which eventually activates the 'effector' members of the caspase family such as caspase-3 leading to cellular destruction. (Blatt and Glick, 2001). Other studies have further shown that up-regulation of Fas-L expression in neurons consistently occurs as an early step in apoptosis induced by injury or stress (Raoul et al., 2000). The control of induction of Fas-L expression is achieved by at least two pathways: the first involves Jun amino-terminal kinase (JNK); (Le Niculescu et al., 1999), the second protein kinase B (PKB or Akt) (Brunet et al., 1999).

Basically, the $p75^{NTR}$ is widely expressed in the brain during neonatal development, but its expression in the adult brain was found to be restricted to a few neuronal populations such as the basal forebrain and striatal cholinergic neurons (Friedman, 2010). A transgenic mouse expressing the $p75^{NTR}$ intracellular domain was generated. Interestingly this animal model showed that the ability of $p75^{NTR}$ to cause apoptosis is restricted not only to certain neuronal types, but also to certain temporal windows during the life of these neurons (Majdan et al., 1997). However, many pathological conditions such as neurodegenerative diseases, ischemia, seizures and traumatic injuries have been shown to induce this receptor, suggesting their important role in mediating apoptotic responses of postmitotic well differentiated neurons (Roux et al., 1999).

Recently, the expression of $p75^{NTR}$ under these pathological conditions was reported to be controlled by inflammatory cytokines such as IL-1β and TNF-α that have been found to be up-regulated prior to $p75^{NTR}$ induction been and governed by cell type specific (Choi and Friedman 2009). The first reports on the apoptotic effect of $p75^{NTR}$ came from in vitro studies on neuronal cell lines (Barrett and Bartlett, 1994; Bredesen and Rabizadeh, 1997). In response to proneurotrophins, this receptor can form complexes with the co-receptor sortilin to mediate many diverse cellular functions such as survival, apoptosis, and axonal growth, depending on the cellular context (Volosin et al., 2006). Confirming such apoptotic role, the studies demonstrated that the pro-nerve growth factor (proNGF) extracted from brain lysates of human Alzheimer's disease patients, or from rat brains following seizures, elicited death of cultured neurons through stimulation of $p75^{NTR}$ (Pedraza and others 2005; Volosin et al., 2006).

The intracellular signal transduction pathway of $p75^{NTR}$ is unlikely to be the same as for Fas since $p75^{NTR}$ has a different death domain (Liepinsh et al., 1997). In this context, the transcription factor nuclear factor kappa B (NFκ-B) is implicated to modulate neuronal apoptosis as an additional putative $p75^{NTR}$-associated signal-transduction element (Taglialatela et al., 1997, 1998). Interestingly, a novel caspase-dependent that does not involve the death domain, but Bcl-2 has recently been described (Coulson et al., 1999).

IV.2. Mitochondria Mediated-Pathway

Special attention is placed on mitochondria as organelles that, on the one hand, act as bioenergetic battery to the cell but, on the other, drive cellular signaling leading to various types of death outcome (Nagley et al., 2010). Mitochondria have an intimate regulatory role in programmed cell death that could be mediated via release of proapoptotic agents and/or disruption of cellular energy by partial failure of the respiratory chain complexes, alteration in intracellular Ca^{2+} homeostasis, regulating the activation of caspases, and increased releasing of reactive oxygen species (ROS) (Kowaltowski et al., 2004)The central position of mitochondria in apoptotic cascade give rise to upstream signaling of mitochondria includes interactions between Bcl-2 family members that together with impairment of function(s) sustained directly by mitochondria, culminate to activate downstream signals. Both upstream and downstream signaling of mitochondria, seem to exhibit cross-talk. In other words, the progression in one may influence simultaneously progression of the other in a positive or negative manner (Nagley et al., 2010). Under normal physiological conditions, 1–2% of molecular oxygen consumed by mammalian cells is metabolized to ROS via electron leakage from the mitochondrial electron transport chain. Therefore, mitochondria are a major source of endogenous ROS in human and animal cells (Liu et al., 2009). This pathway has been elucidated to be triggered directly in mitochondria by a variety of cellular stressors such as ROS, Ca^{2+}, nitric oxide and peroxynitrite, respiratory chain inhibitors, ganglioside D3 (GD3), arachidonate, fatty acids, bilirubin, bile salts, β-amyloid protein, etc.) (Waldmeier, 2003). Increased levels of oxidative radicals, particularly in the presence of increased intramitochondrial Ca^{2+}, can induce apoptosis by causing cross linking of protein thiols in the mitochondrial inner membrane with further stimulation of generation of ROS. Also, ROS induces peroxidation of cardiolipin, a phospholipid that is impeded only in the mitochondrial inner membrane, resulting in alteration of fluidity and other biophysical properties of mitochondrial membranes mainly a fall in the mitochondrial membrane potential ($\Delta\Psi_M$), and induce a decrease in the activity of respiratory chain complex III (Paradies et al., 2001) and complex IV (Paradies et al., 2000). Recently, complex I deficit was found not only to exert direct effect on cytochrome *c* release and consequent cell death, but rather it sensitizes neurons to mitochondrial-dependent apoptosis in response to the cell death agonist Bax through mitochondrial oxidative damage (Perier et al., 2005). In neurons, ROS were reported to also elicit p53 activation with subsequent upregulation of the BH3-only protein, that translocate to the mitochondria and activate the proapoptotic Bax under these conditions (Nagley et al., 2010). Thus, such reciprocal relationship between the mitochondria and the reactive oxygen species (ROS) creates a vicious circle of apoptotic machinery (Tatton and Olanow,1999). However, such relationship must be justified according to the severity of oxidative stress insult. In response to low intensity oxidative stress, neurons trigger the intrinsic apoptotic pathway. While, under higher levels of oxidative stress, activation of alternate cell death pathways such as necrosis or autophagy, is a more likely outcome, where high ROS levels in itself are found to inactivate caspases or lead to a decrease in ATP production due to severe mitochondrial damage (Nicholls et al., 2008). In contrast to the extrinsic pathway, the intrinsic signals usually operate by triggering the release of proteins from the intermembrane space of the mitochondria to the cytosol (Olson and Kornbluth, 2001). Physiologically, cytochrome c (cyt c) exists in the mitochondrial inter-membrane space where it acts in the electron transport chain to transfer electrons from complex III to complex IV (Vaux, 2010). The initial key step in this pathway is the release of cytochrome *c* into the cytosol. The mechanisms by which cytochrome *c* release in mitochondrial apoptosis are supposed to be caused by disruption of the outer mitochondrial membrane (OMM) and/or by the mitochondrial permeability transition (MPT), that is controlled by a voltage- and Ca^{2+}-sensitive megapore, referred to as the permeability transition pore (PTP) (Blatt and Glick, 2001). The integrity of the outer mitochondrial membrane is mainly controlled by the balance between different Bcl-2 family members. On triggering apoptosis, the pro-apoptotic Bcl-2 family members (Bad, Bax, Bid) are activated, delocalized from cytosol or the cytoplasmic surface of OMM to the mitochondria where they induce the release of cytochrome *c* (and other proteins such as apoptosis-inducing factor, AIF) from the intermembrane space. On the other hand, the anti-apoptotic proteins (Bcl-2, Bcl-xL) that reside on OMM, work to prevent cytochrome *c* release from mitochondria, and thereby protect cells from death (Kluck et al., 1997; Yang et al., 1997). Subsequent events involve cytochrome *c* binding to the adaptor protein Apaf-1 (apoptotic protease activating factor-1) forming a high molecular mass cytoplasmic complex referred to as the "apoptosome" (Acehan et al., 2002). recruitment of caspase-9 in this complex and then proteolytically activated (Lossi and Merighi, 2003) On molecular basis, binding of cytochrome *c* to the central apoptotic regulator Apaf-1, promotes its oligomerization and activation. The active Apaf-1 in turn promotes oligomerization and activation of caspase 9 in a reaction r equiring the ATPase activity. Caspase 9 subsequently activates the 'effector' caspases such as 3, 6, and 7 which ultimately lead to cell death (Olson and Kornbluth, 2001). The role of cytochrome *c* is further supplemented by novel insights into a causal role of caspase-independent apoptosis signaling pathway, such as the mitochondrial release of apoptosis inducing factor (AIF), in delayed neuronal death after cerebral ischemia. Other pro-apoptotic proteins released from the mitochondria are well recognized; the first one is termed Smac (second mitochondrial activator of caspases)/Diablo was found to translocate into the cytosol in cells undergoing apoptosis (Du et al., 2000). Another one is the HtrA2/Omi that acts biologically as a chaperone and serine protease that re-folds or degrades misfolded mitochondrial proteins, and has the ability to bind to IAPs, but its over-expression increased neuronal sensitivity to apoptosis due to presence of a serine protease in the cytoplasm, rather than HtrA2/Omi having a physiological role as an IAP regulator (Hegde et al., 2003) Recently, more attention has been paid for the mitochondrial dynamics which is an essential function in maintaining cell viability. In neurons, mitochondria may need to travel long distances from the cell body to dendritic and axonal processes, where they serve in maintaining calcium buffering and cycling of reserve pool synaptic vesicles. The regulation of mitochondrial dynamics requires a delicate balance of two opposing processes referred to as fusion and fission. Mitochondrial fusion is a fundamental cellular process, involves coordinated fusion of both outer and inner membranes of adjacent mitochondria and is essential for embryonic development. The fission process also appears to play important roles in proper mitochondrial localization, prevention of the improper segregation of mitochondria in dividing cells and is necessary for proper recruitment of mitochondria to nerve terminals. (Chan, 2006). Disturbed balance of mitochondrial dynamics causes severe mitochondrial behavioral change that leads to cellular perturbation and apoptosis in many neurodegenerative diseases. Recent studies found decreased expression levels of mitochondrial fusion proteins, including MFN-1, MFN-2 while levels of mitochondrial fission protein FIS1 is increased (Du and Yan, 2010) , indicating that mitochondrial fission could be involved in induction of programmed cell death by a mechanism thought to enhance permeabilization of the mitochondrial outer membrane and the subsequent apoptotic machinery, while fusion process may play a protective role in apoptosis. (Chan, 2006).

IV.3. Endoplasmic Reticulum Mediated-Pathway

In neurons, as in all eukaryotic cells, the endoplasmic reticulum (ER) is important for the synthesis, folding and post-translational modification of transmembrane and secreted proteins (Ron and Walter, 2007). Also, it serves as the major cellular Ca^{2+} store and the site of biosynthesis of steroids, cholesterol and other lipids, thus playing a crucial role in organelle biogenesis and signaling (Chang et al., 2006). ER stress is caused by a perturbation of any of these major functions of the ER, and accumulation of unfolded proteins in the ER lumen is a shared hallmark of disturbance of these functions (Nagley et al., 2010). Impairment of protein folding in the endoplasmic reticulum (ER) triggers the the combined actions of signal transduction pathways named the unfolded protein response (UPR) (Schroder, 2008), At least three distinct signalling pathways are involved, and each one is mediated by ER integral membrane proteins that sense the protein folding status in the lumen and transmit this information to the cytosol. These UPR sensor proteins are activating transcription factor 6 (ATF6), PRK-like ER kinase (PERK), and inositol-requiring kinase 1 (IRE1) (Urano et al., 2000; Ma et al., 2002; Shen et al., 2005); which either fixes the problem or induces apoptosis through activation of the c-Jun-N-terminal kinase (JNK) pathway, and/or the transcription factor C/EPB homologous protein (CHOP), which is thought to act through downregulation of Bcl-2 (Waldmeier, 2003). Recently, it has been proposed that the balance between cell survival and cell death is controlled by the relative ratio of pro-apoptotic to pro-survival Bcl-2 proteins processing at the ER membrane (Oakes et al., 2006). Tilting such balance towards PCD may take time, perhaps explaining why prolonged UPR is necessary for cell death. An alternative explanation is that the levels of CHOP mediate the transition from pro-survival to pro-apoptosis. CHOP mRNA and in addition the two sets of proteins are relatively unstable compared to other UPR proteins; hence, prolonged ER stress may be required before a critical threshold for apoptosis is reached (Rutkowski et al., 2006). Another factor in ER-mediated apoptosis is disturbed Ca^{2+} homeostasis or signaling, leading to activation of the calcium-responsive cysteine protease m-calpain, which in turn activates a procaspase localized at the external side of the ER and cleaves Bcl-X_L to a proapoptotic, presumably MMP-inducing species. Likewise, the Golgi apparatus may also function as a stress sensor triggering apoptosis, although this remains to be clarified. Golgi membranes harbour death receptors normally occurring at plasma membranes, caspase-2 and a ganglioside synthase capable of converting ceramide into GD3, which can translocate into mitochondria to induce mitochondrial membrane permeabilization (MMP). Finally, there is also some evidence that lysosomal enzymes like, e.g., cathepsins can trigger downstream apoptotic processes. On the other hand, stimulation of cell surface receptors for cytokines, neurotrophins, or growth factors mediate survival signals through the PI3K/protein kinase B (AKT) pathway, which upregulates Bcl-2 and oppose apoptotic signals. In addition, ligation of death receptors can activate nuclear factor-κB (NF-κB), which opposes apoptosis through induction of antiapoptotic Bcl-2 family members and some inhibitor of apoptosis proteins (IAPs). NF-κB is held in check by the inhibitor of NF-κB (IkB), which itself is controlled by a kinase (IkB kinase, IKK) phosphorylating and inactivating IkB (Waldmeier, 2003). PCD may also result from proteolytic activation of specific caspases located in the ER. Caspase-12 in rodents is bound to the ER membrane and may be activated by a similar process to the death receptor pathway. Activated IRE1 recruits TNF receptor-associated factor 2 (TRAF2) from pro-caspase-12, resulting in its activation, leading to subsequent cleavage of caspase-9 and caspase-3 independent of classical activation of the intrinsic mitochondrial pathway (Nagley et al., 2010). In this pathway, the initiator of caspase cascade is caspase-12 that is located on the cytoplasmic side of the ER (Nakagawa et al., 2000). The stress inducers lead the translocation of caspase-7 from the ER surface. Caspase-7 activates procaspase-12 by cleaving its prodomain through interaction. The activated caspase-12 then activates procaspase-9, and the activated caspase-9 in turn activates caspases-3, -6 and -7 (Rao et al., 2001; Rao et al., 2002). In this activation pathway, no Cytochrome *c* is released from mitochondria (Chowdhury et al., 2008).

V. Neuronal Apoptosis and Prospective Therapy

Progressive neuronal loss and neuronal dysfunction are not only key features of several neurodegenerative diseases but even in the natural phenomenon of aging; in which the adequate disease-modifying therapy is lacking for most of them. (Marks and Berg, 1999; Sastry and Rao, 2000; Waldmeier and Tatton, 2004). In terms of drug development for treatment of neurodegenerative disorders, apoptosis intervention opens new era. The role of apoptosis has been established either in those of chronic behavior such as alzheimer's disease (AD), parkinson's disease (PD) and amyotrophic lateral sclerosis (ALS), or in the more acute conditions such as cerebral ischemia, traumatic brain injury (TBI), and spinal cord injury (SCI). As clarified before; programmed cell death could play a "double agent" role where in some circumstances it behaves as "defense mechanism" to remove the affected dysfunctional neurons meanwhile in other conditions it lead to aberrations in the apoptotic machinery in certain neurons and thus cause "death by mistake". In both situations, intervention might be beneficial, by encouraging self-repair mechanisms in the first way and by preventing unwanted cell death in the other way. The concept that apoptosis might be the direct cause of, or intimately contribute to these incurable neuronal disorders, gives rise to the hope for developing promising agents in a rational way to cut off the underlying pathogenesis rather than attenuate or compensate for some symptoms. Many questions must be raised and answered before starting to introduce and investigate a new antiapoptotic agent. The basic question is that; does apoptosis actually account for the underlying pathogenesis? If the answer is yes; so the next question is; which type of apoptotic pathways and molecular machinery are implicated? Although needs much effort, but once the answer for this question is explored; this leads to ask if; are there disease-mimic or -relevant animal models that is suitable to evaluate the developed drugs? (Waldmeier, 2003). Of course further research will undoubtedly elucidate the molecular mechanisms underlying cell cycle regulation of neuronal cell death and degeneration (Becker and Bonni; 2004) that will provide schematic approaches to design therapy that could interfere the apoptotic cascade at target crucial point(s) such as 1) eliminating the apoptosis-triggering factor(s) (e.g. reactive oxygen species ROS, intracellular calcium Ca^{2+}, excitatory neurotransmitters as glutamate....etc), 2) maintaining mitochondrial respiratory chain activity, closure of the PTP and preservation of the mitochondrial membrane potential ($\Delta\Psi_M$), 3) modulation of apoptosis-related protein(s) either by direct binding to them or by modifying their expression and/or activity properties. Such approaches have the advantage of providing benefit regardless of the specific etiologic basis of neurodegeneration (Tatton and Olanow; 1999). However, there is still major dogma that in many circumstances the neuronal cell loss alone does not fully explain the neurological deficits. Thus, the underlying pathogenesis is further viewed as a network of events that ultimately reduce the complexity of interneuronal connections and the capacity of affected neurons to support their electrophysiological and synaptic functions. Such concept is further supported by the observation that neurological deficits become detectable only when 50–70% of the functional neurons have been lost. This could be partially explained by the compensatory neuronal circuit reorganization through expanding their individual connective interactions and functional capacities (Waldmeier and Tatton, 2004) In the way to study aniapoptotic agents; some of the already approved drugs are recently reported to exhibit additional antiapoptotic properties beside their original effects such as rasagiline (Maruyama et al., 2001; Youdim et al., 2001), desmethyldeprenyl (Tatton and Chalmers-Redman, 1996; Tatton et al., 2002) and 2-heptylmethylpropargylamine (Boulton et al., 1998; Zhang et al., 1999). All of them are compounds indicated in PD with qualitatively similar monoamine oxidase-B (MAO-B) inhibitor activity; this may make it difficult to differentiate their neuroprotective effect from the symptomatic one (Waldmeier, 2003). Many other compounds are well investigated for their basic disease-modifying role as antiapoptotics; one is coenzyme Q10, an electron acceptor for mitochondrial complexes I and II and an antioxidant, has been recently investigated as a promising neuroprotective agent for neurodegenerative disorders such as parkinson's disease (Shults, 2003; Abdin and Hamouda, 2008) Also, Riluzole, the glutamate release inhibitor, extends survival of ALS patients (Jackson et al., 2002). Another compound is memantine, acts as non-competitive antagonist of N-methyl-d-aspartate (NMDA) glutamate receptors, for the treatment of moderate to severe AD (Jarvis and Figgitt, 2003). Surprisingly, minocycline, a second-generation tetracycline antibiotic, has been discovered to inhibit ROS release, decrease activity of inducible nitric oxide synthase and reduces the activities of p38-mitogen activated protein kinase as well as caspases-1 and –3, presumably by interfering with their upstream activation, so sometimes referred to as a caspase inhibitor. In addition, it prevents mitochondrial permeability transition-mediated cytochrome *c* release (Zhu, et al., 2002). Owing to its promising effects, minocycline is reported as a candidate in Phase III trial in ALS (Thomas, et al., 2003). Among the newly investigated agents; one namely flupirtine that is launched as a centrally-acting, non-opioid analgesic with muscle relaxant properties. It is proposed to be NMDA receptor antagonist of unknown nature, perhaps through an action on their redox (Osborne et al., 1998). Flupirtine protects neuronal cells from apoptosis induced by glutamate (Perovic et al., 1996), prevent mitochondria from oxidative stress insult (Schlüter et al., 2000), and could upregulates Bcl-2 and increases glutathione levels. It could improve some cases that do not express NMDA receptors, such as ischemia induced apoptosis (Osborne et al., 1998), indicating that it has an antiapoptotic property independent of its NMDA antagonistic effect, which needs to be illuminated. The bioavailable and brain penetrating anti-apoptotic compounds that could reach the clinical stages of exploration is still rather limited and inconclusive. Preclinical projects are conducted to evaluate analogues of some antiapoptotic molecules such as Bcl-2 and glia derived neurotrophic factor (GDNF). Other studies are on going to develop modulators for the apoptotic protease activating factor-1 (Apaf-1), specific inhibitors of caspases 8 and 9 and even broad spectrum caspase inhibitors, and JNK inhibitors (Waldmeier and Tatton, 2004). Considering that proNGF has proapoptotic activity; several studies have shown that infusion of a newly developed anti-proNGF antibodies provided significant protection from neuronal cell death after different types of injury such as corticospinal lesions or seizures by specific recognition of the pro-domain of proNGF (Harrington et al., 2004; Volosin et al., 2008). The glyceraldehyde-3-phosphate dehydrogenase (GAPDH) is a housekeeping glycolytic enzyme but also has several functions that are unrelated to its role in energy production, one of which is its implication in p53-dependent apoptosis (Waldmeier and Tatton, 2004). Recently, CGP3466B was shown to bind directly to GAPDH (Kragten et al., 1998); blocking its additional activity in mediating p53-dependent apoptosis (Chen et al., 1999; Tatton et al., 2000), initiation of oxidative stress, decreasing Bcl-2, and increasing in mitochondrial Bax caused by trophic withdrawal in PC12 cells (Tatton et al., 2002). In the same way; TCH346, a tricyclic propargylamine derivative, prevents apoptosis induced in a variety of cell lines by either; direct binding and stabilization of the dimeric form of GAPDH, or by indirect prevention of its upregulation and nuclear translocation. It also, prevents increased mitochondrial membrane permeability, as indicated by the maintenance of mitochondrial membrane potential (Kragten et al., 1998; Carlile et al., 2000l; Wadia et al., 1998). CPI-1189 was synthesized as a novel antioxidant related to the spin-trapping agent phenyl-N-tert-butylnitrone, but also has been suggested to oppose apoptotic effects of TNF-α. Its antiapoptotic properties are contributed to inhibition of IL-1²-induced p38-MAPK phosphorylation and enhancement of extracellular-signal-related kinase (ERK) activity that in turn known to induce the antiapoptotic Bcl-2. However, the development of CPI-1189 in neurodegenerative diseases was declined and redirected to research area on neuropathic pain. The only compound presently in clinical phase III development is CEP-1347, an orally active staurosporin derivative that inhibits the JNK pathway through inhibition of mixed-lineage kinases (Mucke, 2003). Introducing drugs that selectively inhibit caspases could reduce neuronal apoptosis in neurodegenerative diseases (Waldmeier and Tatton, 2004). The neuronal specific inhibitor of apoptosis (NAIP), a member of the inhibitor of apoptosis protein (IAP) family, potently inhibits caspase-3 and -7 (Maier et al., 2002). It has been found that adenovirally-mediated overexpression of NAIP (Xu et al., 1998), or the X chromosome-linked IAP (XIAP) (Xu et al., 1999), reduces the loss of hippocampal neurons following transient ischemia. As potent and selective non-peptidic inhibitors for caspase-3 and -7; isatin sulfonamides compounds have been described, but they standstill without further evaluation has yet been reported (Concha and Abdel-Meguid, 2002) Pifithrin-α is a synthetic inhibitor of p53-induced transcriptional activation, designed originally to protect non-cancerous cells from genomic stress inflicted by cancer therapy (Komarov et al., 1999). A series of analogs of pifithrin-a with similar or higher potency was recently reported (Zhu et al., 2002). In view of the implication of p53 in several neurodegenerative diseases and because it penetrates the blood–brain barrier (BBB), this compound seems to meet considerable interest. However, pifithrin-α might be expected to promote genomic instability and disturb the cell cycle (Bassi et al., 2002). Thus, its safety for long-term application remains to be determined and its neuronal rescue effect must be justified for preservation of function or not. Nevertheless, it might prove useful for short-term administration in stroke or trauma (Waldmeier, 2003). In this context; although there is respect to antiapoptotic therapy, but there is scarcity of information about long-term effects on the physiological apoptotic processes, that is a problematic issue. Possible scenarios range from the consequences of preserving dysfunctional cells to a shift from apoptosis to necrosis, to promotion of preexisting or outright induction of cancer. It appears reasonable to assume that the corresponding risks vary depending on the pathway(s) with which a drug interferes, and how much downstream in the apoptotic cascade(s) this occurs. For example, the danger of preserving dysfunctional cells or shifting from apoptosis to necrosis might be expected to be greater, the more downstream a drug acts. The fear that cancer might be promoted or induced is based on the fact that in many cancers, key apoptotic regulators including the Bcl-2 family, Apaf-1, and p53 are altered or dysregulated (Bunz, 2001). Nevertheless, there is ample reason to be optimistic about the prospects for antiapoptotic drugs to show therapeutic benefit in one or another neurodegenerative disease. The more we learn about the involved pathways, the better we will be able to design compounds that interfere with disease processes without causing trouble (Waldmeier, 2003).

References

Abdin, AA; Hamouda, HE. Mechanism of the neuroprotective role of coenzyme Q10 with or without L-dopa in rotenone-induced parkinsonism. *Neuropharmacology*. 2008, 55(8):1340-1346.

Acehan, D; Jiang, X; Morgan, DG; Heuser, JE; Wang, X., Akey, CW. Three dimensional structure of the apoptosome. Implications for assembly, procaspase-9 binding, and activation. *Mol. Cell*. 2002, 9:423-432.

Adams, JM; Cory, S. The Bcl-2 protein family: arbiters of cell survival. *Science*. 281:1322-1326.

Aloyz, R; Bamji, S; Pozniak, C; Toma, JG; Atwal, J; Kaplan, DR; Miller, FD. p53 is essential for developmental neuron death as regulated by the trkA and p75 neurotrophin receptors. *J. Cell Biol.* 1998, 143:1691-1703.

Antonsson, B; Conti, F; Ciavatta, AM; Montessuit, S; Lewis, S; Martinou, I; Bernasconi, L; Bernard, A; Mermod, J-J; Mazzei, G; Maundrell, K; Gambale, F; Sadoul, R; Martinou, J-C. Inhibition of Bax channel-forming activity by Bcl-2. *Science*. 1997, 277:370-372.

Appella, E; Anderson, CW. Post-translational modifications and activation of p53 by genotoxic stresses. *Eur. J. Biochem*. 2001, 268:2764-2772.

Baehrecke, EH. How death shapes life during development. *Nat Cell Biol* 2002, 3(10):779-787.

Baffy, G; Miyashita, T; Williamson, JR; Reed, JC. Apoptosis induced by withdrawal of interleukin-3 (IL-3) and an IL-3-dependent hematopoietic cell line is associated with repartitioning of intracellular calcium and is blocked by enforced Bcl-2 oncoprotein production. *J. Biol. Chem*. 1993, 268:6511-6519.

Barde, YA. Neurotrophins: a family of proteins supporting the survival of neurons. *Prog.Clin. Biol. Res.* 1994, 390:45-56.

Barrett, GL; Bartlett, PF. The p75 nerve growth factor receptor mediates survival or death depending on the stage of sensory neuron development. *Proc. Natl. Acad. Sci. USA* 1994, 91(14):6501-6505.

Bassi, L; Carloni, M; Fonti, E; Palma, DLP; Meschini, R; Palitti, F. Pifithrin-alpha, an inhibitor of p53, enhances the genetic instability induced by etoposide (VP16) in human lymphoblastoid cells treated in vitro. *Mutat. Res.* 2002, 499:163-176.

Becker, EB; %5BAuthor%5D"Bonni, A. Cell cycle regulation of neuronal apoptosis in development and disease. *Prog. Neurobiol* 2004, 72(1):1-25.

Bernier, PJ; Parent, A. Bcl-2 protein expression as a marker of neuronal immaturity in postnatal primate brain. *J. Neurosci.* 1998, 18:2486-2497.

Blatt, NB; %5BAuthor%5D"Glick, GD. Signaling pathways and effector mechanisms pre-programmed cell death. *Bio. org Chem.* 2001, 9(6):1371-1384.

Bonfoco, E; Krainc, D; Ankarcrona, M; Nicotera, P; Lipton, SA. Apoptosis and necrosis: two distinct events induced, respectively, by mild and intense insults with N-methyl-D-aspartate or nitric oxide/superoxide in cortical cell cultures. *Proc. Nat. Acad. Sci. USA* 1995, 92:7162-7166.

Boulton, AA; Yu, PH; Davis, BA; Paterson, IA; Li, XM; Juorio, AV; Durden, DA; Dyck, LE. Aliphatic N-methylpropargylamincs: onoaminc oxidasc-B inhibitors and antiapoptotic drugs. *Adv. Pharmacol.* 1998, 42:308-311.

Bredesen, DE; Rabizadeh, S. $p75^{NTR}$ and apoptosis: Trk-dependent and Trk-independent effects. *Trends Neurosci.* 1997, 20(7):287-290.

Bredesen, DE; Rao, RV, Mehlen, P. Cell death in the nervous system. *Nature* 2006, 443(7113):796-802.

Brooks, CL; Gu, W. Ubiquitination, phosphorylation and acetylation: the molecular basis for p53 regulation. *Curr. Opin. Cell Biol.* 2003, 15:164-171.

Brunet, A; Bonni, A; Zigmond, MJ; Lin, MZ; Juo, P; Hu, LS; Anderson, MJ; Arden, KC; Blenis, J; Greenberg, ME. Akt promotes cell survival by phosphorylating and inhibiting a Forkhead transcription factor. *Cell.* 1999, 96(6):857-868.

Budihardjo, I; Oliver, H; Lutter, M; Luo, X; Wang, X. Biochemical pathways of caspase activation during apoptosis. *Annu. Rev. Cell Dev. Biol.* 1999, 15:269-290.

Bulat, N; Widmann, C. Caspase substrates and neurodegenerative diseases. *Brain Res. Bull.* 2009, 80(4-5):251-267.

Bunz, F. Cell death and cancer therapy. *Curr. Op. Pharmacol.* 2001, 1:337-341.

Buss, RR;Sun, W; Oppenheim, RW. Adaptive roles of programmed cell death during nervous system development. *Annu. Rev. Neurosci.* 2006, 29:1-35.

Carlile, GW; Chalmers-Redman, RM; Tatton, NA; Pong, A; Borden, KE; Tatton, WG. Reduced apoptosis after NGF and serum withdrawal: Conversion of tetrameric glyceraldehyde-3-phosphate dehydrogenase to a lycer. *Mol. Pharmacol.* 2000, 57:2-12

Chan, DC. Mitochondria: dynamic organelles in disease, aging, and development. *Cell.* 2006, 125:1241-1252.

Chen, RW; Saunders, PA; Wei, HF; Li, ZW; Seth, P; Chuang, D. Involvement of glyceraldehyde-3-phosphate dehydrogenase (GAPDH) and p53 in neuronal apoptosis: evidence that GAPDH is upregulated by p53. *J. Neurosci.* 1999, 19:9654-9662.

Choi, S; Friedman, W. Inflammatory cytokines IL-1b and TNF-a regulate $p75^{NTR}$ expression in CNS neurons and astrocytes by distinct celltype-specific signalling mechanisms. *ASN Neuro.* 2009, 1(2):e00010.

Chowdhury, I; %5BAuthor%5D"Tharakan, B; %5BAuthor%5D"Bhat, GK. Caspases - an update. *Comp. Biochem. Physiol. Biochem. Mol.Biol.* 2008, 151(1):10-27.

Cohen, GM. Caspases: the executioners of apoptosis. Biochem. J. 1997, 326 (1):1-16.

Concha, NO; Abdel-Meguid, SS. Controlling apoptosis by inhibition of caspases. *Curr. Med. Chem.* 2002, 9:713-726.

Coulson, EJ; Reid, K; Barrett, GL; Bartlett, PF. p75 neurotrophin receptor-mediated neuronal death is promoted by Bcl-2 and prevented by Bcl-xL. *J. Biol. Chem.* 1999, 274:16387-16391.

Courtois, S; de Fromentel CC; Hainaut, P. p53 protein variants: structural and functional similarities with p63 and p73 isoforms. *Oncogene.* 2004, 23:631-638.

Cregan, SP; Dawson, VL; Slack, RS. Role of AIF in caspase-dependent and caspase-independent cell death. *Oncogene.* 2004, 23:2785-2796.

Cregan, SP; MacLaurin, JG; Craig, CG; Robertson, GS; Nicholson, DW; Park, DS; Slack, RS. Bax-dependent caspase-3 activation is a key determinant in p53-induced apoptosis in neurons. *J. Neurosci.* 1999, 19:7860-7869.

Crumrine, RC; Thomas, AL; Morgan, PF. Attenuation of p53 expression protects against focal ischemic damage in transgenic mice. J Cereb Blood Flow Metab. 1994, 14:887-891.

Culmsee, C; Mattson, MP. p53 in neuronal apoptosis. *Biochem. Biophys. Res. Commun.* 2005, 331(3):761-777.

Daugas, E; Susin, SA; Zamzami, N; Ferri, KF; Irinopoulou, T; Larochette, N; Prevost, MC; Leber, B; Andrews, D; Penninger, J; Kroemer, G. Mitochondrio-nuclear translocation of AIF in apoptosis and necrosis. *FASEB J.* 2000, 14:729-739.

Denecker, G; Hoste, E; Gilbert, B; Hochepied, T; Ovaere, P; Lippens, S; Van den Broecke, C; Van Damme, P; 'Herde, K; Hachem, JP; Borgonie, G; Presland, RB; Schoonjans, L; Libert, C; Vandekerckhove, J; Gevaert, K; Vandenabeele, P. Caspase-14 protects against epidermal UVB photodamage and water loss. *Nat. Cell Biol.* 2007, 9(6):666-674.

Du, CY; Fang, M; Li, YC; Li, L; Wang, XD. Smac, a mitochondrial protein that promotes cytochrome c-dependent caspase activation by eliminating IAP inhibition, *Cell.* 2000, 102:33-42.

Du, H; Yan, SS. Mitochondrial medicine for neurodegenerative diseases. *Int. J. Biochem. Cell Biol.* 2010, 42(5):560-572.

Eizenberg, O; Faber-Elman, A; Gottlieb, E; Oren, M; Rotter, V; Schwartz, M. p53 plays a regulatory role in differentiation and apoptosis of central nervous system-associated cells. *Mol. Cell Biol.* 1996, 16:5178-5185.

Ellerby, LM; Ellerby, HM; Park, SM; Holleran, AL; Murphy, AN; Fiskum, G; Kane, DJ; Testa, MP; Kayalar, C; Bredesen, DE. Shift of the cellular oxidation-reduction potential in neural cells expressing Bcl-2. *J. Neurochem.* 1996, 67:1259-1267.

Emerit, J; Edeas, M; Bricaire, F. Neurodegenerative diseases and oxidative stress. *Biomed. Pharmacother* 2004, 58(1):39-46.

Feddersen, RM; Ehlenfeldt, R; Yunis, WS; Clark, HB; Orr, HT. Disrupted cerebellar cortical development and progressive degeneration of Purkinje cells in SV40 T antigen transgenic mice. *Neuron.* 1992, 9:955-966.

Foghsgaard, L; Wissing, D; Mauch, D; Lademann, U; Bastholm, L; Boes, M; Elling, F; Leist, M; Ja¨a¨ttela¨, M. Cathepsin B acts as a dominant execution protease in tumor cell apoptosis induced by tumor necrosis factor. *J. Cell Biol.* 2001, 153:999-1010.

Friedman, WJ. Proneurotrophins, seizures, and neuronal apoptosis. *Neuroscientist.* 2010, 16(3):244-252.
Glucksmann, A. Cell deaths in normal vertebrate ontogeny. Biol Rev. 1951, 26:59-86.
Gonzalez-Garcia, M; Garcia, I; Ding, L; O' Shea S; Boise LH; Thompson CB; Nunez, G. Bcl-x is expressed in embryonic and postnatal neural tissues and functions to prevent neuronal cell death. *Proc. Natl. Acad. Sci. USA*_1995, 92:4304-4308.
Grana, X; Reddy, EP. Cell cycle control in mammalian cells: role of cyclins, cyclin dependent kinases (CDKs), growth suppressor genes and cyclin-dependent kinase inhibitors (CKIs). *Oncogene.* 1995, 11:211-219.
Gross, A; McDonnell, JM; Korsmeyer, SJ. BCL-2 family members and the mitochondria in apoptosis. *Genes Dev.* 1999, 13:1899-1911.
Hardwick, JM. Caspases in Programmed Cell Death. Encyclopedia Cancer. Second Edition. Volume 1. 2002, pp:369-373.
Harper, SJ; LoGrasso, P. Signalling for survival and death in neurones: the role of stress-activated kinases, JNK and p38. Cell Signal. 2001, 13:299-310.
Harrington, AW; Leiner, B; Blechschmitt, C; Arevalo, JC; Lee, R; Mörl, K; Meyer, M; Hempstead, BL; Yoon, SO; Giehl, KM. Secreted proNGF is a pathophysiological death-inducing ligand after adult CNS injury. *Proc. Natl. Acad. Sci. USA* 2004, 101(16):6226-6230.
Hartmann, A; Mouatt-Prigent, A; Vila, M; Abbas, N; Perier, C; Faucheux, BA; Vyas, S; Hirsch, EC. Increased expression and redistribution of the antiapoptotic molecule Bcl-xL in parkinson's disease. *Neurobiol. Dis.* 2002, 10(1):28-32.
Hass, R. Retrodifferentiation and cell death_ *Crit. Rev. Oncog.* 1994, 5(4):359-371.
Hayes, TE; Valtz, NL; McKay, RD. Downregulation of CDC2 upon terminal differentiation of neurons. *New Biol.* 1991, 3:259-269.
Hegde, R; Srinivasula, SM; Zhang, ZJ; Wassell, R; Mukattash, R; Cilenti, L; DuBois, G; Lazebnik, Y; Zervos, AS; Fernandes-Alnemri, T; Alnemri, ES. Identification of Omi/HtrA-2 as a mitochondrial apoptotic serine protease that disrupts inhibitor of apoptosis protein-caspase interaction, *J. Biol. Chem.* 2002, 277:432-438.
Heintz, N. Cell death and the cell cycle: a relationship between transformation and neurodegeneration? *Trends Biochem. Sci.* 1993, 18:157-159.
Hennet, T; Bertoni, G; Richter, C; Peterhans, E. Expression of Bcl-2 protein enhances the survival of mouse fibrosarcoid cells in tumor necrosis factor-mediated cytotoxicity. *Cancer Res.* 1993, 53:1456-1460.
Hetman, M; Cavanaugh, JE; Kimelman, D; Xia, Z. Role of glycogen synthase kinase-3beta in neuronal apoptosis induced by trophic withdrawal. J Neurosci. 2000, 20:2567-2574.
Hockenbery, DM; Oltvai, ZN; Yin, X-M; Milliman, CL; Korsmeyer, SJ. Bcl-2 functions in an antioxidant pathway to prevent apoptosis. *Cell.* 1993, 75:241-251.
Hofseth, LJ; Hussain, SP; Harris, CC. p53: 25 years after its discovery, *Trends Pharmacol. Sci.* 2004, 25:177-181.
Hongisto, V; Smeds, N; Brecht, S; Herdegen, T; Courtney, MJ; Coffey, ET. Lithium blocks the c-Jun stress response and protects neurons via its action on glycogen synthase kinase 3. *Mol. Cell Biol.* 2003, 23:6027-6036.
Honig, LS; Rosenberg, RN. Apoptosis and neurologic disease. *Am. J. Med.* 2000, 108:317-330.

Huang, EJ; Reichardt, LF. Neurotrophins: roles in neuronal development and function. *Annu. Rev. Neurosci.* 2001, 24:677-736.

Hunot, S; Brugg, B; Ricard, D; Michel, PP; Muriel, MP; Ruberg, M; Faucheux, BA; Agid, Y; Hirsch, EC. Nuclear translocation of NF-kappaB is increased in dopaminergic neurons of patients with parkinson disease. *Proc. Natl. Acad. Sci. USA* 1997, 94(14):7531-7536.

Husseman, JW; Nochlin, D; Vincent, I. Mitotic activation: a convergent mechanism for a cohort of neurodegenerative diseases. *Neurobiol. Aging* 2000, 21:815-828.

Jackson, M; Lladó, J; Rothstein, JD. Therapeutic developments in the treatment of amyotrophic lateral sclerosis. *Expert Opin. Investig. Drugs.* 2002, 11:1343-1364.

Jarvis, B; Figgitt, DP. Memantine. *Drugs Aging.* 2003, 20:465-476.

Jeffers JR; Parganas, E; Lee, Y; Yang, C; Wang, J; Brennan, J; MacLean, KH; Han, J; Chittenden, T; Ihle, JN; McKinnon, PJ; Cleveland, JL; Zambetti, GP. PUMA is an essential mediator of p53-dependent and -independent apoptotic pathways, *Cancer Cell.* 2003, 4:321-328.

Joza, N; Susin, SA; Daugas, E; Stanford,WL; Cho, SK; Li, CY; Sasaki, T; Elia, AJ; Cheng, HY; Ravagnan, L; Ferri, KF; Zamzami, N; Wakeham, A; Hakem, R; Yoshida, H; Kong, YY; Mak, TW; Zuniga-Pflucker, JC; Kroemer, G; Penninger, JM. Essential role of the mitochondrial apoptosis-inducing factor in programmed cell death. *Nature.* 2001, 410:549-554.

Kane, DJ; Sarafian, TA; Anton, R; Hahn, H; Butler Gralla, E; Selverstone Valentine, J; Ord, T; Bredesen, DE. Bcl-2 inhibition of neural death: decreased generation of reactive oxygen species. *Science.* 1993, 262:1274-1277.

Kaplan, DR; Miller, FD. Neurotrophin signal transduction in the nervous system. Curr Opin Neurobiol. 2000, 10(3):381-391.

Kaufmann, JA; Bickford, PC; Taglialatela, G. Oxidative-stressdependent up-regulation of Bcl-2 expression in the central nervous system of aged Fisher-344 rats. *J. Neurochem.* 2001, 76:1099-1108.

Kerr, JF; Wyllie, AH; Currie AR. Apoptosis: a basic biological phenomenon with wide-ranging implications in tissue kinetics. *Br. J. Cancer.* 1972, 26(4):239-257.

Kharbanda, S; Pandey, P; Schofield, L; Israels, S; Roncinske, R; Yoshida, K; Bhartia, A; Yuan, Z-M; Saxena, S; Weichselbaum, R; Nalin, C; Kufe, D. Role for Bcl-x_L as an inhibitor of cytosolic cytochrome C accumulation in DNA damage-induced apoptosis. *Proc. Natl. Acad. Sci. USA* 1997, 94:6939-6942.

King, KL; Cidlowski, JA. Cell cycle and apoptosis: common pathways to life and death. *J. Cell Biochem.* 1995, 58:175-180.

Kluck, RM; Bossy-Wetzel, E; Green, DR; Newmeyer, DD. The release of cytochrome *c* from mitochondria: a primary site for Bcl-2 regulation of apoptosis. *Science.* 1997, 275:1132-1136.

Komarov, PG; Komarova, EA; Kondratov, RV; Christov-Tselkov, K; Coon, JS; Chernov, MV; Gudkov, AV. A chemical inhibitor of p53 that protects mice from the side effects of cancer therapy. *Science.* 1999, 285:1733-1737.

Konishi, Y; Lehtinen, M; Donovan, N; Bonni, A. Cdc2 phosphorylation of BAD links the cell cycle to the cell death machinery. *Mol. Cell.* 2002, 9:1005-1016.

Kowaltowski, AJ; Fenton, RG; Fiskum G. Bcl-2 family proteins regulate mitochondrial reactive oxygen production and protect against oxidative stress. *Free Radic Biol. Med.* 2004, 37(11):1845-1853.

Kragten, E; Lalande, I; Zimmermann, K; Roggo, S; Schindler, P; Müller, D; van Oostrum, J; Waldmeier, P; Fürst, P. Glyceraldehyde-3-phosphate dehydrogenase, the putative target of the antiapoptotic compounds CGP 3466 and R-(-)-deprenyl. *J. Biol. Chem.* 1998, 273:5821-5828.

Krajewski, S; Tanaka, S; Takayama, S; Schibler, MJ; Fenton, W; Reed JC. Investigation of the subcellular distribution of the bcl-2 oncoprotein: residence in the nuclear envelope, endoplasmic reticulum, and outer mitochondrial membranes. *Cancer Res.* 1993, 53:4701-4714

Krantic, S; Mechawar, N; Reix, S; Quirion, R. Apoptosis-inducing factor: a matter of neuron life and death. *Prog. Neurobiol.* 2007, (3):179-96.

Labbé, K; Saleh, M. Cell death in the host response to infection. *Cell Death Differ.* 2008, 15(9):1339-49.

Larsen, KE; Sulzer, D. Autophagy in neurons: a review. *Histo Histopathol.* 2002, 17:897-908.

Lee, R; Kermani, P; Teng, KK; Hempstead, BL. Regulation of cell survival by secreted proneurotrophins. *Science.* 2001, 294(5548):1945-1948.

Lee, Y; McKinnon, PJ. ATM dependent apoptosis in the nervous system. *Apoptosis.* 2000, 5:523-529.

Lees, E. Cyclin dependent kinase regulation. Curr. Opin. 1995, *Cell Biol.* 7:773-780.

Lemasters, JJ; Qian, T; He, L; Kim, JS; Elmore, SP; Cascio, WE; Brenner, DA. Role of mitochondrial inner membrane permeabilization in necrotic cell death, apoptosis, and autophagy. *Antioxid Redox Signal.* 2002, 4(5):769-781.

Le-Niculescu, H; Bonfoco, E; Kasuya, Y; Claret, FX; %5BAuthor%5D"Green, DR; %5BAuthor%5D"Karin, M. Withdrawal of survival factors results in activation of the JNK pathway in neuronal cells leading to Fas ligand induction and cell death. *Mol. Cell Biol.* 1999, 19(1):751-763.

Levrero, M; de Laurenzi, V; Constanzo, A; Gong, J; Wang JY; Melino, G. The p53/p63/p73 family of transcription factors: Overlapping and distinct functions. *J. Cell Sci.* 2000, 113:1661-1670.

Liepinsh, E; Ilag, LL; Otting, G; Ibáñez, CF. NMR structure of the death domain of p75 neurotrophin receptor. *EMBO J.* 1997, 16:4999-5005.

Lipton, SA; Nakanishi N. Shakespeare in love - with NMDA receptors? *Nature med.* 1999, 5:270-271.

Lipton, SA; Rosenberg, PA. Excitatory amino acids as a final common pathway for neurologic disorders. *N. Engl. J. Med.* 1994, 330:613-622.

Liu, CY; Lee, CF; Wei, YH. Role of reactive oxygen species-elicited apoptosis in the pathophysiology of mitochondrial and neurodegenerative diseases associated with mitochondrial DNA mutations. *J. Formos Med. Assoc* 2009, 108(8):599-611.

Liu, DX; Greene, LA. 2001. Neuronal apoptosis at the G1/S cell cycle checkpoint. *Cell Tissue Res.* 305, 217-228.

Loeffler, M; Daugas, E; Susin, SA; Zamzami, N; Metivier, D; Nieminen, AL; Brothers, G; Penninger, JM; Kroemer, G. Dominant cell death induction by extramitochondrially targeted apoptosis-inducing factor. *FASEB J.* 2001, 15:758-767.

Lohrum, MA; Vousden, KH. Regulation and function of the p53-related proteins: same family, different rules. *Trends Cell Biol.* 2000, 10:197-202.

Lossi, L; Merighi, A. In vivo cellular and molecular mechanisms of neuronal apoptosis in the mammalian CNS. *Prog. Neurobiol.* 2003, 69(5):287-312.

Ma, K; Vattem, KM; Wek, RC. Dimerization and release of molecular chaperone inhibition facilitate activation of eukaryotic initiation factor-2 kinase in response to endoplasmic reticulum stress. *J. Biol. Chem.* 2002, 277:18728-18735.

Maier, JK; Lahoua, Z; "Gendron, NH; Fetni, R; Johnston, A; Davoodi, J; Rasper, D; Roy, S; Slack, RS; Nicholson, DW; MacKenzie, AE. The neuronal apoptosis inhibitory protein is a direct inhibitor of caspases 3 and 7. *J. Neurosci.* 2002, 22:2035-2043.

Majdan, M; Lachance, C; Gloster, A; Aloyz, R; Zeindler, C; Bamji, S; Bhakar, A; Belliveau, D; Fawcett, J; Miller, FD; Barker, PA. Transgenic mice expressing the intracellular domain of the p75 neurotrophin receptor undergo neuronal apoptosis. *J. Neurosci.* 1997, 17(18):6988-6998.

Malanga, M; Pleschke, JM; Kleczkowska, HE; Althaus, FR. Poly(ADP-ribose) binds to specific domains of p53 and alters its DNA binding functions, *J. Biol. Chem.* 273:11839-11843.

Marks, N; Berg, MJ. Recent advances on neuronal caspases in development and neurodegeneration. *Neurochem. Int.* 1999, 35:195-220.

Martin SJ. Apoptosis. In Delves PJ, Roitt IM, eds. Encyclopedia of Immunology, ed 2. San Diego, Academic Press, 1998, pp 220-227.

Martin, LJ. Neurodegenerative Disorders. *Encyclopedia hum. brain.* 2003, 3:441-463.

Martin, LJ; Al-Abdulla, NA; Brambrink, AM; Kirsch, JR; Sieber, FE; Portera-Cailliau, C. Neurodegeneration in excitotoxicity, global cerebral ischemia, and target deprivation: a perspective on the contributions of apoptosis and necrosis. *Brain Res. Bull.* 1998, 46:281-309.

Maruyama, W; Youdim, MB; Naoi, M. Antiapoptotic properties of rasagiline, N-propargylamine-1(R)-aminoindan, and its optical (S)-isomer, TV1022. *Ann. N. Y. Acad. Sci.* 2001, 939:320-329.

Masoudi, R; Ioannou, MS; Coughlin, MD; Pagadala, P; Neet, KE; Clewes, O; Allen, SJ; Dawbarn, D; Fahnestock, M. Biological activity of nerve growth factor precursor is dependent upon relative levels of its receptors. *J. Biol. Chem.* 2009, 284(27):18424-18433.

Mattson, MP; Duan, WZ. "Apoptotic" biochemical cascades in synaptic compartments: roles in adaptive plasticity and neurodegenerative disorders. *J. Neurosci. Res.* 1999, 58:152-166.

Miller, FD; aplan, DR. Neurotrophin signalling pathways regulating neuronal apoptosis. *Cell Mol. Life Sci.* 2001, 8:1045-1053.

Miller, RL; James-Kracke, M; Sun, GY; Sun, AY. *Oxidative and Res.* 2009, 34(1):55-65.

Miramar, MD; Costantini, P; Ravagnan, L; Saraiva, LM; Haouzi, D; Brothers, G; Penninger, JM; Peleato, ML; Kroemer, G; Susin, SA. NADH oxidase activity of mitochondrial apoptosis-inducing factor. *J. Biol. Chem.* 2001, 276:16391-16398.

Moreira, PI; Zhu, X; Wang, X; Lee, HG; Nunomura, A; Petersen, RB; Perry, G; Smith, MA. Mitochondria: a therapeutic target in neurodegeneration. *Biochim. Biophys. Acta.* 2010, 1802(1), 212-220.

Morgan, DO. Cyclin-dependent kinases: engines, clocks, and microprocessors. *Annu. Rev. Cell Dev. Biol.* 1997, 13:261-291.

Moritani, T; Shrier, D; Wang, H; Fitzgerald, J; Echevarri, J; Hiwatashi, A; Kinoshita, T; Namaguchi, Y; Westesson, P. 211 Excitotoxic Mechanism in Pediatric Brain. *Neurographics.* 2002, Volume 2, Number 1.

Mucke, HA. CEP-1347 (Cephalon). *Idrugs.* 2003, 6:377-383.

Munoz-Pinedo, C; Carrion, AG; Goldstein, JC; Fitzgerald, P; Newmeyer, DD; Green, DR. Different mitochondrial intermembrane space proteins are released during apoptosis in a manner that is co-ordinately initiated but can vary in duration. *Proc. Natl. Acad. USA* 2006, 103:11573-11578.

Murphy, AN; Bredesen, DE; Cortopassi, G; Wang, E; Fiskum, G. Bcl-2 potentiates the maximal calcium uptake capacity of neural cell mitochondria. *Proc. Natl. Acad. Sci. USA* 1996, 93:9893-9898.

Nagley, P; Higgins, GC; Atkin, JD; Beart, PM. Multifaceted deaths orchestrated by mitochondria in neurones. *Biochim. Biophys. Acta* 2010, 1802(1):167-85.

Nagy, Z. Cell cycle regulatory failure in neurones: causes and consequences. *Neurobiol. Aging* 2000, 21:761-769.

Nakagawa, T; Zhu, H; Morishima, N; Li, E; Xu, J; Yankner, BA; Yuan, J. Caspase-12 mediates endoplasmic reticulum specific apoptosis and cytotoxicity by amyloidbeta. *Nature.* 2000, 403:98-103.

Nicholls, DG. Oxidative stress and energy crises in neuronal dysfunction, *Ann. N. Y. Acad. Sci.* 2008, 1147:53-60.

Nicholson, DW. Caspases and Cell Death. *Encyclopedia Biol. Chemist.* Volume 1. Elsevier Inc. 2004, pp:319-327.

Nicotera, P. Cell Death by Apoptosis and Necrosis. *Encyclopedia Biol. Chem.* 2004, 1:352-355.

Oakes, SA; Lin, SS; Bassik, MC. The control of endoplasmic reticulum-initiated apoptosis by the BCL-2 family of proteins. *Curr. Mol. Med.* 2006, 6:99-109.

Olson, M; Kornbluth, S. Mitochondria in apoptosis and human disease. *Curr. Mol. Med.* 2001, 1(1):91-122.

Orrenius, S; Zhivotovsky, B; Nicotera, P. Regulation of cell death: the calcium-apoptosis link. *Nat. Rev. Mol. Cell Biol.* 2003, 4:552-565.

Osborne, NN; Cazevieille, C; Wood, JP; Nash, MS; Pergande, G; Block, F; Kosinski, C; Schwarz, M. Flupirtine, a nonopioid centrally acting analgesic, acts as an NMDA antagonist. *Gen. Pharmacol.* 1998, 30:255-263.

Otera, H; Ohsakaya, S; Nagaura, Z; Ishihara, N; Mihara, K. Export of mitochondrial AIF in response to proapoptotic stimuli depends on processing at the intermembrane space. *EMBO J.* 2005, 24:1375-1386.

Paradies, G; Petrosillo, G; Pistolese, M; Ruggiero, FM. Reactive oxygen species generated by the mitochondrial respiratory chain affect the complex III activity via cardiolipin peroxidation in beef-heart submitochondrial particles. *Mitochondrion.* 2001, 1:151-159.

Paradies, G; Petrosillo, G; Pistolese, M; Ruggiero, FM. The effect of reactive oxygen species generated from the mitochondrial electron transport chain on the cytochrome c oxidase activity and on the cardiolipin content in bovine heart submitochondrial particles. *FEBS Lett.* 2000, 466:323-326.

Patapoutian, A; Reichardt, LF. Trk receptors: mediators of neurotrophin action. *Curr. OpinNeurobiol.* 2001, 11(3):272-280.

Pedraza, CE Podlesniy, P; Vidal, N; Arévalo, JC; Lee, R; Hempstead, B; Ferrer, I; Iglesias, M; Espinet, C. Pro-NGF isolated from the human brain affected by Alzheimer's disease induces neuronal apoptosis mediated by $p75^{NTR}$. *Am. J. Pathol.* 2005, 166(2):533-543.

Perier, C; "Tieu, K; Guégan, C; Caspersen, C; Jackson-Lewis, V; Carelli, V; Martinuzzi, A; Hirano, M; Przedborski, S; %5BAuthor%5D"Vila, M. Complex I deficiency primes Bax-dependent neuronal apoptosis through mitochondrial oxidative damage. *Proc. Natl. Acad. Sci. USA* 2005, 102(52):19126-19131.

Perovic, S; Pialoglou, P; Schroder, HC; Pergande, G; Muller, WE. Flupirtine increases the levels of glutathione and Bc1-2 in hNT (human Ntera/D1) neurons: mode of action of the drug-mediated antiapoptotic effect. *Eur. J. Pharmacol.* 1996, 317:157-164.

Pop, C; Salvesen, GS. Human caspases: activation, specificity, and regulation. *J. Biol. Chem.* 2009, 284(33):21777-21781.

Pozniak, CD; Barnabe-Heider, F; Rymar, VV; Lee, AF; Sadikot, AF; Miller, FD. p73 is required for survival and maintenance of CNS neurons. *J. Neurosci.* 2002, 22:9800-9809.

Pozniak, CD; Radinovic, S; Yang, A; McKeon, F; Kaplan, DR; Miller, FD. An anti-apoptotic role for the p53 family member, p73, during developmental neuron death. *Science.* 2000, 289:304-306.

Raff, MC; Barres, BA; Burne, JF; Coles, HSR; Ishizaki, Y; Jacobson, MD. Programmed cell death and the control of cell survival. *Philos. Trans R Soc. Lond B Biol. Sci.* 1994, 345:265-268.

Raina, AK; Hochman, A; Zhu, XW; Rottkamp, CA; Nunomura, A; Siedlak, SL; Boux, H; Castellani, RJ; Perry, G; Smith, MA. Abortive apoptosis in Alzheimer's disease. *Acta Neuropathol.* 2001, 101:305-310.

Rao, RV; Castro-Obregon, S; Frankowski, H; Schuler, M; Stoka, V; del Rio, G; Bredesen, DE; Ellerby, HM. Coupling endoplasmic reticulum stress to the cell death program. An Apaf-1 independent intrinsic pathway. *J. Biol. Chem.* 2002, 277(24): 21836-21842.

Rao, RV; Hermel, E; Castro-Obregon, S; del Rio, G; Ellerby, LM; Ellerby, HM; Bredesen, DE. Coupling endoplasmic reticulum stress to the cell death program. Mechanism of caspase activation. *J. Biol. Chem.* 2001, 276(36): 33869-33874.

Raoul, C; Henderson, CE; Pettmann, B. Programmed cell death of embryonic motoneurons triggered through the Fas death receptor. *J. Cell Biol.* 1999, 147(5):1049-62.

Raoul, C; Pettmann, B; Henderson, CE. Active killing of neurons during development and following stress: a role for p75(NTR) and Fas? *Curr. Opin. Neurobiol.* 2000, 10(1):111-117.

Reed, JC. Caspases and cytokines: roles in inflammation and autoimmunity. *Adv. Immunol.* 1999, 73:265-299.

Ron, D; Walter, P. Signal integration in the endoplasmic reticulum unfolded protein response. *Nat. Rev. Mol. Cell Biol.* 2007, 8:519-529.

Roux, PP; Colicos, MA; Barker, PA; Kennedy, TE. p75 neurotrophin receptor expression is induced in apoptotic neurons after seizure. *J. Neurosci.* 1999, 19(16):6887-6896.

Rutkowski, DT; Arnold, SM; Miller, CN; Wu, J; Li, J; Gunnison, KM; Mori, K; Sadighi Akha, AA; Raden, D; Kaufman, RJ. Adaptation to ER stress is mediated by differential stabilities of pro-survival and pro-apoptotic mRNAs and proteins. *PLoS Biol.* 2006, 4:e374.

Salvesen, GS; Dixit, VM. Caspases: intracellular signaling by proteolysis. *Cell.* 1997, 91:443-446.

Sastry, PS; Rao, KS. Apoptosis and the nervous system. *J. Neurochem.* 2000, 74(1):1-20.

Schlüter, T; Struy, H; Scho¨nfeld, P. Protection of mitochondrial integrity from oxidative stress by the triaminopyridine derivative flupirtine. *FEBS Lett.* 2000, 481:42-46.

Schroder, M. Endoplasmic reticulum stress responses, *Cell. Mol. Life Sci.* 2008, 6:862-894.

Seth, R; Yang, C; Kaushal, V; Shah, SV; Kaushal, GP. p53-dependent caspase-2 activation in mitochondrial release of apoptosis-inducing factor and its role in renal tubular epithelial cell injury. *J. Biol. Chem.* 2005, 280:31230-31239.

Shahbazian, MD; Orr, HT; Zoghbi, HY. Reduction of Purkinje cell pathology in SCA1 transgenic mice by p53 deletion. *Neurobiol. Dis.* 2001, 8:974-981.

Shen, J; Prywes, R. ER stress signaling by regulated proteolysis of ATF6, Methods. 2005, 35:382-389.

Sherr, CJ; Roberts, JM. CDK inhibitors: positive and negative regulators of G1-phase progression. *Genes. Dev.* 1999, 13:1501-1512.

Shimizu, S; Eguchi, Y; Kamike, W; Funahashi, Y; Mignon, A; Lacronique, V; Matsuda, H; Tsujimoto, Y. Bcl-2 prevents apoptotic mitochondrial dysfunction by regulating proton flux. *Proc. Natl. Acad. Sci. USA* 1998, 95:1455-1459.

Shults, CW. Coenzyme Q10 in neurodegenerative diseases. *Curr. Med. Chem.* 2003, 10:1917-1921.

Soussi, T. The p53 tumor suppressor gene: from molecular biology to clinical investigation. *Ann. N. Y. Acad. Sci.* 2000, 910:121-137.

Sperandio, S; de, BI; Bredesen, DE. An alternative, nonapoptotic form of programmed cell death. *PNAS.* 2000, 97:14376-14381.

Stolzing, A; Grune, T. Neuronal apoptotic bodies: phagocytosis and degradation by primary microglial cells. *FASEB J.* 2004, 18(6):743-5.

Sullivan, PG; Springer, JE; Hall, ED; Scheff, SW. Mitochondrial uncoupling as a therapeutic target following neuronal injury. *J. Bioenerg. Biomembr.* 2004, 36:353-356.

Susin, SA; Lorenzo, HK; Zamzami, N; Marzo, I; Snow, B.E; Brothers, GM; Mangion, J; Jacotot, E; Costantini, P; Loeffler, M; Larochette, N; Goodlett, DR; Aebersold, R; Siderovski, DP; Penninger, JM; Kroemer, G. Molecular characterization of mitochondrial apoptosis-inducing factor. *Nature.* 1999, 397:441-446.

Susin, SA; Zamzami, N; Castedo, M; Hirsch, T; Marchetti, P; Macho, A; Daugas, E; Geuskens, M; Kroemer, G. Bcl-2 inhibits the mitochondrial release of an apoptogenic protease. *J. Exp. Med.* 1996, 184(4):1331-1341.

T.Y. Chang, C.C. Chang, N. Ohgami, Y. Yamauchi, Cholesterol sensing, trafficking, and esterification, *Annu. Rev. Cell Dev. Biol.* 22 (2006) 129-157.

Taglialatela, G; Kaufmann, JA; Trevino, A; Perez-Polo, JR. Central nervous system DNA fragmentation induced by the inhibition of nuclear factor kappa B. *NeuroReport.* 1998, 9:489-493.

Tatton, WG; Chalmers-Redman, RM. Modulation of gene expression rather than monoamine oxidase inhibition: (-)-deprenyl-related compounds in controlling neurodegeneration. *Neurology.* 1996, 47(Suppl. 3):S171-S183.

Tatton, WG; Chalmers-Redman, RM; Elstner, M; Leesch, W; Jagodzinski, FB; Stupak, DP; Sugrue, MM; Tatton, NA. Glyceraldehyde- 3-phosphate dehydrogenase in neurodegeneration and apoptosis signaling. *J. Neural. Transm. Suppl.* 2000, 60:77-100.

Tatton, WG; Chalmers-Redman, RM; Ju, WJ; Mammen, M; Carlile, GW; Pong, AW; Tatton, NA. Propargylamines induce antiapoptotic new protein synthesis in serum- and nerve growth factor (NGF)-withdrawn, NGF-differentiated PC-12 cells. *J. Pharmacol. Exp. Ther.* 2002, 301:753- 764.

Tatton, WG; Olanow, CW. Apoptosis in neurodegenerative diseases: the role of mitochondria. *Biochim. Biophys Acta.* 1999, 1410(2):195-213.

Taylor, RC; Cullen, SP; Martin, SJ. Apoptosis: controlled demolition at the cellular level. *Nat. Rev. Mol. Cell Biol.* 2008, 9:231-241.

Thomas, M; Le, WD; Jankovic, J. Minocycline and other tetracycline derivatives: a neuroprotective strategy in Parkinson's disease and Huntington's disease. *Clin. Neuropharmacol.* 2003, 26:18-23.

Torres-Aleman, I. Serum growth factors and neuroprotective surveillance: focus on IGF-1. *Mol. Neurobiol.* 2000, 21:153-160.

Torres-Aleman, I; Villalba, M; Nieto-Bona, MP. Insulin-like growth factor-I modulation of cerebellar cell populations is developmentally stage-dependent and mediated by specific intracellular pathways. *Neuroscience.* 1998, 83:321-334.

Urano, F; Wang, X; Bertolotti, A; Zhang, Y; Chung, P; Harding, HP; Ron, D. Coupling of stress in the ER to activation of JNK protein kinases by transmembrane protein kinase IRE1. *Science.* 2000, 287:664-666.

Vahsen, N; Cande, C; Dupaigne, P; Giordanetto, F; Kroemer, RT; Herker, E; Scholz, S; Modjtahedi, N; Madeo, F; Le Cam, E; Kroemer, G. Physical interaction of apoptosis-inducing factor with DNA and RNA. *Oncogene.* 2006, 25:1763-1774.

Vanderhaeghen, P; Cheng, HJ. Guidance molecules in axon pruning and cell death. Cold Spring Harb Perspect Biol. 2010, 2(6):a001859.

Vaux, DL. Apoptogenic factors released from mitochondria. *Biochim. Biophys. Acta.* 2010, 14.

Voehringer, DW; McConkey, DJ; McDonnell, TJ; Meyn, RE. Bcl-2 expression causes redistribution of glutathione to the nucleus. *Proc. Natl. Acad. Sci. USA* 1998, 95:2956-2960.

Volosin, M; Song, W; Almeida, RD; Kaplan, DR; Hempstead, BL; Friedman, WJ. Interaction of survival and death signaling in basal forebrain neurons: roles of neurotrophins and proneurotrophins. *J. Neurosci.* 2006, 26(29):7756-7766.

Volosin, M; Trotter, C; Cragnolini, A; Kenchappa, RS; Light, M; Hempstead, BL; Carter, BD; Friedman, WJ. Induction of proneurotrophins and activation of p75NTR-mediated apoptosis via neurotrophin receptor-interacting factor in hippocampal neurons after seizures. *J. Neurosci.* 2008, 28(39):9870-9879.

Wadia, JS; "Chalmers-Redman, RM; Ju, WJ; Carlile, GW; Phillips, JL; Fraser, AD; Tatton, WG. Mitochondrial membrane potential and nuclear changes in apoptosis caused by serum and nerve growth factor withdrawal: Time course and modification by (-)-deprenyl. *J. Neurosci.* 1998, 18:932-947.

Waldmeier, PC. Prospects for antiapoptotic drug therapy of neurodegenerative diseases. *Prog. Neuropsychopharmacol. Biol. Psychiatry.* 2003, 27(2):303-321.

Waldmeier, PC; Tatton, WG. Interrupting apoptosis in neurodegenerative disease: potential for effective therapy? *Drug Discov. Today* 2004, 9(5):210-218.

Wartiovaara, K; Barnabe-Heider, F; Miller, FD; Kaplan, DR. N-myc promotes survival and induces S-phase entry of postmitotic sympathetic neurons. *J. Neurosci.* 2002, 22:815-824.

Wood, KA; Dipasquale, B; Youle, RJ. In situ labeling of granule cells for apoptosis-associated DNA fragmentation reveals different mechanisms of cell loss in developing cerebellum. *Neuron.* 1993, 11:621-632.

Wood, KA; Youle, RJ. The role of free radicals and p53 in neuron apoptosis in vivo. J. *Neurosci.* 1995, 15:5851-5857.

Xiang, H; Hochman, DW; Saya, H; Fujiwara, Schwartzkroin, TPA; Morrison, RS. Evidence for p53-mediated modulation of neuronal viability. *J. Neurosci.* 1996, 16:6753-6765.

Xu, DG; Bureau, Y; McIntyre, DC; Nicholson, DW; Liston, P; Zhu, Y; Fong, WG; Crocker, SJ; Korneluk, RG; Robertson, GS. Attenuation of ischemia-induced cellular and behavioral deficits by X chromosome-linked inhibitor of apoptosis protein overexpression in the rat hippocampus. *J. Neurosci.* 1999, 19:5026-5033.

Xu, DG; Crocker, SJ; Doucet, JP; St-Jean, M; Tamai, K; Hakim, AM; Ikeda, JE; Liston, P; Thompson, CS; Korneluk, RG; MacKenzie, A; Robertson, GS. Elevation of neuronal expression of NAIP reduces ischemic damage in the rat hippocampus. *Nat. Med.* 1997, 3:997-1004.

Yang, J; Liu, X; Bhalla, K; Kim, CN; Ibrado, AM; Cai, J; Peng, TI; Jones, DP; Wang, X. Prevention of apoptosis by Bcl-2: release of cytochrome *c* from mitochondria blocked. *Science.* 1997. 275:1129-1132.

Ye, H; Cande, C; Stephanou, NC; Jiang, S; Gurbuxani, S; Larochette, N; Daugas, E; Garrido, C; Kroemer, G; Wu, H. DNA binding is required for the apoptogenic action of apoptosis inducing factor. *Nat. Struct. Mol. Biol.* 2002, 9:680-684.

Youdim, MB; Wadia, A; Tatton, W; Weinstock, M. The anti-Parkinson drug rasagiline and its cholinesterase inhibitor derivatives exert neuroprotection unrelated to MAO inhibition in cell culture and in vivo. *Ann. N. Y. Acad. Sci.* 2001, 939:450-458.

Yu XY; Sun, L; Liou, C; Zhao, H; Xu, ZQ. Overexpression of death-associated protein kinase is a component in glutamate-induced neuronal apoptosis cascade. *Int. J. Devl. Neuroscience.* 2006, 24:495-603.

Yuan, J; Lipinski, M; Degterev, A. Diversity in the mechanisms of neuronal cell death. Neuron. 2003, 40(2):401-413.

Yuste, VJ; Moubarak, RS; Delettre, C; Bras, M; Sancho, P; Robert, N; d'Alayer, J; Susin, SA. Cysteine protease inhibition prevents mitochondrial apoptosis-inducing factor (AIF) release. *Cell Death Differ.* 2005, 12:1445-1448.

Zhang, D; Berry, MD; Paterson, IA; Boulton, AA. Loss of mitochondrial membrane potential is dependent on the apoptotic program activated: prevention by R-2HMP. *J. Neurosci. Res.* 1999, 58:284-292.

Zhu, S; Stavrovskaya, IG; Drozda, M; Kim, BY; Ona, V; Li, M; Sarang, S; Liu, AS; Hartley, DM; Wu, DC; Gullans, S; Ferrante, RJ; Przedborski, S; Kristal, BS; Friedlander, RM. Minocycline inhibits cytochrome c release and delays progression of amyotrophic lateral sclerosis in mice. *Nature.* 2002, 417:74-78.

Zhu, X; Yu, QS; Cutler, RG; Culmsee, CW; Holloway, HW; Lahiri, DK; Mattson, MP; Greig, NH. Novel p53 inactivators with neuroprotective action: syntheses and pharmacological evaluation of 2-imino-2,3,4,5,6,7-hexahydrobenzothiazole and 2-imino-2,3,4,5,6,7-hexahydrobenzoxazole derivatives. *J. Med. Chem.* 2002, 45:5090-5097.

Zimmermann, KC; Bonzon, C; Green, DR. The machinery of programmed cell death. *Pharmacol. Ther.* 2001, 92:57-70.

Zimmermann, KC; Green, DR. How cells die: apoptosis pathways. *J. Allergy Clin. Immunol* 2001, 108(4 Suppl):S99-103.

Zindy, F; Cunningham, JJ; Sherr, CJ; Jogal, S; Smeyne, RJ; Roussel, MF. Postnatal neuronal proliferation in mice lacking Ink4d and Kip1 inhibitors of cyclin-dependent kinases. *Proc. Natl. Acad. Sci. USA* 1999, 96:13462-13467.

Zipfel, GJ; Babcock, DJ; Lee, JM; Choi, DW. Neuronal apoptosis after CNS injury: the roles of glutamate and calcium. *J. Neurotrauma.* 2000, 17:857-869.

In: Neuronal Cell Apoptosis
Editors: C. J. Schmid and J. L. Wolfe

ISBN: 978-1-61324-633-7

Chapter II

Witnessing the Mechanistic Insights of Neuronal Apoptosis from a Global Transcriptomic View: Comparative Microarray Analysis of Multiple Neurodegenerative Models

***Minghui Jessica Chen*[1], *Jian Ming Jeremy Ng*[1], *Zhao Feng Peng*[3], *and Nam Sang Cheung*[1,4*]**

[1]Menzies Research Institute, University of Tasmania, Hobart, Tasmania, Australia

[2]Key Laboratory of Biogeology and Environmental Geology of the Ministry of Education, China University of Geosciences, Wuhan, P.R. China

[3]Department of Biochemistry, Yong Loo Lin School of Medicine, National University of Singapore, Singapore

[4]School of Life and Environmental Sciences, Deakin University, Australia

Abstract

Apoptosis is the earliest identified mode of cell death characterized by its closely coordinated signaling mechanisms that put the cell to its demise without disruption to its microenvironment. It is the most reported cause of neuronal death seen in the pathogenesis of numerous neurodegenerative disorders. The current focus of this review revolves around the significance of the major biological processes (e.g. oxidative stress, endoplasmic reticulum/lysosomal disruption, mitochondrial respiratory dysfunction and

*Corresponding author: Nam Sang Cheung, email nscheung@utas.edu.au

ubiquitin-proteasome system inhibition) at work in the induction of neuronal apoptosis during neurodegeneration in various central nervous system diseases. To further complement and substantiate the existing findings, comparative microarray analysis was conducted on five neurodegenerative global gene profiles from our laboratory to provide comprehensive temporal elucidation of the implication of these neuropathological mechanisms during neuronal apoptosis at the transcriptional level. These transcriptomic profiles were derived from the application of widely employed pharmacological agents on *in vitro* rodent models to mimic neuropathological hallmarks. They comprise of lactacystin (proteasomal inhibitor), Glu (excitotoxicity inducer), nitric oxide (source of reactive nitrergic and oxygen species), hypochlorous acid (source of reactive chlorinergic and oxygen species) and rotenone (mitochondrial complex I inhibitor) commonly used to model chronic neurodegenerative diseases such as Alzheimer's disease, amyotrophic lateral sclerosis, Parkinson's disease and Down syndrome as well as acute neurological disorders such as stroke and traumatic brain injury. Mediation of neuronal injury and apoptosis by any one agent is not mutually exclusive to any particular neurodegenerative condition, but at most times demonstrates simultaneous involvement across the neurodegenerative disease spectrum. This is an indirect indication of a significant commonality of signaling cascades in the pathogenesis of most neurodegenerative diseases, and that apoptosis plays a major role in the trigger of these downstream mechanisms. Subsequently, this provides the opportunity for identification of novel biological targets whose manipulation would provide beneficial broad-spectrum neurodegenerative disease therapeutic intervention.

Keywords: Apoptosis; Neurodegeneration; Microarray; in vitro models.

Abbreviations

Acad -	acetyl-coenzyme A dehydrogenase
AD -	Alzheimer's disease
AIF -	apoptosis-inducing factor
ALDH -	aldehyde dehydrogenase
ALS -	amyotrophic lateral sclerosis
Anxs -	annexins
Atf4 -	transcription factor 4
ATP -	adenosine triphosphate
Bax -	Bcl2-associated x protein
Bbc3 -	BCL2 binding component 3
BCL2 -	B-cell lymphoma 2
BDNF -	brain-derived neurotrophic factor
Bid -	BH3 interacting domain death agonist
Capn2 -	Ca^{2+}-activated protease calpain-2
Casp3 -	caspase 3
Casp8 -	caspase 8
cGMP -	cyclic guanosine 3',5'-monophosphate
CNS -	central nervous system
CTSB -	cytoplasmic activation of cathepsin B

Cyb5 -	cytochrome b5 reductase
Cyp -	cytochrome p450 family
CYTC -	cytochrome c
Dap -	death associated protein
DHRS8 -	dehydrogenase/reductase 8
DISC -	death-inducing signaling complex
eIF2α -	eukaryotic initiation factor
Endo G -	endonuclease G
ERphagy –	reticulophagy
ETC -	electron transport chain
FAD^+ -	flavin adenine dinucleotides
FADD -	Fas-associated death domain
FasL -	Fas/Fas ligand
GC -	guanylate cyclase
Glu –	glutamate
GluRs –	glutamate receptors
Gpd -	glycerol-3-phosphate dehydrogenase
Gst -	glutathione *S*-transferase
GTP -	guanosine-5'-triphosphate
H_2O_2 -	hydrogen peroxide
HOCl -	hypochlorous acid
Hsps -	heat shock proteins
IAP -	inhibitor of apoptosis
Idh -	isocitrate dehydrogenase
IGF-I -	insulin-like growth factor I
iGluRs -	ionotropic subtype glutamate receptors
IL-1β -	interleukin-1-beta
IL-6 -	interleukin-6
IMS -	intermembrane space
iNOS -	NOS type II
IRE1 -	inositol-requiring enzyme
LC3 -	light chain 3
MA -	marcoautophagy
MPT -	mitochondrial permeability transition
NAD^+ -	nicotinamide adenine dinucleotides
NMDAR agonist -	N-methyl-D-aspartate
NO -	nitric oxide
NO_2^- -	nitrites
NO_3^- -	nitrates
NOS -	nitric oxide synthase
$O_2\bullet^-$ -	superoxide anion
OH• -	hydroxyl radical
OMN -	outer mitochondrialmembrane
OMNP -	outer mitochondrial membrane permeability
$ONOO^-$ -	peroxynitrite
PCD -	programmed cell death

PCD -	Type I - apoptosis
PCD -	Type II - autophagy
PCD -	Type III - necrosis
PD -	Parkinson's disease
PERK -	PKR-like ER kinase
PKG -	protein kinase G
Pmaip1 -	phorbol-12-myristate-13-acetate-induced protein 1
PPARγ -	peroxisome proliferator-activated receptor gamma
PPP -	pentose phosphate pathway
Prx -	peroxiredoxins
PUMA -	p53-unregulated modulator of apoptosis
RNS -	reactive nitrogen species
ROS -	reactive oxygen species
SIAH1 -	seven in absentia homolog 1
tBID -	truncated BID
Tnfrsf1-	tumor necrosis factor receptors
TNF-α -	tumor necrosis factor-alpha
TRAIL -	TNF-related apoptosis-inducing ligand
Trp53 -	transformation-related p53
Ubc -	ubiquitins
UBE1 -	ubiquitin-activating enzyme E1
UBE2 -	ubiquitin-conjugating enzyme E2
UBE3 -	ubiquitin protein ligase E3
Ufd1l -	ubiquitin fusion degradation 1 like
UPR -	unfolded protein response
UPS -	ubiquitin-proteasome system

Neurons are morphologically unique and functionally sophisticated mammalian cells. They possess the characteristic morphology of a cell body, with extensively elongated processes (axons and dendrites) that pose consequential problems for intracellular trafficking. In addition, the plethora of surface ion channels in neurons increases their vulnerability to injury and death through energy deprivation (ATP deficit) or via hyper-excitation by neurotransmitters leading to excitotoxicity by inducing elevation of intracellular and mitochondrial Ca^{2+} as previously mentioned, which forms one of the key determinants of PCD through various mechanisms (Chinopoulos and Adam-Vizi, 2006; Nicholls, 2008). Other crucial factors include ROS generation and presence of mitochondrial permeability transition (MPT) and/or outer mitochondrial membrane permeability (OMMP). Recent substantiation in non-neuronal cells indicates elevated Ca^{2+} in mitochondria to be a crucial signal in ROS-mediated apoptosis (Baumgartner, Gerasimenko, et al., 2009). Finally, neurons function and maintain in an adaptive cellular milieu made up of mainly glial cells such as astrocytes (Pekny and Nilsson, 2005; Maragakis and Rothstein, 2006). As such, neuronal exposure to glial cells-secreted cytokines during the inflammatory phase of injury in the CNS (Pekny and Nilsson, 2005; Hanisch and Kettenmann, 2007) can detrimentally activate extrinsic pathways to cell death.

Mitochondria have been recognized as a key crucial organelle under numerous neurological pathological settings, playing the key regulatory roles in death processes in the affected cells (reviewed extensively in Higgins, Beart, et al., 2010). It is vital to cellular homeostatic and energy-yielding activities. The definition of PCD has broadened over the recent years to incorporate more complex, distinctive cell death signaling cascades. Mitochondrial implication in PCD has been extensively studied (reviewed by Nagley, Higgins, et al., 2010). Other than its prominent, well-explored effect on apoptosis (PCD-Type I), it also influences the progression of programmed necrosis (PCD-Type III), an alternative death pathway signified by the absence of caspase involvement (Clarke, 1990). Autophagy (PCD-Type II), perceived as a routine homeostatic mechanism for renewal and clearance of cellular constituents, has but recently been identified as a mode of PCD, which activation has been reported to lead to death under certain stress conditions (Clarke, 1990; Boland and Nixon, 2006; Mizushima, Levine, et al., 2008). With the expansion in the subtypes of PCD, numerous cell death markers have been employed to distinguish the mechanistic mode of cellular injury or neurodegeneration under various pathological conditions in the nervous system.

To further add complexity to the picture, all these diverse signaling pathways are able to cross-regulate each other at both upstream and downstream of mitochondria. As a result, activation of one pathway may positively or negatively impact the development of other pathways instantaneously. The existence of these cross-talks can result in two cell death scenarios where firstly, in one initially homogeneous population of cultured neurons exposed to a given treatment, a heterogeneous set of responses between different cells of that population is likely to occur; secondly, individual cells may display a subset of markers characteristic of explicitly defined cell death paradigms (e.g. PCD Types I, II or III) (Nagley, Higgins, et al., 2010). The former is reflective of the pathological features observed in ischemic stroke brains subjected to a single cellular insult. In stroke-affected and HI brains, distinct necrotic features are concentrated in the core of the injury site. On the other hand, the penumbra surrounding this core is occupied by cells showing apoptotic features such as caspase activation and nuclear DNA fragmentation (Dirnagl, Iadecola, et al., 1999).

Programmed Cell Death (PCD) in Neurons

Apoptosis

Apoptosis (PCD-Type I), dependent on caspase activation, is composed of twomain pathways, the extrinsic (death receptor pathway) and intrinsic pathways (reviewed extensively by Nagley, Higgins, et al., 2010). The stark difference lies in that the extrinsic pathway can occur independent of mitochondria while theintrinsic pathway is characterized by mitochondrial involvement. In theintrinsic pathway, apoptogenic proteins translocate from the mitochondrial intermembrane space (IMS) to the cytoplasm, leading to caspase activation that directly activates cell death. The occurrence ofOMMP is regulated by the BCL2 protein family, which is divided into pro-apoptotic andanti-apoptotic (pro-survival) subfamilies. The anti-apoptotic BCL2proteins form the bridging intermediaries between the BH3-only proteins andthe pro-apoptotic BCL2 family members, which include BAX and BAK. One postulation of this mechanistic association is that the anti-apoptotic BCL2 proteins inactivate BAX andBAK for as long as the pro-survival proteins remain uninhibited by the BH3-onlyproteins.

The other view is that some BH3-only proteins may also act asfacilitators, directly interacting with BAX and BAK to initiate theiractivation (reviewed by Youle and Strasser, 2008). BAX and BAK, upon activation, changeconformation, oligomerise and form pores in the outer mitochondrialmembrane (OMM). The "pore" is likely to be lipidic in nature, thusallowing transit of proteins from the IMS to the cytosol and vice versa (Smith, Ng, et al., 2008).

The key executioners that re-localize from the IMS to the cytosol (reviewed in Hengartner, 2000; Kroemer, Galluzzi, et al., 2007) include cytochrome c (CYTC), antagonists of inhibitor of apoptosis (IAP), "second mitochondria-derived activator of caspases/direct inhibitor of apoptosis-binding protein with low pI" (Smac/DIABLO) and HTRA2/OMI that activate the caspase-signaling cascade, and the caspase-independent proteins, AIF and endonuclease G (Endo G) (Beart, Lim, et al., 2007; Kroemer, Galluzzi, et al., 2007; Diwakarla, Nagley, et al., 2009; Higgins, Beart, et al., 2009). CYTC forms an apoptosome complex with APAF-1 and ATP in the cytosol to activate caspase-9. Caspase-9 then in turn activates its downstream effector caspase-3, caspase-6 and caspase-7, which consequently lead to chromatin condensation and DNA fragmentation (reviewed in Hengartner, 2000). The re-localization of AIF and Endo G during apoptosis frequently occurs much later than that of CYTC and Smac/DIABLO (Beart, Lim, et al., 2007; Diwakarla, Nagley, et al., 2009). Activation of both AIF and Endo G alone is able to result in nuclear degradation and DNA cleavage independent of caspases (Takano, Tomioka, et al., 2005). As such, in an apoptotic process, AIF and Endo G are believed to be involved in a feed-forward mechanism whereby their redistribution from mitochondria is triggered by downstream caspase activation (reviewed in Hansen and Nagley, 2003).

BID, another BH3-only protein accountable for the Bax activation during OMMP in neurons, is also regulated by p53 (Desagher, Osen-Sand, et al., 1999). For BID to become functional, it requires cleavage by active caspase-8 or caspase-2 to form truncated BID (tBID) (Konig, Rehm, et al., 2007; Niizuma, Endo, et al., 2008). During focal cerebral ischemia and oxygen/glucose deprivation in neurons, tBid presence is critical for apoptosis to proceed (Plesnila, Zinkel, et al., 2001; Yin, Luo, et al., 2002). Once migrated to the mitochondria, tBid participates in the direct redistribution of AIF to the nucleus in cultured neurons (Culmsee, Zhu, et al., 2005; Landshamer, Hoehn, et al., 2008). However, it is recently reported full length BID is involved in inducing caspase-independent cell death in neurons, implying the BID truncation is not crucial for the activation of the cell death signaling cascade and it may have an important role in AIF redistribution during cell death in the absence of caspase activity (Ward, Rehm, et al., 2006; Konig, Rehm, et al., 2007).

Autophagy

Autophagy is part of a cell's routine cellular homeostatic process to remove molecules and organelles via lysosomal clearance/degradation pathway (Nagley, Higgins, et al., 2010). Its importance in CNS is gaining increasing recognition (Jaeger and Wyss-Coray, 2009). Autophagy can be classified into several pathway subtypes (reviewed in Todde, Veenhuis, et al., 2009), namely macroautophagy (MA; most commonly associated with mitochondria and cell death), microautophagy (autophagy without vesicles) and chaperone-mediated autophagy. As such, we will focus on the discussion of MA involvement in neuronal death and the term "Autophagy" will be used to refer to MA. Autophagy requires the formation of a double membrane vesicle called an autophagosome, which is involved in the sequestration and engulfment of organelles or molecules and targeting them for lysosomal clearance. Microtubule-associated protein 1-light chain 3 (LC3), an essential protein component of the autophagosome membrane, has been adopted as an autophagic biomarker for the localization studies of these vesicles (Kabeya, Mizushima, et al., 2000). There is a widely accepted belief that autophagy is essential for the homeostatic maintenance of functional organelles, such as mitochondria due to the post-mitotic nature of neurons (Terman, Gustafsson, et al., 2006). Due to the lack of cellular division in neurons, autophagy is essential for turnover of dysfunctional mitochondria and protein aggregates (i.e. β-amyloid or α-synuclein in AD and PD, respectively), believed to be detrimental to neuronal survival (reviewed in Klionsky, 2006; Terman, Gustafsson, et al., 2006). There are at least two situations in which autophagy come into play to induce neuronal death: in the first instance, when the autophagic process is insufficient to remove the toxic protein aggregates, as possibly occurs in AD and PD (reviewed in Kroemer and Levine, 2008); and secondly, autophagy is up-regulated as a direct contributor to cell death as in PCD Type II.

Existence of autophagic neuronal cell death was first proposed when an alternative death pathway independent of apoptosis was activated in sympathetic neurones (Xue, Fletcher, et al., 1999).Subsequent similar studies reported multiple vesicles formation (Gomez-Santos, Ferrer, et al., 2003), increase of autophagic protein levels (Kanno, Ozawa, et al., 2009) and cell death attenuation with autophagic inhibitors (Xue, Fletcher, et al., 1999; Gomez-Santos, Ferrer, et al., 2003). As the autophagic inhibitors such as 3 methyladenine or wortmannin used in these studies are non-specific and can also block apoptotic pathways (Canu, Tufi, et al., 2005), most did not classify the death explicitly as autophagic. Furthermore, cooperative signaling coupling between autophagy and apoptosis makes it even more difficult to distinctively classifycell death as belonging exclusively to a certain PCD type.

Another example of this signaling interplay implicating mitochondrial regulation, autophagic players and cell death is demonstrated by Beclin-1 function. This protein is not only a central component of the autophagic pathway, but also contains a BH3-only domain, which can be negatively regulated by interacting with the anti-apoptotic BCL2 or BCL-xL proteins (Oberstein, Jeffrey, et al., 2007).Interference of these BCL2 family/Beclin-1 interactions by mutation of the BH3-only domain within Beclin-1 led to active autophagic death (Pattingre, Tassa, et al., 2005). Functional importance of Beclin-1 in PCD-Type II is further substantiated by over-expression studies that it induces massive autophagy, which progresses to cell death (Pattingre, Tassa, et al., 2005). Suppression of the autophagic process by Beclin-1/BCL2(BCL-XL) interactions is alleviated through competitive disruption by BH3-only proteins such as BAD (Maiuri, Criollo, et al., 2007). With the establishment of these molecular associations, Beclin-1 involvement in autophagy is much more prominent than in apoptosis. This is because the current perspective is that the crosstalk is unidirectional (Maiuri, Criollo, et al., 2007; Ciechomska, Goemans, et al., 2009), such that the BCL2 family regulates autophagy but Beclin-1 does not affect apoptosis. However, these interactions do not necessarily lead to cell death but can be beneficial to cellular homeostasis in other conditions such as starvation. For instance, during starvation in HeLa cells, ABT737 (a BH3-only peptide mimetic) triggered mitophagy rather than reticulophagy (also known as ERphagy), yet only ER-localized BCL2/BCL-xL (not mitochondrially-targeted BCL2/BCL-xL) was capable of inhibiting starvation-evoked autophagy, implying a significant importance of ER-stress/autophagic crosstalk (Maiuri, Le Toumelin, et al., 2007).

Further pathway interlocks between the apoptotic and autophagic signaling cascades can be illustrated via the roles of death signaling molecules such as the autophagy protein ATG5, as well as the BH3-only proteins BNIP3 and BIK. ATG5, an autophagy protein essential for autophagosome formation, can undergo truncation by calpains, transforming it to an effective apoptotic inducer (Mizushima, Yamamoto, et al., 2001). Once cleaved, ATG5 is able to translocatefreely from the cytosol to mitochondria. There, it binds to BCL-xL, triggering CYTC redistribution and downstream caspase activation (Yousefi, Perozzo, et al., 2006). Hypoxia in HEK293 cells triggers BNIP3 up-regulation resulting in cell death (Azad, Chen, et al., 2008). BNIP3 over-expression promotes autophagy, while knockdown diminishes hypoxia-induced autophagy and cell death. Similarly, silencing of Beclin-1 and ATG5 under the same cellular condition also inhibits hypoxia-induced cell death, indicating that all three proteins participate in the same autophagic cell death cascade. This postulation is further verified by BIK over-expression -induced cell death in $BCL2^{-/-}$ murine embryonic fibroblasts, which could be abrogated by autophagic inhibitors and knockdown of Beclin-1 and ATG5 (Rashmi, Pillai, et al., 2008). Autophagy triggered by relevant BCL2 family members during cell death may be used as an alternative resort to provide an additional supply of energy to fulfill the apoptotic process.

Programmed Necrosis

Severe, massive cellular or tissue damages frequently lead to unregulated, accidental necrosis, when cells are subjected to drastic cellular ionic and osmotic perturbations leading to rapid swelling with consequential gross cellular debilitation (Nicotera and Melino, 2004; Blomgren, Leist, et al., 2007; Boujrad, Gubkina, et al., 2007; Gill and Perez-Polo, 2008; Kroemer, Galluzzi, et al., 2009). This aberrant cytotoxic outcome differs from programmed necrosis (PCD-Type III), which is dependent on distinct regulated pathways (distinct from both apoptosis and autophagy). Cells undergoing PCD-Type III though possessing some morphological resemblance to truly necrotic cellsundergo a discrete loss of permeability of the plasma membrane (Boujrad, Gubkina, et al., 2007). In both cases, mitochondrial energisation and ATP production were also inhibited.

Similarly to PCD-Type I and PCD-Type II, mitochondria are also a centre component to cell death regulation in programmed necrosis. However, as opposed to PCD-Type I, PCD-Type III does not rely on caspase activation (Kroemer, Galluzzi, et al., 2009). Instead, programmed necrosis triggers the MPT and is dependent on AIF and Endo G re-localization (Boujrad, Gubkina, et al., 2007; Higgins, Beart, et al., 2009). Calpains, a family of Ca^{2+}-dependent cysteine proteases, frequently form the hallmark of PCD-Type III (Golstein and Kroemer, 2007; Kroemer, Galluzzi, et al., 2009). While calpains have been implicated in all PCD subtypes in an apoptosis-necrosis continuum (e.g. calpains activation during apoptosis in motor neurons)(Momeni and Kanje, 2006; Samantaray, Ray, et al., 2006; Sribnick, Matzelle, et al., 2007), their involvement in programmed necrosis has been especially highlighted due to elevation of intracellular Ca^{2+} in programmed necrosis (Pang, Bondada, et al., 2003; Yamashima, Tonchev, et al., 2003; Diwakarła, Mercer, et al., 2009). Calpains are categorized into tissue-specific isoforms (n-calpains), and two constitutively and ubiquitously expressed isozymes calpain I (μ-calpain) and calpain II (m-calpain), that are activated in vitro by micromolar and millimolar concentrations of Ca^{2+}, respectively(Polster, Basanez, et al.,

2005). Unlike caspases, calpains do not recognize specific substrate cleavage sites and its activity is tightly modulated by cofactors, such as the endogenous inhibitor, calpastatin, and phospholipids (Huang and Reichardt, 2001).

Cross-regulation between caspase-dependent and caspase-independent PCD pathways is evident from observations that calpastatin can be cleaved by caspase-3 to facilitate calpains activation in PCD-Type I (Porn-Ares, Samali, et al., 1998; Wang, Posmantur, et al., 1998) and that calpains too are able to activate endogenous caspases such as caspase-3, -7, -8 and -9 (Chua, Guo, et al., 2000). It is demonstrated in a hypoglycemia model that neuronal death is dependent on both MPT and redistribution of CYTC and AIF triggered by both caspase-3 and calpain activation (Ferrand-Drake, Zhu, et al., 2003). Calpain I is directly accountable for the redistribution of AIF from intact mitochondria, and calpains inhibition is able to suppress AIF release, emphasizing calpains role in caspase-independent cell death (Polster, Basanez, et al., 2005). As such, AIF-dependent death (in the absence of caspase activation) associating with calpains activation has been classified as programmed necrosis (Boujrad, Gubkina, et al., 2007). This emphasis is further highlighted by the significant reduction of damage induced by KA-mediated excitotoxicity *in vivo* upon calpastatin over-expression, demonstrated to be caused by a decrease in cleavage of BID to tBID and the redistribution of AIF and Endo G from mitochondria (Takano, Tomioka, et al., 2005).

Major Biological Processes Associated with Neuronal Apoptosis

Excitotoxicity and Calcium Ion (Ca^{2+}) Influx

Excitotoxicity is one of the major mechanisms of cell death in numerous CNS diseases including stroke, brain trauma, epilepsy and chronic neurodegenerative disorders such as AD and PD (Wang and Qin, 2010). The "excitotoxicity" theory was first formulated in 1969 by Dr. Olney as an undesirable cytotoxic side effect of excessive or prolonged activation of glutamate (Glu) receptors (GluRs), particularly the ionotropic subtype (iGluRs) (Olney, 1969). Much research efforts have been attributed to the mechanistic elucidation of excitotoxicity with current understanding summarized as follow: iGluRs over-stimulation triggers multiple adverse effects comprising impaired intracellular calcium ion homeostasis, organellar dysfunctions, elevated nitric oxide (NO) and reactive oxygen species (ROS) production, unregulated persistent activation of proteases and kinases, and transcriptional activation of pro-death transcription factors and immediate early genes (Wang and Qin, 2010). It has been demonstrated that in the event of excitotoxic neuronal death, all three subtypes of iGluRs (NMDARs, AMPARs and KARs) are actively involved, with the NMDARs playing a major role due to its abundant expression and highest Ca^{2+} permeability (Hara and Snyder, 2007). Upon excess Glu application in experimental models, heterogeneity of neurodegenerative signaling cascades is observed. In neuronal cultures, an apoptotic-autophagic-necrotic continuum cell death is induced depending on the severity of NMDA insult (Berman and Murray, 1997). *In vivo*, cell death morphology is determined by GluRs subunit composition in neurons (Lev, Moreno, et al., 1995).

Ca^{2+} influx was demonstrated to be an early event essential for triggering of Glu excitotoxicity (Choi, 1985). It was shown that in Ca^{2+}-rich extracellular solution, Glu excitotoxicity in neuronal cultures was potentiated, whereas a Ca^{2+}-free extracellular solution attenuated neurodegeneration. Further studies verified that the induction of Ca^{2+} influx by NMDARs, but not the Ca^{2+} load, was important in the GluRs-mediated neurodegenerative process (Choi, 1987; Tymianski, Charlton, et al., 1993), and that lower Ca^{2+} influxes via NMDARs evoked higher lethality as compared to higher Ca^{2+} influxes via other Ca^{2+}-permeable channels (Sattler, Charlton, et al., 1998). Majority of elevated cytosolic Ca^{2+} is sequestered into mitochondria, which can lead to mitochondrial toxicity resulting in metabolic acidosis and free radical generation (Thayer and Wang, 1995; Wang and Thayer, 1996).

Organellar Destabilization

Mitochondrial Dysfunction

The mitochondrial network plays a vital role in the supply of cellular energy currency in the form of adenosine triphosphate (ATP) to ensure the proper functioning of a variety of metabolic processes within a cell. Simpler molecules resulting from the cellular cyclic processing of macromolecular nutrients transfer electrons to carrier proteins such as nicotinamide and flavin adenine dinucleotides (NAD^+ and FAD^+) producing NADH and $FADH_2$, which transfer the electrons to the electron transport chain (ETC) localized at the inner mitochondrial membrane (Saraste, 1999). Due to the constitutive cyclic fluctuation of the redox status between ETC enzymatic protein complexes with consequent high consumption of cellular oxygen in the oxidative phosphorylation process, mitochondria are assumed to be the main cellular producers of ROS (Orrenius, Gogvadze, et al., 2007). Escaping electrons from the ETC can potentially reduce oxygen to form the highly reactive $O_2\bullet^-$, which can undergo a further Fenton reaction to generate hydroxyl radical (OH•) and hydrogen peroxide (H_2O_2), which similarly can cause detrimental cellular damages (Boveris, Oshino, et al., 1972).

As a result of this pivotal physiological function of mitochondria, which if not properly managed can have adverse effects on cell survival, mitochondrial functionality has been proposed to be a crucial regulator and indicator of cellular homeostasis. Indeed, decline in mitochondrial functionality has been closely linked to increasing the age of mammalians. This age-correlated respiratory chain deficiency is especially prevalent in only a subset of mammalian tissues, such as heart, skeletal muscle, colonic crypts and neurons (Dufour, Terzioglu, et al., 2008).

A recent study by Dufour, et al., 2008 demonstrated that the co-existence of functional respiratory chain-deprived and normal neurons accelerated the neurodegenerative process of the adjacent normal cells through a trans-neuronal signaling mechanism.

Lysosomal Rupture

Calpain activation has been reported to associate with lysosomal rupture leading to the death of post-ischemic CA1 neurons (Yamashima, Tonchev, et al., 2003). A "calpain-cathepsin hypothesis" was formulated by (Yamashima, Kohda, et al., 1998) on the basis of the experimental paradigm of global brain ischemia in primates. The calpain-cathepsin cascade mechanism of cell death involves Ca^{2+} mobilization through the uptake of extracellular Ca^{2+} and/or the release from internal Ca^{2+} stores. Ca^{2+} mobilization can lead to the activation of calpains, which induces lysosomal rupture, possibly aided by ROS. The released lysosomal proteases, mainly the cathepsins, will then degrade the cell constituent proteins, ultimately leading to cell death. Cytoplasmic activation of cathepsin B (CTSB) upon lysosomal rupture mediates activation of pro-inflammatory caspase-1 and -11 upon focal cerebral ischemic induction and $Aß_{42}$-induced neurotoxicity (Benchoua, Braudeau, et al., 2004; Gan, Ye, et al., 2004). However, recent finding by (Mueller-Steiner, Zhou, et al., 2006) suggested CTSB by its endogenous proteolytic activity reduced amyloid plaque accumulation through increased protein turnover.

Endoplasmic Reticulum (ER) Stress

ER, with a pivotal pleiotropic physiological role in cellular biogenesis, metabolism, signaling and survival, is also a vital homeostatic organellar regulator of cellular stress (Travers, Patil, et al., 2000). It is the site for the proper synthesis, folding and post-translational modification of cellular proteins (Ron and Walter, 2007) as well as the production of steroids, cholesterol and other lipids (Chang, Chang, et al., 2006). It also serves as a major intracellular Ca^{2+} ion store (Verkhratsky, 2005).

Presence of ER stress has been reported in AD (Hoozemans, Veerhuis, et al., 2005), PD (DeGracia and Montie, 2004) and ischemic stroke (Kitao, Imai, et al., 2007). ER stress, characterized by the accumulation of unfolded proteins in the ER lumen, is frequently manifested upon presence of oxidative stress. This stress induction can occur upon perturbation of any of ER cellular functions, i.e. via protein oxidation, disturbance of Ca^{2+} signaling, and alteration of the homeostatic redox balance (Chakravarthi, Jessop, et al., 2006; Gorlach, Klappa, et al., 2006). An intimate communicative, functional coupling relationship between ER and mitochondria has also been established on the basis of these cellular functions. One instance would be the maintenance of Ca^{2+} equilibrium, crucial for the proper functioning of both organelles (Csordas, Renken, et al., 2006). Mitochondria act as an emergency Ca^{2+} store upon a sudden transient surge in cytosolic Ca^{2+} level, to buffer the ER against any functional disruption. Furthermore, several members of the B-cell lymphoma 2 (BCL2) family prominent for their roles in regulation of mitochondrial-mediated apoptosis also seem to participate in ER-induced cell death and Ca^{2+} signaling between the ER and mitochondria (Breckenridge, Germain, et al., 2003; Rao, Ellerby, et al., 2004; Gorlach, Klappa, et al., 2006; Szegezdi, Logue, et al., 2006; Wu and Kaufman, 2006). Initiation of ER stress has been demonstrated to occur upon mitochondrial energy deficits (Flores-Diaz, Higuita, et al., 2004; Xu, Liu, et al., 2004).

Extensive ER damage can trigger cell death via the production of unfolded proteins, the release of Ca^{2+} into the cytoplasm or altered redox homeostasis (Breckenridge, Germain, et al., 2003) resulting in either classical programmed cell death (PCD) or other mitochondrial cell death pathways (Jimbo, Fujita, et al., 2003). As such, dysfunctional Ca^{2+} regulation arising from ER stress and increased molecular oxidative damage further potentiates activation of a programmed necrotic pathway involving calpains, forming a positive feedback regulatory loop (Nakagawa and Yuan, 2000; Crocker, Smith, et al., 2003).

Oxidative Stress

Neuronal oxidative stress is a prominent phenomenon frequently observed in chronic neurodegenerative disorders such as AD (Sultana and Butterfield, 2009) and PD (Jenner, 2007) and acute neurological disorders such as stroke (Niizuma, Endo, et al., 2009). Oxidative stress results from the aberrant disruption of the physiological balance between the pro-oxidants and anti-oxidants in favour of the former (Lakhan, Kirchgessner, et al., 2009). It is characterized by a substantial rise in intracellular reactive oxygen species (ROS) level that is disruptive to cellular homeostatic balance through infliction of DNA, protein and lipid damages, release of Ca^{2+} from intracellular stores and chemotaxis,which explains the accumulation of oxidized protein aggregates seen in numerous neuropathological conditions (Lennon, Martin, et al., 1991; Liu, Hsu, et al., 1996; Evans and Cooke, 2004; Halliwell, 2006). Severe oxidative stress triggers cell death through necrosis while at moderate levels can induce apoptosis (Lennon, Martin, et al., 1991; Liu, Hsu, et al., 1996; Evans and Cooke, 2004).

Oxidative stress is a major patho-physiological mechanism characterized by the rise in ROS/reactive nitrogen species (RNS) is significantly enhanced through numerous cell death-aggravating cascades e.g. mitochondrial inhibition, Ca^{2+} overload, and inflammation (Coyle and Puttfarcken, 1993). Sources, both intrinsic and extrinsic, contribute to the increase in oxidative load. Mitochondria, as the main cellular site for energy production, offer a rich primary source of ROS through its respiratory electron transport chain, which constantly undergoes fluctuations in redox states (Halliwell, 2006). Other secondary cellular processes contributing to ROS production include lipid peroxidation, cation-associated Fenton reactions, NO-mediated protein nitrosylation and matrix enzymatic interactions (Chinopoulos and Adam-Vizi, 2006; Halliwell, 2006; Zundorf, Kahlert, et al., 2009).

The superoxide anion ($O_2\bullet^-$), a highly reactive primary radical, through xanthine oxidase and subsequently responsible for H_2O_2 formation and finally OH• generation (a short-lived but most reactive type of oxygen species), thus activatesa chain reaction cascade of ROS production after leakage from the mitochondrial electron transport chain (Lakhan, Kirchgessner, et al., 2009). H_2O_2 is lipid-soluble and readily tranverses cell membranes. Concurrently, $O_2\bullet^-$ crosses cell membrane via anionic channels (Kontos, 2001). On the other hand, nitric oxide (NO), a water- and lipid-soluble free radical that is produced from L-arginine by nitric oxide synthase (NOS), forms the basis for RNS formation. Elevated NOS type -I and -III activities have been observed in neurons and vascular endothelium respectively in several neuropathological conditions. As time progresses, elevated NOS type II (iNOS) activity is also detected in other cell types within the brain including glia and infiltrating neutrophils (Lakhan, Kirchgessner, et al., 2009).

Nitric Oxide (NO) Generation

An early study by Dawson(Dawson, et al., 1991) demonstrated the abrogation of Glu-induced neurodegeneration *in vitro* upon application of NO synthase (NOS) inhibitors, suggesting a prominent role of NO in Glu-induced excitotoxicity. NO once generated, can react with the heme group of guanylate cyclase (GC), triggering a conformational change in GC and the catalysis of guanosine-5'-triphosphate (GTP) to cyclic guanosine 3',5'-monophosphate (cGMP)(Ignarro, 1991) and protein phosphorylation. GC activation is believed to be the main nitrergic signal transduction pathway. cGMP acts as a second messenger that activates protein kinase G (PKG) 1 and 2, with the former involved in intracellular Ca^{2+} control and the latter regulating anionic influx, e.g. Cl^- ion (French, Bijman, et al., 1995; Lau, Kong, et al., 2003).This transduction pathway can affect a broad range of proteins directly, e.g. phosphodiesterases of cyclic nucleotides and indirectly, e.g. protein kinase A, thus increasing the level of cAMP with activation of proteins involved in the cAMP downstream pathway (Ono and Trautwein, 1991; Whalin, Scammell, et al., 1991).

Furthermore, NO is also capable of affecting other cellular signaling pathways independent of GC activation. As NO itself is highly reactive due to its thermodynamic instability, it undergoes various chemical reactions with gaseous molecules, anions and ROS readily to form nitrites (NO_2^-), nitrates (NO_3^-) and peroxynitrite ($ONOO^-$). NO reacts quickly with $O_2\bullet^-$ to form $ONOO^-$ to avoid its elimination by the antioxidant systems. $ONOO^-$ has an action radius of 100μm and an even shorter half-life of 1-2s, tending to generate multiple toxic products in its degradation (Beckman, Beckman, et al., 1990; Whiteman, Ketsawatsakul, et al., 2002). During the process of these chemical reactions, intermediate products such as ROS and other free radicals are produced. These nitrergic intermediate and end products can induce modifications of cellular molecules (lipids, proteins and DNA) through oxidation (Poon, Calabrese, et al., 2004), nitration (Souza, Daikhin, et al., 1999) or nitrosylation (Stamler, Toone, et al., 1997). For proteins, these modifications have special importance as they can modify protein conformation and affect their physiological functions. NO-mediated nitrosylation can alter the normal activity of a wide array of proteins by inducing conformational changes, which can hinder protein phosphorylation leading to a loss of function and/or gain-of-function, or affect signal transduction pathways of the growth factors (Cassina, Hodara, et al., 2000; Jonnala and Buccafusco, 2001; Newman, Hoffman, et al., 2002; Amici, Lupidi, et al., 2003). NO itself can also induce reversible inhibition of cytochrome c oxidase, thus resulting in decreasing oxygen consumption and impeding energetic metabolism. Interestingly, NO also promotes mitochondrial synthesis via enhancing the transcriptional activity of peroxisome proliferator-activated receptor gamma (PPARγ) coactivator-1(Scarpulla, 2002; Nisoli, Clementi, et al., 2003).

Recently, a novel cell death pathway implicating GAPDH and seven in absentia homolog 1 (SIAH1) has been linked to NO (Hara, Agrawal, et al., 2005). GAPDH is an omnipresent housekeeping enzyme that, under physiological conditions, participates in glycolysis. However, with increasing intracellular levels, NO can nitrosylate GAPDH allowing it to bind SIAH1, an ubiquitin ligase. The newly formed duo can then translocate to the nucleus by virtue of the SIAH1 nuclear translocation domain and enhances p300/CBP-associated acetylation of nuclear proteins (Sen, Hara, et al., 2008). The downstream activation of nuclear proteins including p53 causes pyknotic nuclei and morphological features suggestive of apoptosis.

Free Radical Generation

Free radical formation has been frequently observed in CGN (Lafon-Cazal, Pietri, et al., 1993) and cortical cultures (Dugan, Sensi, et al., 1995; Reynolds and Hastings, 1995) upon excitotoxic stimulus. Pharmacological application of various antioxidant compounds including nitrone-based scavengers, free radical spin traps and 21-aminosteroids/lazaroids and/or over-expression of anti-oxidative enzyme overexpressing SOD showed successful attenuation of Glu and ischemia-induced neurotoxicity (Gonzalez-Zulueta, Ensz, et al., 1998). It is believed that rise in intracellular Ca^{2+} via Ca^{2+} influxis responsible for the trigger of free radical production (Dykens, 1994) particularly by mitochondria. Mitochondria exposed to escalating Ca^{2+} and Na^{+} concentrations activate a feed-forward system enhancingfree radical production, which can be abolished by removal of extracellular Ca^{2+} but not NOS antagonists (Dugan, Sensi, et al., 1995). Similar dependence on NMDARs-mediated Ca^{2+} entry was substantiated by findings from Reynolds and Hastings (Reynolds and Hastings, 1995). In summary, it can be deduced that free radical generation in mitochondria is a secondary event to Ca^{2+} influx via NMDARs. Subsequently, these free radicals, particularly $O_2\bullet^{-}$, can interact with other radicals, e.g. NO to form powerful oxidants (Huie and Padmaja, 1993).

Neuroinflammation

The CNS has been misunderstood over the years to be an immune-privileged organ. However recent studies have uncovered the CNS to be engaging in substantial crucial, bi-directional crosstalk with the immune system. Emerging data from neuro-pathological studies further revealed that inflammatory cells participate in tissue remodeling after brain injury. Post-ischemic neuroinflammatory modulations cause dysfunction of the blood-brain barrier, cerebral edema, and neuronal cell death (Lakhan, Kirchgessner, et al., 2009).

Microglial cells, as the resident macrophages of the brain, play a critical role as phagocytic scavengers to the immuno-competent CNS. Microglia can evolve into phagocytes secreting a variety of substances, many of which are cytotoxic and/or cytoprotective. These activated microglial cells are also capable of releasing several pro-inflammatory cytokines such as tumor necrosis factor-alpha (TNF-α), interleukin-1-beta (IL-1β) and interleukin-6 (IL-6), as well as other potential cytotoxic molecules including NO, ROS, and prostanoids (Lucas, Rothwell, et al., 2006).Similarly to microglia, astrocytes can also secreteinflammatory factors such as cytokines, chemokines, and NO (Swanson, Ying, et al., 2004). Interestingly, microglia may induce neuroprotection by secreting neurotrophic molecules such as brain-derived neurotrophic factor (BDNF), insulin-like growth factor I (IGF-I), and several other growth factors.

Caspase Activation

Caspases, a family of well-studied cysteine proteases notable for their role in classical apoptosis, was first demonstrated to be implicated in persistent excitotoxic injury in cerebrocortical (Tenneti, D'Emilia, et al., 1998) and cerebellar cultures (Du, Bales, et al.,

1997). Pre-treatment with caspase inhibitors showed neuroprotective efficacy in NMDA-mediated neurodegeneration. However, Ca^{2+} influx and mitochondrial dysfunction were not inhibited, indicating caspase activation occurs downstream of these events. Other than promoting cleavage of cytosolic proteins, nuclear caspase activation facilitates ICAD (a protein inhibiting the activity of caspase-3-activated DNase) truncation, resulting in DNA fragmentation and cell death (Enari, Sakahira, et al., 1998).

Calpains Involvement

Escalating intracellular Ca^{2+} level activates Ca^{2+}-dependent cytoplasmic proteases namely calpains. Proteolytic activation of one calpain isoform, μ-calpain, is necessary for the cleavage and release of apoptosis-inducing factor (AIF) from mitochondria (Polster, Basanez, et al., 2005). Calpain/AIF pathway importance in excitotoxicity is further underscored by neuronal culture studies subjected to oxygen–glucose deprivation showing that calpain suppression attenuated AIF translocation and subsequent neuronal death (Cao, Xing, et al., 2007). Mitochondrial AIF re-localization to the nucleus causes chromatin condensation, DNA fragmentation and cell death (Susin, Lorenzo, et al., 1999). An alternative route to excitotoxicity is via activation of PARP-1, a nuclear DNA repair enzyme, which incidentally also triggers the release of AIF and which involvement is highlighted by cortical cultures derived from PARP-1 knockout mice exhibiting reduced AIF translocation and neurodegeneration after N-methyl-D-aspartate (NMDAR agonist) treatment (Yu, Wang, et al., 2002).

Novel Insights into Neuronal Apoptosis Based on Multiple Neuropathy-Causative Agents

Several apoptosis-associated patho-physiological mechanisms such as oxidative stress, neuroinflammation, cell cycle re-entry and excitotoxicity are not exclusive to any one neurodegenerative disease or neurological disorder, but are frequently found to occur concurrently in several neuropathies. Particularly, oxidative stress can be considered to be a universal patho-physiological phenomenon consistently observed in chronic neurodegenerative disorders such as AD (Sultana and Butterfield, 2009) and PD (Jenner, 2007) and acute neurological disorders such as stroke (Niizuma, Endo, et al., 2009). Even though this phenomenon has been ubiquitously detected in post-mortem brains of these neurological disorders through its consequential effects, the significance of its implication during neuronal death progression remains unclear. As such, comparative microarray analysis of specific neuropathy-representing models becomes an invaluable avenue to decipher the pathological mechanisms upstream and/or downstream of neuronal apoptosis. From there, novel insights into the signaling transduction pathways modulated upon its occurrence would form the foundation for a screening platform in the identification of potential universal biological targets useful in the area of therapeutic management.

Adopting cultured murine primary cortical neurons as the basis of the *in vitro* model, focus on the commonality of the signaling pathways regulated upon individual 24h time-course neuronal treatments with five well-characterized apoptosis-causative agents: HOCl, NO, lactacystin, rotenone and Glu at their respective EC_{50} concentrations as previously published from our laboratory (Chen, Yap, et al., 2006; Yap, Whiteman, et al., 2006; Cheung, Peng, et al., 2007; Peng, Chen, et al., 2008; Yap, Chen, et al., 2010; Choy, Chen, et al., 2011). They are most appropriately selected as they either have been implicated in the pathogenesis of several neurological dysfunctions, or are well-represented agents of apoptosis-implicating neurodegenerative models through reproduction of pathological morphological characteristics of the disease state.

Involvement of HOCl has been documented in numerous inflammatory and oxidative strain–related diseases namely atherosclerosis (Hazell, Arnold, et al., 1996), cystic fibrosis (Kettle, Chan, et al., 2004) and neurodegenerative disorders such as PD (Choi, Pennathur, et al., 2005) and AD (Green, Mendez, et al., 2004). As compared to other fields of disease study, the role of HOCl in neurodegenerative disorders is still lacking. In the mammalian brain, HOCl is produced predominantly by concerted activations of NADPH oxidase and myeloperoxidase (MPO) in activated microglia and infiltrated neutrophils and monocytes during inflammatory pathological conditions (Bianca, Dusi, et al., 1999; Gonzalez-Scarano and Baltuch, 1999). However, expression of NADPH oxidase and MPO in neurons has also been reported (Noh and Koh, 2000; Tammariello, Quinn, et al., 2000; Green, Mendez, et al., 2004; Vallet, Charnay, et al., 2005). In PD, an up-regulation of MPO with a corresponding elevation of HOCl-modified proteins has been demonstrated in the ventral midbrain of post-mortem brains and disease mouse models. Further, in AD brains, 3-chlorotyrosine, a biomarker of HOCl, existed at three-fold that of control brains (Green, Mendez, et al., 2004).

Similarly, excessive endogenous NO production has been identified in neurological disorders linked to oxidative stress such as ischemia (Cuzzocrea, Riley, et al., 2001), ALS(Cookson and Shaw, 1999), AD (Good, Werner, et al., 1996), and PD (Good, Hsu, et al., 1998). The undesirable downstream effects of deregulated NO presence have been previously mentioned. Abundant presence of NO-originated intermediates and byproducts induces dysregulated modifications of cellular molecules (lipids, proteins and DNA) through oxidation (Butterfield, 1997), nitration (Souza, Daikhin, et al., 1999) or nitrosylation (Stamler, Toone, et al., 1997).

Rotenone, a specific mitochondrial complex I inhibitor, is well known for its ability to mimic the pathological characteristics of PD (e.g. presence of Lewy bodies) within dopaminergic neurons. Mitochondrial complex I is involved in the superoxide formation during physiological respiration and its dysfunction is denoted as a hallmark feature of PD identified in the substantia nigra par compacta of disease patients (Lin and Beal, 2006; Mancuso, Coppede, et al., 2006). It has been further suggested that partial inactivation of the complex activity observed in PD would enhance ROS production as compared to that of full inhibition, aggravating the extent of cellular damages (Pitkanen and Robinson, 1996; Votyakova and Reynolds, 2001).

Pharmacological proteasomal inhibitor, lactacystin, can trigger oxidative stress indirectly via the promotion of aberrant, toxic protein buildup as a result of UPS dysfuction, and thus induce apoptosis in cultured mouse cortical neurons (Cheung, Choy, et al., 2004; Yew, Cheung, et al., 2005). With specific regard to the CNS, proteasomal inhibition has been linked to the cellular toxicity and pathology observed in the brain during normal aging, in neurodegenerative diseases such as PD and AD, as well as cerebral ischemia (Keller, Huang, et al., 2000; Ding and Keller, 2001; Grune, Jung, et al., 2004; Keller, Dimayuga, et al., 2004).Incidentally, proteasomal inhibition is most severe in the brain region that exhibits the largest amount of pathology (Keller, Hanni, et al., 2000; McNaught, Olanow, et al., 2001).Proteasomal inhibition is also reported to mediate deleterious alterations in cell cycle regulation, inflammatory processes and protein aggregation and trigger the cell death pathway (Rockwell, Yuan, et al., 2000; Demasi and Davies, 2003; Rideout, Wang, et al., 2003; Yew, Cheung, et al., 2005).Recently, studies have shown that the inhibition of proteasome using a proteasomal inhibitor such as PS-341 induced apoptosis through the induction of ER stress-reactive oxygen species (Fribley, Zeng, et al., 2004).

Glu is the only physiological excitatory neurotransmitter agonist for GluRs activation, be it metabotropic or ionotropic. As such, Glu–mediated neuronal death comprises two concurrent branching signaling components mediated by ionotropic and metabotropic GluRs respectively.However, during the neuropathological state when marked elevation of extracellular Glu was released by damaged neurons and astrocytes, Glu not only activates iGluRs resulting in excitotoxicity, but also mGluRs, all in all contributing to neuronal death.

Gene functional classification and biological process clustering were performed on the global transcriptomic profile of individual models. Enriched biological processes were generated and compared for identification of common pathways. This prevents the potential of eliminating potential genes related to a particular significant pathway that were not universally significantly regulated in all models, and which could otherwise be filtered off if selection of genes common to all models was to be done. Furthermore, this approach is plausible as it allows a more affirmative conclusion of the overall regulatory trend of the signaling pathways through evaluation of all genes encoding for members of a particular signaling cascade.

Response to Oxidative Stress

Oxidative Stress-Induced Expression via Nrf2

Oxidative stress is a result of several intrinsic and extrinsic factors. Mitochondria,which provide cells with the energy for various cellular activities are a rich intrinsic origin of ROS. For any energy generated via the electron transport chain, which involves a series of redox reactions, a small percentage of ROS is produced (Halliwell, 2006). Variousprocesses within the cell that also lead to ROS formation include lipid peroxidation, metal ion-associated Fenton reactions, NO-mediated protein nitrosylation and matrix enzymatic interactions(Chinopoulos and Adam-Vizi, 2006; Halliwell, 2006). Oxidative stress in general elicits both pro-survival and pro-death cellular responses. In particular, pro-survival response can be witnessed at either transcriptional or post-translational modification levels.

Nrf2, a pro-survival transcription factor primarily localized in the nucleus, is ubiquitously expressed in a wide variety of tissue and cell types(McMahon, Itoh, et al., 2001). The neuroprotective role of Nrf2 has been demonstrated in *in vitro* and *in vivo* models of neuropathy(Johnson, Johnson, et al., 2008), amidst its assertive anti-cancerous and anti-inflammatory functions in tumor growth and tissue injuries (Ramos-Gomez, Kwak, et al.,

2001; Reddy, 2009).It is involved in the transcriptional activation of ARE-containing genes encoding detoxifying enzymes and cytoprotective antioxidant proteins,e.g. superoxide dehydrogenase, Nqo1, and glutathione *S*-transferase (Gst) upon oxidative and electrophilic stresses (Zhao, Moore, et al., 2007).Under physiological basal conditions, induction of Nrf2 target gene expression is inhibited by Nrf2 retention in the cytoplasm by its interacting inhibitory partner, KEAP1, which targets it for degradation by the UPS (reviewed in Kaspar, Niture, et al., 2009). In the event of cellular stress, rise in intracellular ROS and electrophiles induce the activation of several protein kinases, including PKC, ERK, MAPK, and PKR-like ER kinase (PERK) (Buckley, Marshall, et al., 2003). These protein kinases induce the phosphorylation of Nrf2, releasing it from KEAP1 sequestration and facilitating its translocation into the nucleus where it binds to ARE sequence, leading to transcriptional activation of antioxidant enzymes (Huang, Nguyen, et al., 2000; Nguyen, Huang, et al., 2000; Zipper and Mulcahy, 2000; Bloom and Jaiswal, 2003; Buckley, Marshall, et al., 2003; Cullinan, Zhang, et al., 2003).

Individual analytical results from the time-course global expression profiles of HOCl-, NO-, rotenone-, lactacystin- and Glu-treated neurons revealed an elevated expression of several oxidative stress-responsive neuroprotective genes that are often activated by Nrf2. Nrf2-targeted pro-survival genes,e.g. Pkc, activating transcription factor 4 (Atf4), Gst subunits, Hmox1 and Hsps(categorized under Endoplasmic reticulum stress via unfolded protein response in Table1) demonstrated consistent transcriptional up-regulation with concomitant Nrf2 elevated gene expression in all five oxidative stressor models. Because proteins encoded by Nrf2-induced transcriptional target genes also play prominent roles in the unfolded protein response (UPR) triggered upon ER stress, it is reasonable to postulate that an intimate regulatory feedback relationship exists between processes that are activated during cellular oxidative and ER tensions.

HSPS and Molecular Chaperones

Oxidative damages to existing cellular proteins can also be mitigated at the post-translational stage through the mobilization and activation of HSPs and metal ion molecular chaperones. HSPs and molecular chaperones are crucial in the facilitation of the refolding of misfolded proteins to avoid their aggregation and accumulation in the cell (Meriin and Sherman, 2005). They cooperate with the ubiquitin-proteasome system (UPS) to prevent intracellular buildup of aberrant toxic proteins, and regulate the degradation of excess cytoplasmic proteins. HSPs confer cellular protection mainly through two mechanisms: firstly, as mentioned, HSPs act as molecular chaperones to ensure correct formation and maintenance of native conformation of cytosolic proteins and stabilization of actin filaments; secondly, suppression of the pro-apoptotic member of the BCL2 family, BID, by HSP70 (HSPA8) and HSP27 (HSPB8) prevents cytochrome c (CYTC) release (reviewed in Franklin, Krueger-Naug, et al., 2005).

Our comparative gene expression profiling study demonstrated a generally similar elevation in gene expression trend of several Hsps such as Hspb1, Hsp2, Hspa8, Hsp9a and Serpinh1 (Hsp47) (categorized under Endoplasmic reticulum stress via unfolded protein response in Table 1).

Table 1. Selected enriched biological processes consisting of significantly up-regulated genes common to four oxidative stressors-induced neuronal injury models

Biological Processes	Oxidative Stressors				
	HOCl	NOC-18	Rotenone	Lactacystin	Glu
Response to oxidative Stress Oxidative stress-induced expression via Nrf2	↑ Nrf2, Gst (alpha)	↑ Nrf2, Gst (alpha), Hmox1, Atf4, Pkc, Fos, Jun, Ugt	↑Nrf2, Gst (alpha), Hmox1, Atf4, Fos, Ugt	↑Nrf2, Gst (alpha), Hmox1, Atf4, Fos, Ugt	↑Nrf2, Hmox1, Gst (alpha), Fos, Ugt
Heat shock proteins (Hsps) and molecular chaperones	↑ApoE, Nqo1, Txnl2, Prdx(1, 6)	↑ Nqo1, Ptgs2, Npn3, Prdx (1, 6)	↑Nqo1, Gab1 / Gab3, Ptgs2, Npn3, Prdx6, Sesn3	↑Npn3, Txnl1, Ptgs2 ↓ApoE, Prdx (2, 3, 5, 6)	↑ ApoE, Nqo1, Gab1, Npn3, Prdx (1, 6), Mt (1, 2, 3), Sesn3
Glutathione (GSH) anti-oxidant pathway	↑ Gst (alpha, mu, pi), mGst1, Gsr1, Gcl	↑Gst (alpha, mu, theta), mGst1, Gpx1, Gss, Gsr1, Gcl, Idh2 ↑	↑Gst (alpha, mu, pi), mGst1, Gpx1 / Gpx7, Gss, Idh2	↑Gst (alpha), Gsr1, Gcl ↓Gst (mu, theta, zeta), mGst, Idh2	↑Gpx1, Gst (alpha, mu, theta), mGst1, Idh2
Endoplasmic reticulum (ER) stress via unfolded protein response (UPR)	↑Serpinh1 ↑↓Hspa1a, Hsp (ca, cb), Hsp105, Chaperonin 10, Hsp (a2, a8), Txdc4 ↑↓	↑Hspa8, Hspb8, Hspa9a, Hsp14, Serpinh1, Txdc4, Perk, Derl1, Ero1l, Herpud1	↑Hsp (a1, a8), Hspa9a, Txdc4, Ern2	↑Hspa1a, Hsp (a5, a8), Hspa9a, Hspb1, Hspd1, Hspe1, Hsp (ca, cb), Chaperonin 10, Hsp105, Serpinh1, Perk, Ero1l, Herpud1	↑Hsp (a2, a8), Serpinh1, Perk
Ubiquitin-Proteasome System (UPS)	↑Ube2a, ↑↓Ube2e, Usp (5, 19, 22), Psma (1, 7), Psmd8	↑Usp (2, 36) ↓Ube2c, Ube2n, Ube3c, Psma1	↓Ube1x, Ube2d, Ube2n, Ube3b, Psma (1, 3)	↑Ubc, Ube2g2, Ube3a, Ufd1l, Usp (8, 16, 40), Psma (1, 7), Psmb (3, 7), Psmc1 ↑↓Usp (3, 30), Psmb1, Psmc (2, 3, 4, 5) ↓Usp (29, 36, 43), Psmb10	↑Usp3 ↑↓Psma1 ↓Ube1x, Ube2n

Table 1. (Continued)

Biological Processes	Oxidative Stressors				
	HOCl	NOC-18	Rotenone	Lactacystin	Glu
Mitochondrial respiratory chain	↑Acad, Cyp7b1, Cyp4v3 ↑↓Gpd2, ETC complex-I (Ndufa3), ETC complex-III (Uqcr), ETC complex-IV (Cox4i1, Cox7b, Cox7c)	↑Acad, Idh2, ETC complex-IV (Cox6a2, Cox7c) ↓ ETC complex-IV (Cox7a1), Gpd1	↑Acad, Idh2, ETC complex-IV (Cox7b), Cyp1b1, Cyp4v3, Cyp4a14, Cyb5r1, Cyb5r2 ↓ ETC complex-I (Nduf), ETC complex-II (Sdhb, Sdhd), ETC complex-III (Uqcr), ETC complex-IV (Cox5b, Cox6a2)	↑Acad, Cyp26b1, Cyp4b1, Cyp7b1, Etfb ↓ ETC complex-I (Ndufa6), ETC complex-III (Uqcr), ETC complex-IV (Cox5b, Cox6a2, Cox6b, Cox6c, Cox7b), Idh2, Cyb561, Cyb5	↑Acad, Idh2, Etf (a, d), ETC complex-IV (Cox6b, Cox8b), ETC complex-IV (Cox6a)
Calcium ion binding and homeostasis	↑Trpc4, S100A (1, 6, 10, 11), Fah, Cib2 AnxA (2, 3) ↑↓Gpd2, Lman	↑Trpm7, S100A (1, 4), S100b, Atp2c, Pkca, Fk506bp (9, 10), Lamc1, AnxA (2, 3, 4, 5) ↑	↑S100A (1, 6, 10, 11), Fk506bp (9, 10), AnxA (2, 3, 5)	↑S100A (10, 13), AnxA2 ↓Fk506bp (2, 4, 8)	↑S100A (6, 11), FK506bp (9, 10), Calml4, Lamb2, AnxA (2, 3, 5), Capn2
Programmed cell death	↑Bax, Bad ↑↓Casp (3, 8, 9), Cytc	↑Tnfrsf (1a, 12a), Pawr, Dap, Btg, Siva, Puma, Noxa ↑↓ Dapk2	↑Tnfrsf (1a, 10b, 12a), Traf1, Pawr, Dap, Casp8, Fas	↑Fas, Puma, Noxa	↑Casp6, Dap, Pawr, Aif ↑↓Tnfrsf12a

The genes demonstrated in the table were chosen based on the criteria of having at least ±1.5 fold transcriptional expression in a minimum of one times-point and passed statistical testing, one-way ANOVA, $p < 0.05$ in each oxidative stressor transcriptomic profile. Cultured mouse primary cortical neurons (dissected from brains of gestation day 15-16 Swiss white albino embryos; Cheung, et al., 1996) were treated with 250μM HOCl, 0.5mM NOC-18, 10nM rotenone and 250μM Glu respectively and RNA collected over a 24h timeframe. Individual transcriptomic profile was subjected to functional-gene ontology classification using the online bioinformatics database DAVID 2008. Selected biological processes (shown in table below) were generated through gene enrichment and statistically validated by gene-term enrichment score through modified Fisher's exact test and Benjamini correction (Huang, et al., 2009; Dennis, et al., 2003). Up (↑) and/or Down (↓) arrow after gene symbol denotes the overall regulatory trend over a 24h time-course transcriptomic study for each oxidative stressor model, i.e. a specific gene which show early up-regulation at the initial phase, with subsequent down-regulation at a later time are indicated with opposed arrows (↑↓).

The same phenomenon is observed for peroxiredoxins (Prx1 and Prx6), prostaglandin-endoperoxide synthase 2 (Ptgs2; also known as cyclo-oxygenase 2) and neoplastic progression 3 (Npn3; also known as sulfiredoxin) and members of the metal ion chaperones, metallothionein family (Mt1, Mt2 and Mt3) in all models with the exception of lactacystin.

Glutathione (GSH) Anti-Oxidant Pathway

Further to the cytoprotective effects offered by the HSPs and chaperones, the GSH pathway also serves to be a major effective mean to reduce oxidative stress via sequestration of NO, $ONOO^-$ and ROS. Genes related to GSH biosynthesis induced in the presence of oxidative stress acts as a cellular defense mechanism. During oxidative stress, increased $NADP^+$ concentration will lead to activation of glucose-6-phosphate dehydrogenase, enzyme that catalyses the first rate-limiting step in the oxidative branch of pentose phosphate pathway (PPP), and subsequently stimulation of PPP (Ursini, Parrella, et al., 1997).Studies have showed that stimulation of PPP in neurons and astrocytes conferred protection against ROS-induced toxicity by producing NADPH (Ben-Yoseph, Boxer, et al., 1996).Using GSH and NADPH as co-factors, detoxification enzymes such as GST, NQO1, mitochondrial aldehyde dehydrogenase (ALDH), and Aldh dehydrogenase/reductase 8 (DHRS8) help to detoxify quinones and maintain redox balance within the cells.As such, for efficient detoxification upon oxidative stress, it would be beneficial to increase expressionof these enzymes together with their co-factors, which include GSH and NADPH. Hsps genes were also regulated in a GSH-dependent manner (Calabrese, Scapagnini, et al., 2003).Fratelli, Goodwin, et al. (2005) showed that HSP40 and HSP70 are strongly induced by GSH depletion.

Indeed, initiation and persistent activation of the GSH anti-oxidative pathway was reflected in microarray analyses, whereby there was a clear and concerted transcriptional up-regulation trend of its pathway members (Gcl, Gst, Gsr and Gpx1) and associated detoxifying enzymes (Nqo1 and Npn3) and Hsps (Serpinh1 and Hspa8) in HOCl, NO, rotenone and Glu models (Table 1). Intriguingly, lactacystin-mediated neuronal injury demonstrated a mixed transcriptional response within the different Gst subunits.

Altogether, the transcriptional up-regulation of detoxification enzymes, antioxidant proteins and Hsps upon oxidative stress is a result of activation by Nrf2 via the ARE motif found in their promoters followingalteration of thiol redox state (Wasserman and Fahl, 1997). Thus, the coordinate up-regulation of these genes can have a synergistic effect in the maintenance of GSH levels as well as detoxification of reactive intermediates. As seen, Nrf2-induced pathway plays a vital role in promoting cell survival by fighting oxidative insults.

Endoplasmic Reticulum (ER) Stress via Unfolded Protein Response (UPR)

ER, with a pivotal pleiotropic physiological role in cellular biogenesis, metabolism, signaling and survival, is also a vital homeostatic organellar regulator of cellular stress(Travers, Patil, et al., 2000). It is the site for the proper synthesis, folding and post-translational modification of cellular proteins (Ron and Walter, 2007), as well as the production of steroids, cholesterol and other lipids (Chang, Chang, et al., 2006). It also serves as a major intracellular Ca^{2+} store (Verkhratsky, 2005).

Presence of ER stress has been reported in AD (Hoozemans, Veerhuis, et al., 2005), PD (Kitao, Imai, et al., 2007) and ischemic stroke (DeGracia and Montie, 2004). ER stress, characterized by the accumulation of unfolded proteins in the ER lumen, is frequently inflicted upon presence of oxidative stress. This stress induction can occur upon perturbation of any of ER cellular functions, i.e. via protein oxidation, disturbance of Ca^{2+} signaling, and alteration of the homeostatic redox balance (Gorlach, Klappa, et al., 2006). An intimate communicative, functional coupling relationship between ER and mitochondria has also been established on the basis of these cellular functions. One instance would be the maintenance of Ca^{2+} equilibrium, crucial for the proper functioning of both organelles (Csordas, Renken, et al., 2006). Mitochondria act as an emergency Ca^{2+} store upon a sudden transient surge in cytosolic Ca^{2+} level, to buffer the ER against any functional disruption. Furthermore, several members of the Bcl2 family prominent for their roles in regulation of mitochondrial-mediated apoptosis also participate in ER-induced cell death and Ca^{2+} signaling between the ER and mitochondria (Rao, Ellerby, et al., 2004; Gorlach, Klappa, et al., 2006). Initiation of ER stress has been demonstrated to occur upon mitochondrial energy deficits (Flores-Diaz, Higuita, et al., 2004).

Upon ER stress, UPR is triggered to restore homeostatic balance (Schroder and Kaufman, 2005). UPR serves as a double edge sword inducing resultant pro- and anti-apoptotic effects. For the start where oxidative stress induced cellular damage is less severe and within ER tolerable limits, UPR activates signaling pathways that involve kinases residing in ER namely PERK and inositol requiring enzyme (IRE1) as well as transcription factors Atf 4 and 6, to alleviate ER burden and promote cell survival. Mitigation of ER stress is achieved via two cellular routes.Firstly, increased expression of ER chaperones via IRE1-induced expression of Xbp1, a transcription factor that activates transcription of genes coding for proteins needed for the ER protein folding and processing reactions (Schroder, 2008), allows rapid correction and elimination of unfolded and/or misfolded proteins. Secondly, PERK-mediated inhibitory phosphorylation of the eukaryotic initiation factor (eIF2α) prevents accumulation of freshly synthesized, unassembled proteins by shutting down cellular protein translation machineries(Harding, Zhang, et al., 1999).

However, once ER stress is too much for cells to handle whereby homeostasis cannot be restored, UPR activates mechanisms that summon cell death. Programmed cell death pathways resulting from ER stress responses can be mitochondria dependent and/or independent (Li, Lee, et al., 2006). Prolonged and extensive ER damage causes apoptosis onset through the production of unfolded proteins or the release of Ca^{2+} into the cytoplasm (Rao, Ellerby, et al., 2004). ER stress can activate its resident protease, caspase-12 and Atf6-induced CHOP expression, causing mitochondrial membrane permeabilization which eventually leads to either classical apoptosis or other mitochondrial cell death pathways (van der Sanden, Houweling, et al., 2003). In addition, the Ca^{2+} release from ER can activate calpains, which are normally kept inactive by their endogenous inhibitor calpastatin, eventually leading to programmed necrosis(Wang, 2000). Excessive calpain activation in vitro can lead to non-specific degradation of constituent proteins, including cytoskeletal proteins, and growth factor receptors.

Considerable amount of ER stress is evident from the gene expression profiles of all five oxidative stressor models. Momentous up-regulation of genes such as Atf4, Nrf2 and Perk, indicates UPR activation (Table 1). Possibly, augmentation of UPR-related genes expression might be a consequence of Nrf2 transcriptional activation, as mentioned earlier. Moreover, ER-stress inducible chaperones (Txdc4 and Hsps) and pro-apoptotic proteins (Puma and Noxa) (discussed in details in Cell Death section) also showed transcriptional up-regulation. As a whole, it is clear that ER stress, which is experienced by cells during oxidative tension, fires UPR to remove oxidized proteins and maintain proper intracellular ionic environment.

Ubiquitin-Proteasome System (UPS)

The UPS is important for the degradation and removal of unwanted and/or misfolded proteins in all eukaryotes (Schroder and Kaufman, 2005). It is a temporally controlled and tightly regulated process whereby compromises in its function and efficiency have been linked to aging(reviewed in Gaczynska, Osmulski, et al., 2001) and various neurodegenerative disorders where aberrant protein aggregates and inclusions were observed (Mandel, Grunblatt, et al., 2005). A down-regulation of genes encoding proteasome subunits in the substantia nigra pars compacta of PD brain has been reported by (Mandel, Grunblatt, et al., 2005), indicating the maintenance of functional proteasomal expression isinevitable for neuronal survival under stress conditions. UPS can be segregated into two distinct processes which starts off with ubiquitinylation of the proteins mediated by three enzymes namely, ubiquitin-activating enzyme E1 (UBE1), ubiquitin-conjugating enzyme E2 (UBE2) and ubiquitin protein ligase E3 (UBE3), followed by the targeting of the ubiquitin-linked protein/polypeptide to the proteasome for clearance.

Time dependent regulatory fashion of UPS-related genes as seen in the gene expression profiles of all oxidative stressor models, with the exception of HOCl and lactacystin, suggested the occurrence of an UPS dysfunction in which genes encoding for ubiquitinylation enzymes and certain proteasome subunits (Table 1) were mostly down-regulated. These genes include those encoding for the proteasome subunits (Psma1 and Psma3), Ube1 (Ube1x), Ube2 (Ube2c, Ube2a, Ube2d2 and Ube2n) and Ube3 (Ube3b and Ube3c). In contrary, deubiquitinylating enzymes like ubiquitin-specific peptidases (Usp2, Usp3, Usp22, Usp36), which were significantly up-regulated further affirm the decreased efficiency of UPS function.

An early initial up-regulation of UPS genes that eventually falters with time is especially prominent in HOCl-mediated neuronal injury (Table 1), suggestive of a regulatory feedback mechanism in an attempt to reverse the detrimental effects of proteasomal inhibition in the event of a proteasome dysfunction-induced cell death (Yew, Cheung, et al., 2005). This mirrors a partial regulatory trend of the UPS genes (Usp3, Usp30, Psmb1, Psmc[2-5]) in lactacytin-mediated neuronal injury.

On the other hand, the remaining genes were up-regulated in lactacystin profile.These include genes that encode for the proteasome subunits (Psma1, Psma7, Psmb3 and Psmb7), ubiquitins (Ubc), ubiquitin-conjugating enzyme E2 (Ube2g2), ubiquitin protein ligase E3 (Ube3a) and ubiquitin fusion degradation 1 like (Ufd1l). This is in consistent with report made by (Meiners, Heyken, et al., 2003)whereby proteasomal inhibition triggered a transient and concerted up-regulation of all 26S proteasome subunit mRNAs and enhanced synthesis of all proteasomal subunits that resulted in an increase in the number of proteasomes.

Mitochondrial Respiratory Chain/Non-Respiratory Chain Enzymes

The mitochondrial network plays a vital role in the supply of cellular energy currency in the form of ATP to ensure the proper functioning of for a variety of metabolic processes within a cell. Simpler molecules resulting from the cellular cyclic processing of macromolecular nutrients transfer electrons to carrier proteins such as NAD+ and FAD+ producing NADH and $FADH_2$, which transfer the electrons to the ETC localized at the inner mitochondrial membrane (Saraste, 1999).

Due to the constitutive cyclic fluctuation of the redox status between ETC enzymatic protein complexes with consequent high consumption of cellular oxygen in the oxidative phosphorylation process, mitochondria are assumed to be the main cellular producers of ROS (Orrenius, Gogvadze, et al., 2007).

Escaping electrons from the ETC can potentially reduce oxygen to form the highly reactive free radical superoxide anion, which can undergo further Fenton reaction to generate hydroxyl radical and hydrogen peroxide which similarly can cause detrimental cellular damages (Boveris, Oshino, et al., 1972).

As a result of this pivotal physiological function of mitochondria which if not properly managed can result in adverse effects on cell survival. Mitochondrial functionality has been proposed to be a crucial regulator and indicator of cellular homeostasis. Indeed, decline in mitochondrial functionality has been closely linked to the increasing age of mammalians. This age-correlated respiratory chain deficiency is especially prevalent in only a subset of mammalian tissues, such as heart, skeletal muscle, colonic crypts and neurons (Dufour, Terzioglu, et al., 2008).

A recent study by Dufour, et al.(2008) demonstrated that the co-existence of functional respiratory chain-deprived and normal neurons accelerated the neurodegenerative process of the adjacent normal cells through a trans-neuronal signaling mechanism.

As inferred from DNA microarray analysis, genes accounting for mitochondria related enzymes that are involved in oxidative phosphorylation and ETC processes were substantially regulated.

Main players of the mitochondrial ETC for examples, complexes I to IV, were mainly down-regulated across five oxidative stressor models, with the rotenone treatment being especially prominent with all four complexes being transcriptionally affected (Table 1).In particular, mitochondrial complex IV demonstrated significant transcriptional regulation across all five oxidative stressor models.

Interestingly, other mitochondrial independent microsomal electron transferring pathways like the cytochrome p450 family (Cyp), and cytochrome b5 reductase (Cyb5) were up-regulated. This might signify an attempt to counterbalance energy shortage (Table 1). Likewise, proteins that are involved in oxidative phosphorylation such asglycerol-3-phosphate dehydrogenase (Gpd), acetyl-coenzyme A dehydrogenase (Acad) and isocitrate dehydrogenase (Idh) also demonstrated elevated gene expression.

Calcium Ion Binding and Homeostasis

Because Ca^{2+}-activated protein families can either elicit cell survival or its demise, proper and delicate handling of calcium ion homeostasis is of critical importance. Briefly, these opposing consequences are facilitated by Ca^{2+} released from organelles via different calcium channel subtypes during stressful times. This calcium release will then stimulate Ca^{2+}-dependent proteins that are localized at specific areas within cells. To counteract this unexpected and temporal increase in intracellular calcium due to oxidative stress, cells possess an intrinsic fail-safe system,guarded and enforced by the mitochondria (reviewed in Chinopoulos and Adam-Vizi, 2006).Here, mitochondria respond and regulate cellular Ca^{2+} signals by buffering Ca^{2+}content and provide cells with energy in response to elevated intracellular Ca^{2+}concentration (Santo-Domingo and Demaurex, 2010). However, sustained leakage of calcium ions to surroundings due to prolonged ER stress and increased ROS production could result in over activation of calpains and interfere with normal cellular signaling cascades that eventually cause cell death(Crocker, Smith et al. 2003).

In all five oxidative stressor models, evaluation of transcriptomic profiles disclosed significantelevated gene expression of proteins involved in calcium ion binding and homeostasis that included Ca^{2+}-activated S100A, FK506 Ca^{2+}-binding protein families and annexins (Anxs) (Table 1). In addition, a transcriptional up-regulation of the Ca^{2+}-activated protease calpain-2 (Capn2), prominently implicated in programmed necrosis (discussed in detail in Cell Death), is observed in the Glu-mediated neuronal injury model. This is a probable indication of the presence of atypical, high intracellular Ca^{2+} level, which enhances Ca^{2+} signaling resulting in an increased demand for elevated expression of its Ca^{2+}-dependent members.

Cell Death

It is evident from the microarray analysis that the apoptotic mechanisms, both mitochondrial-dependent and -independent, were at work in the facilitation of neuronal death in the four distinct oxidative stressor models. However, the up-regulated set of PCD genes differ across models even though the genes encode for proteins involving in the same pathway. Transcriptional up-regulation of the main players in mitochondrial-dependent pathway, comprising of cytochrome c (Cytc), caspase 3 (Casp3), BH3 interacting domain death agonist (Bid), Bcl2-associated x protein (Bax) and apoptosis-inducing factor (Aif), and that of the extrinsic apoptotic pathway, including the tumor necrosis factor receptors (Tnfrsf1), Fas, caspase 8 (Casp8) and death associated protein (Dap), were detected (Table 1).

Intrinsic Mitochondrial-Dependent Apoptosis

Activation of the intrinsic mitochondrial apoptotic pathway as a response to elevatedROS production and mitochondrial DNA damage facilitates outer membranepermeabili-zation and mitochondrial-cytoplasmic translocation of CYTC, AIF, or SMAC/DIABLO, which induce downstream caspase-dependent or -independent cytosolic signaling events (Ryter, Kim, et al.,

2007). In the event of the occurrence of former, CYTC interacts with APAF1 to form the apoptosome complex, which further recruits and activates the zymogen of CASP9. The fully assembled apoptosome then proceeds to induce cleavage of downstream effector CASP3 and CASP7. In form of positive feedback mechanism, activated CASP3 promotesCASP2 and CASP6 activation that further promoteCASP9 processing(Slee, Harte, et al., 1999). Concurrently, the antagonistic effect of Smac/Diablo on the inhibitors of apoptotic proteins (IAPs) enhances caspase activation. On the other hand, AIF mediates caspase-independent signaling through cytoplasmic-nuclear translocation to mediate nuclear chromatin condensation and DNA fragmentation (Susin, Lorenzo, et al., 1999). From Table 1, it is apparent that HOCl and Glu –mediated neuronal injuries engaged the mitochondrial-dependent apoptotic pathway as demonstrated by the transcriptional up-regulation of Cytc, Bax, Bad, Casp (3, 6 and 9) and Aif in both models.

Extrinsic Tumor Necrosis Factor Receptors (TNFRs)-Mediated Apoptosis

The extrinsic apoptotic pathway is triggered by the activation of death receptors present on the exterior of plasma membrane through ligand-receptor interactions, which result in the eventual activation of the initiator caspases through protein-protein interactions. These death receptors include Fas, TNFRSF, and TNF-related apoptosis-inducing ligand (TRAIL) receptors (Ashkenazi and Dixit, 1999), of which the former two demonstrated transcriptional up-regulation in all five oxidative stressor models (Table 1). The Fas/Fas ligand (FasL) pathway is one of the extensively studied mechanisms in extrinsically stimulated death receptors-mediated apoptosis (reviewed in Circu and Aw, 2010). With binding of Fas/FasL, the Fas-associated death domain (FADD) and procaspase 8 are recruited to form the death-inducing signaling complex (DISC), which is then endocytosed (Watanabe, Morita, et al., 1988). The accumulation of cytoplasmic DISC after its release from the endosomes and detachment from the receptors promotes activation of the initiator CASP8 (Li, Lee, et al., 2006). The relative degree of CASP8 activation determines the types of downstream signaling courses being induced: substantial CASP8 activation can directly activate CASP3, whereas low CASP8 activation adopts intrinsic mitochondrial-dependent pathway as previously to activate CASP3 (Barnhart, Alappat, et al., 2003). The extrinsic pathways mediated by the plasma membrane integrated death receptors (TNFRs) play a major role in cell death promotion in all five oxidative stressor models by demonstrating transcriptional up-regulation over the 24h gene profiling period (Table 1).

p53-Mediated Apoptosis

p53 (alternatively known as transformation-related p53 [Trp53]) is a well-known transcription factor for its physiologic role in guarding the integrity of the DNA genome by keeping the cell cycle process in check. However, recent discoveries have shifted the research attention onto its role in cell death induction through activation of ER stress-responsive apoptotic genes such as p53-unregulated modulator of apoptosis (PUMA; BCL2 binding component 3 [Bbc3]) and phorbol-12-myristate-13-acetate-induced protein 1 (NOXA; phorbol-12-myristate-13-acetate-induced protein 1 [Pmaip1]) (Yu and Zhang, 2003) as well as its non-transcriptional-dependent mediation of cell death through its non-nuclear localization (Yee and Vousden, 2005). p53 has also been demonstrated to associate with and inactivate the anti-apoptotic/pro-survival members of the BCL2 superfamily such as BCL2 and BCL-xL (Erster and Moll, 2005).

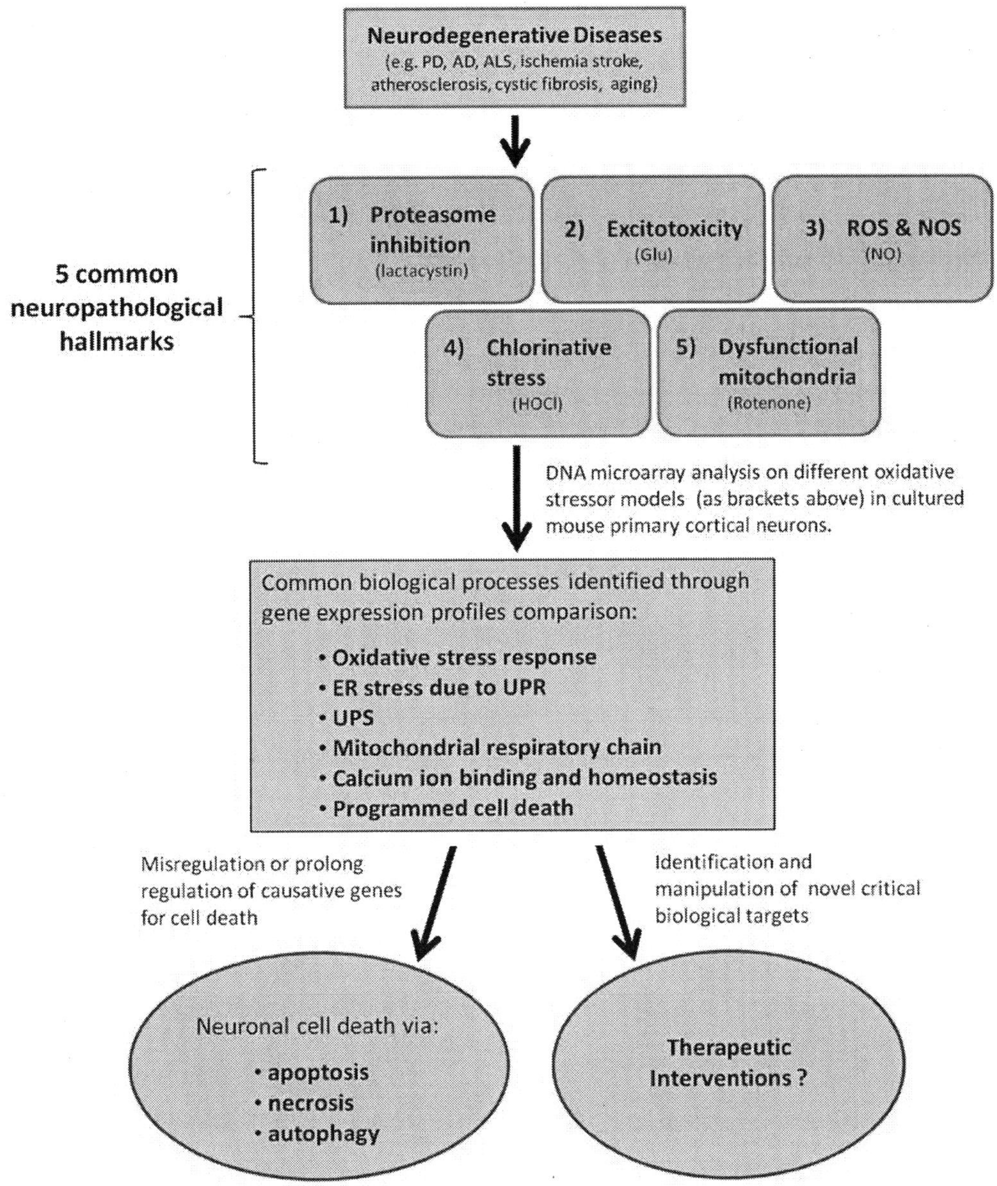

Figure 1. highlights 5 common neuropathological hallmarks associated with various neurodegenerative diseases. DNA microarray experiments were done on mouse primary cortical neurons treated with different causative agents to mimic these common hallmarks in vitro. Comparative analysis of gene expression profiles obtained from each oxidative stressor model identified several commonly regulated genes that are associated with a few critical biological processes (see Table 1) that when misregulated, could result in neuronal cell death. Conversely with results from DNA microarray analysis, it is possible to identify novel biological targets associated with neurodegeneration. From which, researchers could derive methods to appropriately manipulate them in the search for effective therapeutic interventions.

On the other hand, PUMA and NOXA are BH3-containing proteins have the ability to displace BCL2 and BCL-xL inhibitory interactions with the pro-apoptotic proteins BAX and/or BAK (also members of the BCL2 superfamily), thus leading to the latter duo activationwhich causes mitochondrial outer membrane permeabilization and release of apoptogenic factors such as CYTC (Kim, Tu, et al., 2009). Interestingly, BCL-xL has been reported to be involved in the inhibition of cytoplasmic p53 activity, and that transcriptional activation of Puma by nuclear p53 is required for the PUMA protein to release cytoplasmic p53 from BCL-xL to activate BAX (Chipuk, Bouchier-Hayes, et al., 2005).Only in NO and lactacystin-mediated neuronal death models did PUMA and NOXA demonstrated significant transcriptional up-regulation, indicative of the activation of p53 signaling pathway induced by ER stress (Table 1).

Summary of the Comparative Microarray Analysis Across the Five Oxidative Stressors Models

Transcriptional analysis though surface deep, is a powerful screening method that could be effectively used to look for crucial biological hints related to the onset and regulation of neuronal cell death pathways in response to excessive oxidative insult. Notably, it is almost impossible to separately distinguish different forms of PCD as they are interconnected via several regulatory feedback loops as discussed. In addition, with each player having concurrent functions in different modes, this further complicates the entire PCD process. However what is interesting is that, because PCD pathways are intertwined as seen, it might be possible to identify a singular pivotal biological target that is commonly involved in different modes of cell death and upon its inhibition could effectively prevent or delay cell death. The take home message is that neuronal PCD should not be viewed as a simple singular process but one that is intricate and involves various complex pathways. .

In conclusionas summarized in Figure 1, comparative analysis of the global transcriptomic profiles of HOCl, NO, rotenone, lactacystin and Glu, all of which are prominent oxidative stressors well-represented of neurological disorders such as AD, PD and ischemic stroke models, enable us to achieve novel insights into neuronal oxidative stress. This would greatly facilitate downstream experimental works via identification of novel biological targets which singular effects through manipulative means (e.g. suppression or over-expression) would prove to be effective in the impediment of neurodegeneration in a variety of neuropathies. The enriched biological processes discussed here centralized and highlighted the significant implication of mitochondria and other organelles including ER, in a multi-process aggravation of neuronal injury. Furthermore, in the context of most oxidative stress-related neurodegenerative diseases such as AD, PD, ALS and even in ischemic stroke, which are also recognized as protein aggregation disorders, the presence of these aggregates provide an additional aspect aggravating the cellular stresses leading to neurodegeneration and cell death. Even though studies at the cellular level have limitations in terms of understanding the detailed pathophysiological process leading to disease in humans or animal models of particular diseases, they do provide the key underpinnings of molecular and cellular pathways, on the basis of which the behaviours of cells, tissues and organs in the more complex milieu of the mammal can be deciphered.

References

Amici, M., G. Lupidi, et al. (2003). "Peroxynitrite-induced oxidation and its effects on isolated proteasomal systems." *Free Radic. Biol. Med.* 34(8): 987-996.

Ashkenazi, A. and V. M. Dixit (1999). "Apoptosis control by death and decoy receptors." *Curr. Opin. Cell Biol.* 11(2): 255-260.

Azad, M. B., Y. Chen, et al. (2008). "Hypoxia induces autophagic cell death in apoptosis-competent cells through a mechanism involving BNIP3." *Autophagy* 4(2): 195-204.

Barnhart, B. C., E. C. Alappat, et al. (2003). "The CD95 type I/type II model." *Semin. Immunol.* 15(3): 185-193.

Baumgartner, H. K., J. V. Gerasimenko, et al. (2009). "Calcium elevation in mitochondria is the main Ca2+ requirement for mitochondrial permeability transition pore (mPTP) opening." *J. Biol. Chem.* 284(31): 20796-20803.

Beart, P. M., M. L. Lim, et al. (2007). "Hierarchical recruitment by AMPA but not staurosporine of pro-apoptotic mitochondrial signaling in cultured cortical neurons: evidence for caspase-dependent/independent cross-talk." *J. Neurochem.* 103(6): 2408-2427.

Beckman, J. S., T. W. Beckman, et al. (1990). "Apparent hydroxyl radical production by peroxynitrite: implications for endothelial injury from nitric oxide and superoxide." *Proc. Natl. Acad. Sci. USA* 87(4): 1620-1624.

Ben-Yoseph, O., P. A. Boxer, et al. (1996). "Assessment of the role of the glutathione and pentose phosphate pathways in the protection of primary cerebrocortical cultures from oxidative stress." *J. Neurochem.* 66(6): 2329-2337.

Benchoua, A., J. Braudeau, et al. (2004). "Activation of proinflammatory caspases by cathepsin B in focal cerebral ischemia." *J. Cereb. Blood Flow Metab.*24(11): 1272-1279.

Berman, F. W. and T. F. Murray (1997). "Domoic acid neurotoxicity in cultured cerebellar granule neurons is mediated predominantly by NMDA receptors that are activated as a consequence of excitatory amino acid release." *J. Neurochem.* 69(2): 693-703.

Bianca, V. D., S. Dusi, et al. (1999). "beta-amyloid activates the O-2 forming NADPH oxidase in microglia, monocytes, and neutrophils. A possible inflammatory mechanism of neuronal damage in Alzheimer's disease." *J. Biol. Chem.* 274(22): 15493-15499.

Blomgren, K., M. Leist, et al. (2007). "Pathological apoptosis in the developing brain." *Apoptosis* 12(5): 993-1010.

Bloom, D. A. and A. K. Jaiswal (2003). "Phosphorylation of Nrf2 at Ser40 by protein kinase C in response to antioxidants leads to the release of Nrf2 from INrf2, but is not required for Nrf2 stabilization/accumulation in the nucleus and transcriptional activation of antioxidant response element-mediated NAD(P)H:quinone oxidoreductase-1 gene expression." *J. Biol. Chem.* 278(45): 44675-44682.

Boland, B. and R. A. Nixon (2006). "Neuronal macroautophagy: from development to degeneration." *Mol. Aspects Med.* 27(5-6): 503-519.

Boujrad, H., O. Gubkina, et al. (2007). "AIF-mediated programmed necrosis: a highly regulated way to die." *Cell Cycle* 6(21): 2612-2619.

Boveris, A., N. Oshino, et al. (1972). "The cellular production of hydrogen peroxide." *Biochem. J.* 128(3): 617-630.

Breckenridge, D. G., M. Germain, et al. (2003). "Regulation of apoptosis by endoplasmic reticulum pathways." *Oncogene* 22(53): 8608-8618.

Buckley, B. J., Z. M. Marshall, et al. (2003). "Nitric oxide stimulates Nrf2 nuclear translocation in vascular endothelium." *Biochem. Biophys. Res. Commun.* 307(4): 973-979.

Butterfield, D. A. (1997). "beta-Amyloid-associated free radical oxidative stress and neurotoxicity: implications for Alzheimer's disease." *Chem. Res. Toxicol.* 10(5): 495-506.

Calabrese, V., G. Scapagnini, et al. (2003). "Redox regulation of heat shock protein expression in aging and neurodegenerative disorders associated with oxidative stress: a nutritional approach." *Amino Acids* 25(3-4): 437-444.

Canu, N., R. Tufi, et al. (2005). "Role of the autophagic-lysosomal system on low potassium-induced apoptosis in cultured cerebellar granule cells." *J. Neurochem.* 92(5): 1228-1242.

Cao, G., J. Xing, et al. (2007). "Critical role of calpain I in mitochondrial release of apoptosis-inducing factor in ischemic neuronal injury." *J. Neurosci.* 27(35): 9278-9293.

Cassina, A. M., R. Hodara, et al. (2000). "Cytochrome c nitration by peroxynitrite." *J. Biol. Chem.*275(28): 21409-21415.

Chakravarthi, S., C. E. Jessop, et al. (2006). "The role of glutathione in disulphide bond formation and endoplasmic-reticulum-generated oxidative stress." *EMBO Rep.* 7(3): 271-275.

Chang, T. Y., C. C. Chang, et al. (2006). "Cholesterol sensing, trafficking, and esterification." *Annu. Rev. Cell Dev. Biol.* 22: 129-157.

Chen, M. J., Y. W. Yap, et al. (2006). "Early induction of calpains in rotenone-mediated neuronal apoptosis." *Neurosci. Lett.* 397(1-2): 69-73.

Cheung, N. S., M. S. Choy, et al. (2004). "Lactacystin-induced apoptosis of cultured mouse cortical neurons is associated with accumulation of PTEN in the detergent-resistant membrane fraction." *Cell Mol. Life Sci.* 61(15): 1926-1934.

Cheung, N. S., Z. F. Peng, et al. (2007). "Hydrogen sulfide induced neuronal death occurs via glutamate receptor and is associated with calpain activation and lysosomal rupture in mouse primary cortical neurons." *Neuropharmacology* 53(4): 505-514.

Chinopoulos, C. and V. Adam-Vizi (2006). "Calcium, mitochondria and oxidative stress in neuronal pathology. Novel aspects of an enduring theme." *FEBS J.* 273(3): 433-450.

Chipuk, J. E., L. Bouchier-Hayes, et al. (2005). "PUMA couples the nuclear and cytoplasmic proapoptotic function of p53." *Science* 309(5741): 1732-1735.

Choi, D. K., S. Pennathur, et al. (2005). "Ablation of the inflammatory enzyme myeloperoxidase mitigates features of Parkinson's disease in mice." *J. Neurosci.* 25(28): 6594-6600.

Choi, D. W. (1985). "Glutamate neurotoxicity in cortical cell culture is calcium dependent." *Neurosci. Lett.* 58(3): 293-297.

Choi, D. W. (1987). "Ionic dependence of glutamate neurotoxicity." *J. Neurosci.*7(2): 369-379.

Choy, M. S., M. J. Chen, et al. (2011). "Up-regulation of endoplasmic reticulum stress-related genes during the early phase of treatment of cultured cortical neurons by the proteasomal inhibitor lactacystin." *J. Cell Physiol.* 226(2): 494-510.

Chua, B. T., K. Guo, et al. (2000). "Direct cleavage by the calcium-activated protease calpain can lead to inactivation of caspases." *J. Biol. Chem.* 275(7): 5131-5135.

Ciechomska, I. A., G. C. Goemans, et al. (2009). "Bcl-2 complexed with Beclin-1 maintains full anti-apoptotic function." *Oncogene* 28(21): 2128-2141.

Circu, M. L. and T. Y. Aw (2010). "Reactive oxygen species, cellular redox systems, and apoptosis." *Free Radic. Biol. Med.* 48(6): 749-762.

Clarke, P. G. (1990). "Developmental cell death: morphological diversity and multiple mechanisms." *Anat. Embryol.* (Berl) 181(3): 195-213.

Cookson, M. R. and P. J. Shaw (1999). "Oxidative stress and motor neurone disease." *Brain Pathol.* 9(1): 165-186.

Coyle, J. T. and P. Puttfarcken (1993). "Oxidative stress, glutamate, and neurodegenerative disorders." *Science* 262(5134): 689-695.

Crocker, S. J., P. D. Smith, et al. (2003). "Inhibition of calpains prevents neuronal and behavioral deficits in an MPTP mouse model of Parkinson's disease." *J. Neurosci.* 23(10): 4081-4091.

Csordas, G., C. Renken, et al. (2006). "Structural and functional features and significance of the physical linkage between ER and mitochondria." *J. Cell Biol.* 174(7): 915-921.

Cullinan, S. B., D. Zhang, et al. (2003). "Nrf2 is a direct PERK substrate and effector of PERK-dependent cell survival." *Mol. Cell Biol.* 23(20): 7198-7209.

Culmsee, C., C. Zhu, et al. (2005). "Apoptosis-inducing factor triggered by poly(ADP-ribose) polymerase and Bid mediates neuronal cell death after oxygen-glucose deprivation and focal cerebral ischemia." *J. Neurosci.* 25(44): 10262-10272.

Cuzzocrea, S., D. P. Riley, et al. (2001). "Antioxidant therapy: a new pharmacological approach in shock, inflammation, and ischemia/reperfusion injury." *Pharmacol. Rev.* 53(1): 135-159.

Dawson, V. L., T. M. Dawson, et al. (1991). "Nitric oxide mediates glutamate neurotoxicity in primary cortical cultures." *Proc. Natl. Acad. Sci. USA* 88(14): 6368-6371.

DeGracia, D. J. and H. L. Montie (2004). "Cerebral ischemia and the unfolded protein response." *J. Neurochem.* 91(1): 1-8.

Demasi, M. and K. J. Davies (2003). "Proteasome inhibitors induce intracellular protein aggregation and cell death by an oxygen-dependent mechanism." *FEBS Lett.* 542(1-3): 89-94.

Desagher, S., A. Osen-Sand, et al. (1999). "Bid-induced conformational change of Bax is responsible for mitochondrial cytochrome c release during apoptosis." *J. Cell Biol.* 144(5): 891-901.

Ding, Q. and J. N. Keller (2001). "Proteasomes and proteasome inhibition in the central nervous system." *Free Radic. Biol. Med.* 31(5): 574-584.

Dirnagl, U., C. Iadecola, et al. (1999). "Pathobiology of ischaemic stroke: an integrated view." *Trends Neurosci.* 22(9): 391-397.

Diwakarla, S., L. D. Mercer, et al. (2009). "GABAergic striatal neurons exhibit caspase-independent, mitochondrially mediated programmed cell death." *J. Neurochem.* 109 Suppl 1: 198-206.

Diwakarla, S., P. Nagley, et al. (2009). "Differential insult-dependent recruitment of the intrinsic mitochondrial pathway during neuronal programmed cell death." *Cell Mol. Life Sci.* 66(1): 156-172.

Du, Y., K. R. Bales, et al. (1997). "Activation of a caspase 3-related cysteine protease is required for glutamate-mediated apoptosis of cultured cerebellar granule neurons." *Proc. Natl. Acad. Sci. USA* 94(21): 11657-11662.

Dufour, E., M. Terzioglu, et al. (2008). "Age-associated mosaic respiratory chain deficiency causes trans-neuronal degeneration." *Hum. Mol. Genet.* 17(10): 1418-1426.

Dugan, L. L., S. L. Sensi, et al. (1995). "Mitochondrial production of reactive oxygen species in cortical neurons following exposure to N-methyl-D-aspartate." *J. Neurosci.* 15(10): 6377-6388.

Dykens, J. A. (1994). "Isolated cerebral and cerebellar mitochondria produce free radicals when exposed to elevated CA2+ and Na+: implications for neurodegeneration." *J. Neurochem.* 63(2): 584-591.

Enari, M., H. Sakahira, et al. (1998). "A caspase-activated DNase that degrades DNA during apoptosis, and its inhibitor ICAD." *Nature* 391(6662): 43-50.

Erster, S. and U. M. Moll (2005). "Stress-induced p53 runs a transcription-independent death program." *Biochem. Biophys. Res. Commun.* 331(3): 843-850.

Evans, M. D. and M. S. Cooke (2004). "Factors contributing to the outcome of oxidative damage to nucleic acids." *Bioessays* 26(5): 533-542.

Ferrand-Drake, M., C. Zhu, et al. (2003). "Cyclosporin A prevents calpain activation despite increased intracellular calcium concentrations, as well as translocation of apoptosis-inducing factor, cytochrome c and caspase-3 activation in neurons exposed to transient hypoglycemia." *J. Neurochem.* 85(6): 1431-1442.

Flores-Diaz, M., J. C. Higuita, et al. (2004). "A cellular UDP-glucose deficiency causes overexpression of glucose/oxygen-regulated proteins independent of the endoplasmic reticulum stress elements." *J. Biol. Chem.* 279(21): 21724-21731.

Franklin, T. B., A. M. Krueger-Naug, et al. (2005). "The role of heat shock proteins Hsp70 and Hsp27 in cellular protection of the central nervous system." *Int. J. Hyperthermia* 21(5): 379-392.

Fratelli, M., L. O. Goodwin, et al. (2005). "Gene expression profiling reveals a signaling role of glutathione in redox regulation." *Proc. Natl. Acad. Sci. USA* 102(39): 13998-14003.

French, P. J., J. Bijman, et al. (1995). "Isotype-specific activation of cystic fibrosis transmembrane conductance regulator-chloride channels by cGMP-dependent protein kinase II." *J. Biol. Chem.* 270(44): 26626-26631.

Fribley, A., Q. Zeng, et al. (2004). "Proteasome inhibitor PS-341 induces apoptosis through induction of endoplasmic reticulum stress-reactive oxygen species in head and neck squamous cell carcinoma cells." *Mol. Cell Biol.* 24(22): 9695-9704.

Gaczynska, M., P. A. Osmulski, et al. (2001). "Caretaker or undertaker? The role of the proteasome in aging." *Mech. Ageing Dev.* 122(3): 235-254.

Gan, L., S. Ye, et al. (2004). "Identification of cathepsin B as a mediator of neuronal death induced by Abeta-activated microglial cells using a functional genomics approach." *J. Biol. Chem.* 279(7): 5565-5572.

Gill, M. B. and J. R. Perez-Polo (2008). "Hypoxia ischemia-mediated cell death in neonatal rat brain." *Neurochem. Res.* 33(12): 2379-2389.

Golstein, P. and G. Kroemer (2007). "Cell death by necrosis: towards a molecular definition." *Trends Biochem. Sci.* 32(1): 37-43.

Gomez-Santos, C., I. Ferrer, et al. (2003). "Dopamine induces autophagic cell death and alpha-synuclein increase in human neuroblastoma SH-SY5Y cells." *J. Neurosci. Res.* 73(3): 341-350.

Gonzalez-Scarano, F. and G. Baltuch (1999). "Microglia as mediators of inflammatory and degenerative diseases." *Annu. Rev. Neurosci.* 22: 219-240.

Gonzalez-Zulueta, M., L. M. Ensz, et al. (1998). "Manganese superoxide dismutase protects nNOS neurons from NMDA and nitric oxide-mediated neurotoxicity." *J. Neurosci.* 18(6): 2040-2055.

Good, P. F., A. Hsu, et al. (1998). "Protein nitration in Parkinson's disease." *J. Neuropathol. Exp. Neurol.* 57(4): 338-342.

Good, P. F., P. Werner, et al. (1996). "Evidence of neuronal oxidative damage in Alzheimer's disease." *Am. J. Pathol.* 149(1): 21-28.

Gorlach, A., P. Klappa, et al. (2006). "The endoplasmic reticulum: folding, calcium homeostasis, signaling, and redox control." *Antioxid. Redox Signal* 8(9-10): 1391-1418.

Green, P. S., A. J. Mendez, et al. (2004). "Neuronal expression of myeloperoxidase is increased in Alzheimer's disease." *J. Neurochem.* 90(3): 724-733.

Grune, T., T. Jung, et al. (2004). "Decreased proteolysis caused by protein aggregates, inclusion bodies, plaques, lipofuscin, ceroid, and 'aggresomes' during oxidative stress, aging, and disease." *Int. J. Biochem. Cell Biol.* 36(12): 2519-2530.

Halliwell, B. (2006). "Oxidative stress and neurodegeneration: where are we now?" *J. Neurochem.* 97(6): 1634-1658.

Hanisch, U. K. and H. Kettenmann (2007). "Microglia: active sensor and versatile effector cells in the normal and pathologic brain." *Nat. Neurosci.* 10(11): 1387-1394.

Hansen, T. M. and P. Nagley (2003). "AIF: a multifunctional cog in the life and death machine." *Sci. STKE* 2003(193): PE31.

Hara, M. R., N. Agrawal, et al. (2005). "S-nitrosylated GAPDH initiates apoptotic cell death by nuclear translocation following Siah1 binding." *Nat. Cell Biol.* 7(7): 665-674.

Hara, M. R. and S. H. Snyder (2007). "Cell signaling and neuronal death." *Annu. Rev. Pharmacol. Toxicol.* 47: 117-141.

Harding, H. P., Y. Zhang, et al. (1999). "Protein translation and folding are coupled by an endoplasmic-reticulum-resident kinase." *Nature* 397(6716): 271-274.

Hazell, L. J., L. Arnold, et al. (1996). "Presence of hypochlorite-modified proteins in human atherosclerotic lesions." *J. Clin. Invest.* 97(6): 1535-1544.

Hengartner, M. O. (2000). "The biochemistry of apoptosis." *Nature* 407(6805): 770-776.

Higgins, G. C., P. M. Beart, et al. (2009). "Oxidative stress triggers neuronal caspase-independent death: endonuclease G involvement in programmed cell death-type III." *Cell Mol. Life Sci.* 66(16): 2773-2787.

Higgins, G. C., P. M. Beart, et al. (2010). "Oxidative stress: emerging mitochondrial and cellular themes and variations in neuronal injury." *J. Alzheimers Dis.* 20 Suppl 2: S453-473.

Hoozemans, J. J., R. Veerhuis, et al. (2005). "The unfolded protein response is activated in Alzheimer's disease." *Acta Neuropathol.* 110(2): 165-172.

Huang, E. J. and L. F. Reichardt (2001). "Neurotrophins: roles in neuronal development and function." *Annu. Rev. Neurosci.* 24: 677-736.

Huang, H. C., T. Nguyen, et al. (2000). "Regulation of the antioxidant response element by protein kinase C-mediated phosphorylation of NF-E2-related factor 2." *Proc. Natl. Acad. Sci. USA* 97(23): 12475-12480.

Huie, R. E. and S. Padmaja (1993). "The reaction of no with superoxide." *Free Radic. Res. Commun.* 18(4): 195-199.

Ignarro, L. J. (1991). "Signal transduction mechanisms involving nitric oxide." *Biochem. Pharmacol.* 41(4): 485-490.

Jaeger, P. A. and T. Wyss-Coray (2009). "All-you-can-eat: autophagy in neurodegeneration and neuroprotection." *Mol. Neurodegener.*4: 16.

Jenner, P. (2007). "Oxidative stress and Parkinson's disease." *Handb. Clin. Neurol.* 83: 507-520.

Jimbo, A., E. Fujita, et al. (2003). "ER stress induces caspase-8 activation, stimulating cytochrome c release and caspase-9 activation." *Exp. Cell Res.* 283(2): 156-166.

Johnson, J. A., D. A. Johnson, et al. (2008). "The Nrf2-ARE pathway: an indicator and modulator of oxidative stress in neurodegeneration." *Ann. N. Y. Acad. Sci.* 1147: 61-69.

Jonnala, R. R. and J. J. Buccafusco (2001). "Inhibition of nerve growth factor signaling by peroxynitrite." *J. Neurosci. Res.* 63(1): 27-34.

Kabeya, Y., N. Mizushima, et al. (2000). "LC3, a mammalian homologue of yeast Apg8p, is localized in autophagosome membranes after processing." *EMBO J.* 19(21): 5720-5728.

Kanno, H., H. Ozawa, et al. (2009). "Spinal cord injury induces upregulation of Beclin 1 and promotes autophagic cell death." *Neurobiol. Dis.* 33(2): 143-148.

Kaspar, J. W., S. K. Niture, et al. (2009). "Nrf2:INrf2 (Keap1) signaling in oxidative stress." *Free Radic. Biol. Med.* 47(9): 1304-1309.

Keller, J. N., E. Dimayuga, et al. (2004). "Autophagy, proteasomes, lipofuscin, and oxidative stress in the aging brain." *Int. J. Biochem. Cell Biol.* 36(12): 2376-2391.

Keller, J. N., K. B. Hanni, et al. (2000). "Impaired proteasome function in Alzheimer's disease." *J. Neurochem.* 75(1): 436-439.

Keller, J. N., F. F. Huang, et al. (2000). "Oxidative stress-associated impairment of proteasome activity during ischemia-reperfusion injury." *J. Cereb. Blood Flow Metab.* 20(10): 1467-1473.

Kettle, A. J., T. Chan, et al. (2004). "Myeloperoxidase and protein oxidation in the airways of young children with cystic fibrosis." *Am. J. Respir. Crit. Care Med.* 170(12): 1317-1323.

Kim, H., H. C. Tu, et al. (2009). "Stepwise activation of BAX and BAK by tBID, BIM, and PUMA initiates mitochondrial apoptosis." *Mol. Cell* 36(3): 487-499.

Kitao, Y., Y. Imai, et al. (2007). "Pael receptor induces death of dopaminergic neurons in the substantia nigra via endoplasmic reticulum stress and dopamine toxicity, which is enhanced under condition of parkin inactivation." *Hum. Mol. Genet.* 16(1): 50-60.

Klionsky, D. J. (2006). "Neurodegeneration: good riddance to bad rubbish." *Nature* 441(7095): 819-820.

Konig, H. G., M. Rehm, et al. (2007). "Full length Bid is sufficient to induce apoptosis of cultured rat hippocampal neurons." *BMC Cell Biol.* 8: 7.

Kontos, H. A. (2001). "Oxygen radicals in cerebral ischemia: the 2001 Willis lecture." *Stroke* 32(11): 2712-2716.

Kroemer, G., L. Galluzzi, et al. (2007). "Mitochondrial membrane permeabilization in cell death." *Physiol. Rev.* 87(1): 99-163.

Kroemer, G., L. Galluzzi, et al. (2009). "Classification of cell death: recommendations of the Nomenclature Committee on Cell Death 2009." *Cell Death Differ.* 16(1): 3-11.

Kroemer, G. and B. Levine (2008). "Autophagic cell death: the story of a misnomer." *Nat. Rev. Mol. Cell Biol.* 9(12): 1004-1010.

Lafon-Cazal, M., S. Pietri, et al. (1993). "NMDA-dependent superoxide production and neurotoxicity." *Nature* 364(6437): 535-537.

Lakhan, S. E., A. Kirchgessner, et al. (2009). "Inflammatory mechanisms in ischemic stroke: therapeutic approaches." *J. Transl. Med.* 7: 97.

Landshamer, S., M. Hoehn, et al. (2008). "Bid-induced release of AIF from mitochondria causes immediate neuronal cell death." *Cell Death Differ.* 15(10): 1553-1563.

Lau, K. L., S. K. Kong, et al. (2003). "cGMP stimulates endoplasmic reticulum Ca(2+)-ATPase in vascular endothelial cells." *Life Sci.* 73(16): 2019-2028.

Lennon, S. V., S. J. Martin, et al. (1991). "Dose-dependent induction of apoptosis in human tumour cell lines by widely diverging stimuli." *Cell Prolif.* 24(2): 203-214.

Lev, S., H. Moreno, et al. (1995). "Protein tyrosine kinase PYK2 involved in Ca(2+)-induced regulation of ion channel and MAP kinase functions." *Nature* 376(6543): 737-745.

Li, J., B. Lee, et al. (2006). "Endoplasmic reticulum stress-induced apoptosis: multiple pathways and activation of p53-up-regulated modulator of apoptosis (PUMA) and NOXA by p53." *J. Biol. Chem.* 281(11): 7260-7270.

Lin, M. T. and M. F. Beal (2006). "Mitochondrial dysfunction and oxidative stress in neurodegenerative diseases." *Nature* 443(7113): 787-795.

Liu, P. K., C. Y. Hsu, et al. (1996). "Damage, repair, and mutagenesis in nuclear genes after mouse forebrain ischemia-reperfusion." *J. Neurosci.* 16(21): 6795-6806.

Lucas, S. M., N. J. Rothwell, et al. (2006). "The role of inflammation in CNS injury and disease." *Br. J. Pharmacol.*147 Suppl 1: S232-240.

Maiuri, M. C., A. Criollo, et al. (2007). "BH3-only proteins and BH3 mimetics induce autophagy by competitively disrupting the interaction between Beclin 1 and Bcl-2/Bcl-X(L)." *Autophagy* 3(4): 374-376.

Maiuri, M. C., G. Le Toumelin, et al. (2007). "Functional and physical interaction between Bcl-X(L) and a BH3-like domain in Beclin-1." *EMBO J.* 26(10): 2527-2539.

Mancuso, M., F. Coppede, et al. (2006). "Mitochondrial dysfunction, oxidative stress and neurodegeneration." *J. Alzheimers Dis.* 10(1): 59-73.

Mandel, S., E. Grunblatt, et al. (2005). "Gene expression profiling of sporadic Parkinson's disease substantia nigra pars compacta reveals impairment of ubiquitin-proteasome subunits, SKP1A, aldehyde dehydrogenase, and chaperone HSC-70." *Ann. N. Y. Acad. Sci.* 1053: 356-375.

Maragakis, N. J. and J. D. Rothstein (2006). "Mechanisms of Disease: astrocytes in neurodegenerative disease." *Nat. Clin. Pract. Neurol.* 2(12): 679-689.

McMahon, M., K. Itoh, et al. (2001). "The Cap'n'Collar basic leucine zipper transcription factor Nrf2 (NF-E2 p45-related factor 2) controls both constitutive and inducible expression of intestinal detoxification and glutathione biosynthetic enzymes." *Cancer Res.* 61(8): 3299-3307.

McNaught, K. S., C. W. Olanow, et al. (2001). "Failure of the ubiquitin-proteasome system in Parkinson's disease." *Nat. Rev. Neurosci.* 2(8): 589-594.

Meiners, S., D. Heyken, et al. (2003). "Inhibition of proteasome activity induces concerted expression of proteasome genes and de novo formation of Mammalian proteasomes." *J. Biol. Chem.* 278(24): 21517-21525.

Meriin, A. B. and M. Y. Sherman (2005). "Role of molecular chaperones in neurodegenerative disorders." *Int. J. Hyperthermia* 21(5): 403-419.

Mizushima, N., B. Levine, et al. (2008). "Autophagy fights disease through cellular self-digestion." *Nature* 451(7182): 1069-1075.

Mizushima, N., A. Yamamoto, et al. (2001). "Dissection of autophagosome formation using Apg5-deficient mouse embryonic stem cells." *J. Cell Biol.* 152(4): 657-668.

Momeni, H. R. and M. Kanje (2006). "Calpain inhibitors delay injury-induced apoptosis in adult mouse spinal cord motor neurons." *Neuroreport* 17(8): 761-765.

Mueller-Steiner, S., Y. Zhou, et al. (2006). "Antiamyloidogenic and neuroprotective functions of cathepsin B: implications for Alzheimer's disease." *Neuron* 51(6): 703-714.

Nagley, P., G. C. Higgins, et al. (2010). "Multifaceted deaths orchestrated by mitochondria in neurones." *Biochim. Biophys. Acta* 1802(1): 167-185.

Nakagawa, T. and J. Yuan (2000). "Cross-talk between two cysteine protease families. Activation of caspase-12 by calpain in apoptosis." *J. Cell Biol.* 150(4): 887-894.

Newman, D. K., S. Hoffman, et al. (2002). "Nitration of PECAM-1 ITIM tyrosines abrogates phosphorylation and SHP-2 binding." *Biochem. Biophys. Res. Commun.* 296(5): 1171-1179.

Nguyen, T., H. C. Huang, et al. (2000). "Transcriptional regulation of the antioxidant response element. Activation by Nrf2 and repression by MafK." *J. Biol. Chem.* 275(20): 15466-15473.

Nicholls, D. G. (2008). "Oxidative stress and energy crises in neuronal dysfunction." *Ann. N. Y. Acad. Sci.* 1147: 53-60.

Nicotera, P. and G. Melino (2004). "Regulation of the apoptosis-necrosis switch." *Oncogene* 23(16): 2757-2765.

Niizuma, K., H. Endo, et al. (2009). "Oxidative stress and mitochondrial dysfunction as determinants of ischemic neuronal death and survival." *J. Neurochem.* 109 Suppl 1: 133-138.

Niizuma, K., H. Endo, et al. (2008). "The PIDDosome mediates delayed death of hippocampal CA1 neurons after transient global cerebral ischemia in rats." *Proc. Natl. Acad. Sci. USA* 105(42): 16368-16373.

Nisoli, E., E. Clementi, et al. (2003). "Mitochondrial biogenesis in mammals: the role of endogenous nitric oxide." *Science* 299(5608): 896-899.

Noh, K. M. and J. Y. Koh (2000). "Induction and activation by zinc of NADPH oxidase in cultured cortical neurons and astrocytes." *J. Neurosci.* 20(23): RC111.

Oberstein, A., P. D. Jeffrey, et al. (2007). "Crystal structure of the Bcl-XL-Beclin 1 peptide complex: Beclin 1 is a novel BH3-only protein." *J. Biol. Chem.* 282(17): 13123-13132.

Olney, J. W. (1969). "Brain lesions, obesity, and other disturbances in mice treated with monosodium glutamate." *Science* 164(880): 719-721.

Ono, K. and W. Trautwein (1991). "Potentiation by cyclic GMP of beta-adrenergic effect on Ca2+ current in guinea-pig ventricular cells." *J. Physiol.* 443: 387-404.

Orrenius, S., V. Gogvadze, et al. (2007). "Mitochondrial oxidative stress: implications for cell death." *Annu. Rev. Pharmacol. Toxicol.* 47: 143-183.

Pang, Z., V. Bondada, et al. (2003). "Calpain facilitates the neuron death induced by 3-nitropropionic acid and contributes to the necrotic morphology." *J. Neuropathol. Exp. Neurol.* 62(6): 633-643.

Pattingre, S., A. Tassa, et al. (2005). "Bcl-2 antiapoptotic proteins inhibit Beclin 1-dependent autophagy." *Cell* 122(6): 927-939.

Pekny, M. and M. Nilsson (2005). "Astrocyte activation and reactive gliosis." *Glia* 50(4): 427-434.

Peng, Z. F., M. J. Chen, et al. (2008). "Proteasome inhibition: an early or late event in nitric oxide-induced neuronal death?" *Nitric Oxide* 18(2): 136-145.

Pitkanen, S. and B. H. Robinson (1996). "Mitochondrial complex I deficiency leads to increased production of superoxide radicals and induction of superoxide dismutase." *J. Clin. Invest.* 98(2): 345-351.

Plesnila, N., S. Zinkel, et al. (2001). "BID mediates neuronal cell death after oxygen/ glucose deprivation and focal cerebral ischemia." *Proc. Natl. Acad. Sci. USA* 98(26): 15318-15323.

Polster, B. M., G. Basanez, et al. (2005). "Calpain I induces cleavage and release of apoptosis-inducing factor from isolated mitochondria." *J. Biol. Chem.* 280(8): 6447-6454.

Poon, H. F., V. Calabrese, et al. (2004). "Free radicals: key to brain aging and heme oxygenase as a cellular response to oxidative stress." *J. Gerontol A Biol. Sci. Med. Sci.* 59(5): 478-493.

Porn-Ares, M. I., A. Samali, et al. (1998). "Cleavage of the calpain inhibitor, calpastatin, during apoptosis." *Cell Death Differ.* 5(12): 1028-1033.

Ramos-Gomez, M., M. K. Kwak, et al. (2001). "Sensitivity to carcinogenesis is increased and chemoprotective efficacy of enzyme inducers is lost in nrf2 transcription factor-deficient mice." *Proc. Natl. Acad. Sci. USA* 98(6): 3410-3415.

Rao, R. V., H. M. Ellerby, et al. (2004). "Coupling endoplasmic reticulum stress to the cell death program." *Cell Death Differ.* 11(4): 372-380.

Rashmi, R., S. G. Pillai, et al. (2008). "BH3-only protein BIK induces caspase-independent cell death with autophagic features in Bcl-2 null cells." *Oncogene* 27(10): 1366-1375.

Reddy, P. H. (2009). "Amyloid beta, mitochondrial structural and functional dynamics in Alzheimer's disease." *Exp. Neurol.* 218(2): 286-292.

Reynolds, I. J. and T. G. Hastings (1995). "Glutamate induces the production of reactive oxygen species in cultured forebrain neurons following NMDA receptor activation." *J. Neurosci.* 15(5 Pt 1): 3318-3327.

Rideout, H. J., Q. Wang, et al. (2003). "Cyclin-dependent kinase activity is required for apoptotic death but not inclusion formation in cortical neurons after proteasomal inhibition." *J. Neurosci.* 23(4): 1237-1245.

Rockwell, P., H. Yuan, et al. (2000). "Proteasome inhibition in neuronal cells induces a proinflammatory response manifested by upregulation of cyclooxygenase-2, its accumulation as ubiquitin conjugates, and production of the prostaglandin PGE(2)." *Arch. Biochem. Biophys.* 374(2): 325-333.

Ron, D. and P. Walter (2007). "Signal integration in the endoplasmic reticulum unfolded protein response." *Nat. Rev. Mol. Cell Biol.* 8(7): 519-529.

Ryter, S. W., H. P. Kim, et al. (2007). "Mechanisms of cell death in oxidative stress." *Antioxid. Redox. Signal* 9(1): 49-89.

Samantaray, S., S. K. Ray, et al. (2006). "Calpain activation in apoptosis of motoneurons in cell culture models of experimental parkinsonism." *Ann. N. Y. Acad. Sci.* 1074: 349-356.

Santo-Domingo, J. and N. Demaurex (2010). "Calcium uptake mechanisms of mitochondria." *Biochim. Biophys. Acta* 1797(6-7): 907-912.

Saraste, M. (1999). "Oxidative phosphorylation at the fin de siecle." *Science* 283(5407): 1488-1493.

Sattler, R., M. P. Charlton, et al. (1998). "Distinct influx pathways, not calcium load, determine neuronal vulnerability to calcium neurotoxicity." *J. Neurochem.* 71(6): 2349-2364.

Scarpulla, R. C. (2002). "Nuclear activators and coactivators in mammalian mitochondrial biogenesis." *Biochim. Biophys. Acta* 1576(1-2): 1-14.

Schroder, M. (2008). "Endoplasmic reticulum stress responses." *Cell Mol. Life Sci.* 65(6): 862-894.

Schroder, M. and R. J. Kaufman (2005). "ER stress and the unfolded protein response." *Mutat. Res.* 569(1-2): 29-63.

Sen, N., M. R. Hara, et al. (2008). "Nitric oxide-induced nuclear GAPDH activates p300/CBP and mediates apoptosis." *Nat. Cell Biol.* 10(7): 866-873.

Slee, E. A., M. T. Harte, et al. (1999). "Ordering the cytochrome c-initiated caspase cascade: hierarchical activation of caspases-2, -3, -6, -7, -8, and -10 in a caspase-9-dependent manner." *J. Cell Biol.* 144(2): 281-292.

Smith, D. J., H. Ng, et al. (2008). "The mitochondrial gateway to cell death." *IUBMB Life* 60(6): 383-389.

Souza, J. M., E. Daikhin, et al. (1999). "Factors determining the selectivity of protein tyrosine nitration." *Arch. Biochem. Biophys.* 371(2): 169-178.

Sribnick, E. A., D. D. Matzelle, et al. (2007). "Direct evidence for calpain involvement in apoptotic death of neurons in spinal cord injury in rats and neuroprotection with calpain inhibitor." *Neurochem. Res.* 32(12): 2210-2216.

Stamler, J. S., E. J. Toone, et al. (1997). "(S)NO signals: translocation, regulation, and a consensus motif." *Neuron* 18(5): 691-696.

Sultana, R. and D. A. Butterfield (2009). "Oxidatively modified, mitochondria-relevant brain proteins in subjects with Alzheimer disease and mild cognitive impairment." *J. Bioenerg. Biomembr.* 41(5): 441-446.

Susin, S. A., H. K. Lorenzo, et al. (1999). "Molecular characterization of mitochondrial apoptosis-inducing factor." *Nature* 397(6718): 441-446.

Swanson, R. A., W. Ying, et al. (2004). "Astrocyte influences on ischemic neuronal death." *Curr. Mol. Med.* 4(2): 193-205.

Szegezdi, E., S. E. Logue, et al. (2006). "Mediators of endoplasmic reticulum stress-induced apoptosis." *EMBO Rep.* 7(9): 880-885.

Takano, J., M. Tomioka, et al. (2005). "Calpain mediates excitotoxic DNA fragmentation via mitochondrial pathways in adult brains: evidence from calpastatin mutant mice." *J. Biol. Chem.* 280(16): 16175-16184.

Tammariello, S. P., M. T. Quinn, et al. (2000). "NADPH oxidase contributes directly to oxidative stress and apoptosis in nerve growth factor-deprived sympathetic neurons." *J. Neurosci.*20(1): RC53.

Tenneti, L., D. M. D'Emilia, et al. (1998). "Role of caspases in N-methyl-D-aspartate-induced apoptosis in cerebrocortical neurons." *J. Neurochem.* 71(3): 946-959.

Terman, A., B. Gustafsson, et al. (2006). "Mitochondrial damage and intralysosomal degradation in cellular aging." *Mol. Aspects Med.* 27(5-6): 471-482.

Thayer, S. A. and G. J. Wang (1995). "Glutamate-induced calcium loads: effects on energy metabolism and neuronal viability." *Clin. Exp. Pharmacol. Physiol.* 22(4): 303-304.

Todde, V., M. Veenhuis, et al. (2009). "Autophagy: principles and significance in health and disease." *Biochim. Biophys. Acta* 1792(1): 3-13.

Travers, K. J., C. K. Patil, et al. (2000). "Functional and genomic analyses reveal an essential coordination between the unfolded protein response and ER-associated degradation." *Cell* 101(3): 249-258.

Tymianski, M., M. P. Charlton, et al. (1993). "Source specificity of early calcium neurotoxicity in cultured embryonic spinal neurons." *J. Neurosci.*13(5): 2085-2104.

Ursini, M. V., A. Parrella, et al. (1997). "Enhanced expression of glucose-6-phosphate dehydrogenase in human cells sustaining oxidative stress." *Biochem. J.* 323 (Pt 3): 801-806.

Vallet, P., Y. Charnay, et al. (2005). "Neuronal expression of the NADPH oxidase NOX4, and its regulation in mouse experimental brain ischemia." *Neuroscience* 132(2): 233-238.

van der Sanden, M. H., M. Houweling, et al. (2003). "Inhibition of phosphatidylcholine synthesis induces expression of the endoplasmic reticulum stress and apoptosis-related protein CCAAT/enhancer-binding protein-homologous protein (CHOP/GADD153)." *Biochem. J.* 369(Pt 3): 643-650.

Verkhratsky, A. (2005). "Physiology and pathophysiology of the calcium store in the endoplasmic reticulum of neurons." *Physiol. Rev.* 85(1): 201-279.

Votyakova, T. V. and I. J. Reynolds (2001). "DeltaPsi(m)-Dependent and -independent production of reactive oxygen species by rat brain mitochondria." *J. Neurochem.* 79(2): 266-277.

Wang, G. J. and S. A. Thayer (1996). "Sequestration of glutamate-induced Ca2+ loads by mitochondria in cultured rat hippocampal neurons." *J. Neurophysiol.* 76(3): 1611-1621.

Wang, K. K. (2000). "Calpain and caspase: can you tell the difference?" *Trends Neurosci.* 23(1): 20-26.

Wang, K. K., R. Posmantur, et al. (1998). "Caspase-mediated fragmentation of calpain inhibitor protein calpastatin during apoptosis." *Arch. Biochem. Biophys.* 356(2): 187-196.

Wang, Y. and Z. H. Qin (2010). "Molecular and cellular mechanisms of excitotoxic neuronal death." Apoptosis.

Ward, M. W., M. Rehm, et al. (2006). "Real time single cell analysis of Bid cleavage and Bid translocation during caspase-dependent and neuronal caspase-independent apoptosis." *J. Biol. Chem.* 281(9): 5837-5844.

Wasserman, W. W. and W. E. Fahl (1997). "Comprehensive analysis of proteins which interact with the antioxidant responsive element: correlation of ARE-BP-1 with the chemoprotective induction response." *Arch. Biochem. Biophys.* 344(2): 387-396.

Watanabe, T., I. Morita, et al. (1988). "Preventive effect of MCI-186 on 15-HPETE induced vascular endothelial cell injury in vitro." Prostaglandins Leukot Essent Fatty Acids 33(1): 81-87.

Whalin, M. E., J. G. Scammell, et al. (1991). "Phosphodiesterase II, the cGMP-activatable cyclic nucleotide phosphodiesterase, regulates cyclic AMP metabolism in PC12 cells." *Mol. Pharmacol.* 39(6): 711-717.

Whiteman, M., U. Ketsawatsakul, et al. (2002). "A reassessment of the peroxynitrite scavenging activity of uric acid." *Ann. N. Y. Acad. Sci.* 962: 242-259.

Wu, J. and R. J. Kaufman (2006). "From acute ER stress to physiological roles of the Unfolded Protein Response." *Cell Death Differ.* 13(3): 374-384.

Xu, W., L. Liu, et al. (2004). "Nitric oxide induces coupling of mitochondrial signalling with the endoplasmic reticulum stress response." *Nat. Cell Biol.* 6(11): 1129-1134.

Xue, L., G. C. Fletcher, et al. (1999). "Autophagy is activated by apoptotic signalling in sympathetic neurons: an alternative mechanism of death execution." *Mol. Cell Neurosci.* 14(3): 180-198.

Yamashima, T., Y. Kohda, et al. (1998). "Inhibition of ischaemic hippocampal neuronal death in primates with cathepsin B inhibitor CA-074: a novel strategy for neuroprotection based on 'calpain-cathepsin hypothesis'." *Eur. J. Neurosci.* 10(5): 1723-1733.

Yamashima, T., A. B. Tonchev, et al. (2003). "Sustained calpain activation associated with lysosomal rupture executes necrosis of the postischemic CA1 neurons in primates." *Hippocampus* 13(7): 791-800.

Yap, Y. W., M. J. Chen, et al. (2010). "Temporal transcriptomic profiling reveals cellular targets that govern survival in HOCl-mediated neuronal apoptosis." *Life Sci.* 87(15-16): 457-467.

Yap, Y. W., M. Whiteman, et al. (2006). "Hypochlorous acid induces apoptosis of cultured cortical neurons through activation of calpains and rupture of lysosomes." *J. Neurochem.* 98(5): 1597-1609.

Yee, K. S. and K. H. Vousden (2005). "Complicating the complexity of p53." *Carcinogenesis* 26(8): 1317-1322.

Yew, E. H., N. S. Cheung, et al. (2005). "Proteasome inhibition by lactacystin in primary neuronal cells induces both potentially neuroprotective and pro-apoptotic transcriptional responses: a microarray analysis." *J. Neurochem.* 94(4): 943-956.

Yin, X. M., Y. Luo, et al. (2002). "Bid-mediated mitochondrial pathway is critical to ischemic neuronal apoptosis and focal cerebral ischemia." *J. Biol. Chem.* 277(44): 42074-42081.

Youle, R. J. and A. Strasser (2008). "The BCL-2 protein family: opposing activities that mediate cell death." *Nat. Rev. Mol. Cell Biol.* 9(1): 47-59.

Yousefi, S., R. Perozzo, et al. (2006). "Calpain-mediated cleavage of Atg5 switches autophagy to apoptosis." *Nat. Cell Biol.* 8(10): 1124-1132.

Yu, J. and L. Zhang (2003). "No PUMA, no death: implications for p53-dependent apoptosis." *Cancer Cell* 4(4): 248-249.

Yu, S. W., H. Wang, et al. (2002). "Mediation of poly(ADP-ribose) polymerase-1-dependent cell death by apoptosis-inducing factor." *Science* 297(5579): 259-263.

Zhao, J., A. N. Moore, et al. (2007). "Enhancing expression of Nrf2-driven genes protects the blood brain barrier after brain injury." *J. Neurosci.* 27(38): 10240-10248.

Zipper, L. M. and R. T. Mulcahy (2000). "Inhibition of ERK and p38 MAP kinases inhibits binding of Nrf2 and induction of GCS genes." *Biochem. Biophys. Res. Commun.* 278(2): 484-492.

Zundorf, G., S. Kahlert, et al. (2009). "alpha-Ketoglutarate dehydrogenase contributes to production of reactive oxygen species in glutamate-stimulated hippocampal neurons in situ." *Neuroscience* 158(2): 610-616.

In: Neuronal Cell Apoptosis
Editors: C. J. Schmid and J. L. Wolfe
ISBN: 978-1-61324-633-7

Chapter III

Death of Neuronal and Glial Cells Induced by Photodynamic Treatment: Ultrastructural Changes and Signaling Mechanisms

A. B. Uzdensky*
Department of Biophysics and Biocybernetics
and A.B. Kogan Research institute for Neurocybernetics,
Southern Federal University, Rostov-on-Don, Russia

Abstract

Photodynamic therapy (PDT), an inducer of oxidative stress, is used for treatment of cancer including brain tumors. To study PDT effect on neurons and glial cells, we used a simple model object - an isolated crayfish mechanoreceptor consisting of only two neurons enveloped by glial cells. PDT abolished neuronal activity and caused necrosis of neurons and necrosis and apoptosis of glial cells. Percent of necrotic neurons and glial cells progressively increased after PDT. At the ultrastructural level, the swelling of mitochondria, dictyosomes and endoplasmic reticulum was observed in neurons and glial cells after PDT. At 1 hour postirradiation these alterations were expanded throughout the cells. Neurons lost the "tigroid" morphology that was a result of segregation of the cytoplasm by Nissl bodies enriched with ER, mitochondria, ribosomes and polysomes, which are involved in protein synthesis and transport along neurites. The structures involved in glia-to-neuron (neuronal submembrane cisterns, glial protrusions, double-wall vesicles), intra-glial (tubular lattices) and intra-neuronal transport (Nissl bodies, Golgi apparatus, microtubular bundles) were also impaired at the earlier stages of the stretch receptor damage. Local inactivation of a neuron body by a focused laser beam enhanced glial apoptosis but not necrosis that was induced by following

* Institute of Neurocybernetics, 194/1 Stachky ave., NII NK, Rostov-on-Don, 344090, Russia. tel: +7-8632-433111, fax: +7-8632-433577. e-mail: auzd@yandex.ru.

photosensitization. Therefore, the stretch receptor neuron prevented glial apoptosis. In order to find molecular signals involved in the neuroglial interactions, we tested neurotrophic factors NGF, BDNF and GDNF. NGF and GDNF protected glial cells from PDT-induced necrosis and apoptosis. Inhibition of protein kinase JNK enhanced PDT-induced apoptosis of glial cells and prevented the anti-apoptotic effect of NGF. Hence, JNK was involved in the NGF-mediated protection of glial cells from PDT-induced apoptosis. Unlike, another MAP kinase, p38, was involved in PDT-induced glial necrosis. Calmodulin, calmodulin-dependent kinase II, phospholipase C, protein kinase C and adenylate cyclase also participated in PDT-induced necrosis of glial cells. Phospholipase C and glycogen synthase kinase also participated in PDT-induced apoptosis of glial cells, whereas adenylate cyclase, JNK, protein kinases A and C played the anti-apoptotic role in glial cells. The involvement of these signaling pathways in responses of neurons and glial cells to PDT suggests that their modulation can modify PDT efficacy. The scheme of participation of diverse inter- and intracellular signaling processes in responses of neurons and glial cells to PDT-induced oxidative stress has been created.

Keywords: photodynamic therapy, neuron, glia, firing, necrosis, apoptosis, signal transduction.

Introduction

Reactive oxygen species (e.g. H_2O_2, $OH^{\bullet}$, or $O_2^{\bullet -}$) and different free radicals are present in small amounts in cells as byproducts of the normal metabolism. However, their level increases dramatically under environmental stress (e.g., ultraviolet or ionizing radiation). These highly reactive oxidants damage cell structures, induce oxidative stress and finally cell death. Oxidative stress is the general reason of a variety of diseases including cancer, heart and brain ischemia and infarction, neurodegenerative diseases, radiation sickness, and aging (Zenkov et al., 2001; Karageuzyan, 2005; Dubinina, 2006; Fatokun et al., 2008; Elahi et al., 2009; Klaunig, 2009; Niizuma et al., 2009). At the same time, it is used in medicine for killing of unwanted pathological cells and destruction of malignant tissues.

In photodynamic therapy (PDT), dye molecules staining the tissue generate highly toxic singlet oxygen ($^1O_2$) under light exposure in the presence of oxygen. $^1O_2$ induces oxidative damage to cellular proteins, peroxidation of membrane lipids and cell death, necrosis or apoptosis. PDT provides effective destruction of tumors, which accumulated selectively photosensitizing dyes (Dougherty et al., 1998; Brown et al., 2004; Castano et al., 2004, Uzdensky, 2010). It is a promising adjuvant method for treatment of brain tumors (Eljamel, 2004; Stylli, Kaye, 2006; Kostron, 2010). However, not only malignant cells but also adjacent normal neurons and glial cells may be damaged during photodynamic treatment of brain tumors. Moreover, peripheral nerve elements in the normal tissues surrounding a tumor are injured during photodynamic therapy. Therefore, thorough experimental studies of photodynamic effect on neuronal and glial cells should precede practical applications of PDT.

The experiments on the animal brain give the averaged data that are not easily and unambiguously interpreted at the cellular level because of numerous interactions of brain neurons with hundreds other neurons and glial cells, difficulties in their identification, in registration of activity of a given neuron, and in identifying glial cells that interact with it.

Using a simpler neuroglial system that consists of a small number of neurons and glial cells is preferable for elucidation of mechanisms of their photodynamic damage. We used the isolated crayfish stretch receptor as a very simple but informative model object for such study. It consists of a single sensory neuron enwrapped by satellite glial cells. Since the association of glial cells with this neuron is evident, this object is very suitable for study of neuroglial interactions that occur in extreme situations. In the present review we summarize the results of the complex electrophysiological, fluorescence-microscopic, pharmacological and ultrastructural study of PDT effect on the mechanoreceptor neuron and surrounding glial cells. The special attention is paid to the structural and signaling aspects of intercellular neuroglial interactions and their role in responses of neurons and glial cells to photodynamic injury.

Isolated Crayfish Stretch Receptor as a Model Object

SRN Features

Two muscle receptor organs are placed bilaterally in each abdominal segment of a crayfish. They consist of a couple of receptor muscles, slowly and rapidly adapting, expanded to the neighboring segments. Correspondingly, two sensory neurons (SRN), slowly and rapidly adapting ones, are mounted on these muscles. The sensory neurons are enwrapped by the multiplayer glial envelope (Figure 1). SRN are the large and flattened neuronal cells with a body size of 50-100 μM (Florey and Florey, 1955). These are very suitable for microscopic study. Another advantage of this object includes its ability to fire with a regular rate during prolonged time (several hours). The firing frequency is proportional to the receptor muscle length. Stretching of this muscle allows setting a desirable firing frequency rate. Small fluctuations of firing frequency (no more than 5-10%) provide registration of tiny initial changes in SRN activity induced by external impacts. SRN is very sensitive to photodynamic treatment (Uzdensky, Mironov, 1999; Uzdensky, 2010).

Recording of Neuronal Activity

Muscle receptor organs of two years old crayfishes *Astacus leptodactylus* from the Don River affluents were isolated according to Florey and Florey (1955). SRN axons were cut at approximately 1 cm from the neuron body. The receptor muscles were isolated together with 4x4 mm carapace pieces. These were placed in a plexiglass chamber equipped with a device for receptor muscle extension, and filled with van Harreveld's saline (mM: NaCl - 205; KCl- 5.4; $NaHCO_3$ - 0.24; $MgCl_2$ - 5.4; $CaCl_2$ - 13.5; pH 7.2-7.4). SRN spikes were derived extracellularly from axons by the glass pipette suction electrodes, amplified (the amplifier UU-90, Institute of Experimental Medicine, Sankt-Petersburg, Russia), digitized (the analog-digital converter L-761, L-Card, Moscow, Russia), and processed by a personal computer using the home-made software, which provided continuous monitoring of a firing frequency level. Experiments were carried out at 25±3 °C.

Photodynamic Treatment

At the beginning of each experiment the initial firing level of SRN was set near 6-12 Hz by application of the appropriate receptor muscle extension. After 30 min control recording of neuronal activity, photosensitizing solution was added into the chamber to a final concentration of 10^{-7} M. The photosensitizer Photosens (NIOPIK, Moscow, Russia) is a mixture of sulphonated aluminum phthalocyanines $AlPcS_n$ (n = 2, 3 or 4; the mean n = 3.1). After 30-min incubation with Photosens, cells were irradiated by the helium-neon laser (633 nm, 0.3 W/cm^2) or by the laser diode (670 nm, 0.4 W/cm^2) until irreversible firing abolition. A laser beam diameter was of 3 mm, so that the neuron soma and the significant part of the axon were irradiated. The radiant power was measured by the laser dosimeter IMO-2N ("Etalon", Volgograd, Russia). Such treatment inhibited and irreversibly abolished neuron activity for 15-25 min. 30-min light exposure used in the experiments was sufficient for irreversible firing abolition of almost all neurons.

Fluorescence-Microscopic Study of Cell Death

In order to visualize dead cells, at certain time intervals after firing abolition 20 μM propidium iodide and 10-20 μM Hoechst 33342 were added into the experimental chamber.

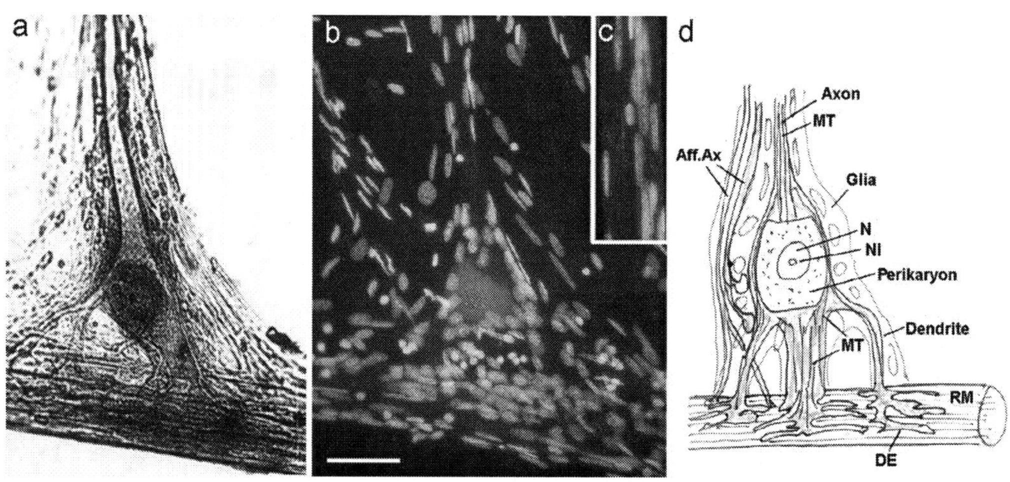

Figure 1. Slowly adapting abdominal stretch receptor of a crayfish. (a) The brightfield microphotograph. (B) and (C): Fluorescent images of the stretch receptor (b) and the fragment of its axon (c) fluorochromed with Hoechst 33342 (blue-green fluorescence) and propidium iodide, which visualizes nuclei of necrotic cells (red fluorescence). (d) The scheme of this preparation. Mechanoreceptor neuron surrounded by glial cells is mounted on the receptor muscle (RM). Its axon and dendrites are filled with numerous microtubules (MT). Dendrite endings (DE) are ramified between muscle fibers and contact with their membranes. N -nucleus, Nl – nucleolus, Aff.Ax – afferent axons. Scale bar on B – 50 μm.

Propidium iodide, a membrane impermeable fluorochrome, imparts red fluorescence to the nuclei of necrotic neuronal and glial cells with the compromised plasma membrane. Hoechst 33342 imparts blue fluorescence to the nuclear chromatin. Fragmented blue nuclei belonged to cells at the late apoptosis stages. In the next 10 min, the preparations were washed with van Harreveld's saline, postfixed with 0.1-0.2% glutaraldehyde, repeatedly washed and mounted in glycerol. Fluorescent images were obtained using fluorescence microscope Lumam-I3 (LOMO, Leningrad, USSR).

The total number of glial nuclei ("glia density") and the percentage of PI-stained nuclei of necrotic glial cells were counted in the predetermined standard fields (10000 μm^2) around SRN soma. Fragmented apoptotic nuclei were counted in the proximal 2 mm axon fragment, where apoptosis was more profound. Standard statistical methods based on the Student's t-test were used for evaluation of the difference between experimental and control values. Data are presented as mean ± SEM.

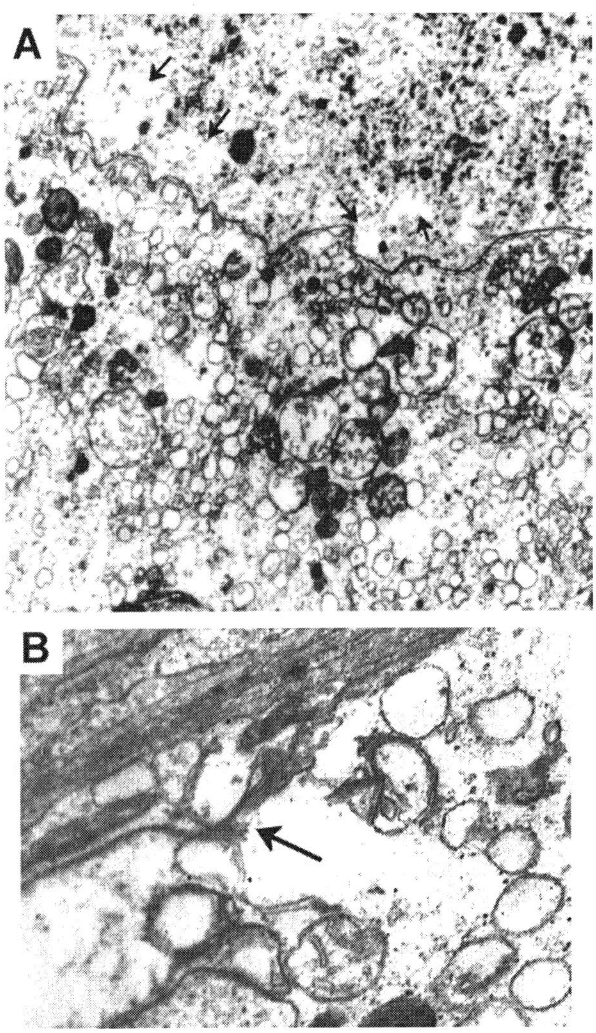

Figure 2. Examples of the main types of neuron response dynamics to photodynamic effect. Thin arrow indicates the moment of the dye application, thick arrow - the irradiation start.

Table 1. Main types of crayfish neuron response dynamics to photodynamic effect

Type of neuron response	Sequence of firing frequency changes
E	Excitation and abrupt cessation of firing
EIE	Excitation, inhibition, new excitation, and abrupt cessation of firing
IE	Transient inhibition, excitation and abrupt cessation of firing
I	Gradual inhibition until complete cessation of firing
EI	Transient excitation, then gradual inhibition until complete cessation of firing
IEI	Inhibition, excitation, repeated inhibition until complete cessation of firing

Pharmacological Analysis of the Possible Involvement of Signaling Proteins in Responses of Glial and Neuronal Cells to Photodynamic Treatment

The inhibitors of various signaling proteins were purchased in Sigma-Aldrich Co (USA), Alomone Co (Israel), or Peprotech (USA). Their concentrations were chosen to be twice lower than the predetermined drug concentrations, which disturbed neuron firing in the darkness for 2-3 h. They were added to the experimental chamber 5 min after application of Photosens, i.e. 25 min before irradiation. Photosensitizer and modifier substances were present in the chamber during irradiation and following incubation.

Transmission Electron Microscopy

At the chosen moment, the stretch receptor preparations (3 in each group) were fixed 30 min with 2.5 % glutaraldehyde dissolved in 0.1 M phosphate buffer (pH 7.2). Then the mechanoreceptor neurons were cut out together with the 1-2 mm pieces of the receptor muscle so that the preparations were T-shaped. This provided their correct orientation in the polymer blocks and identification of SRN in the sections. After the following washing in phosphate buffer, the preparations were incubated 1 h with 1% OsO_4, contrasted with uranyl acetate and lead citrate, dehydrated in a graded series of ethanol and acetone, and embedded into an epoxy resin (Gayer, 1984). Ultra-thin sections were obtained on the ultramicrotome Leica EM UC6 (Leica, FRG) and studied on the electron microscope Tecnai Spirit 12 (Phillips, Holland).

Bioellectric Neuron Response to Photodynamic Treatment

SRN was found to be insensitive to red laser irradiation or action of a photosensitizer staining alone, but very sensitive to their combined action, i.e. to photodynamic effect. We observed various neuron response kinetics with alternating firing acceleration and inhibition phases that were followed by irreversible abolition of neuron activity (Figure 2). The main types of neuron response dynamics are characterized in the Table 1. These depended on the photosensitizer, its concentration and photoirradiation intensity (Uzdensky, Mironov, 1999; Uzdensky et al., 2001). The most typical neuron response to photodynamic effect of 10^{-5} M Photosens or higher was rapid firing frequency increase and following abrupt and irreversible block (Figure 2A). At lower photosensitizer concentrations, additional firing inhibition phases were superimposed upon such response kinetics (Figure 2B,F). Gradual inhibition until irreversible firing abolition (Figure 2C-E) was typical for photosensitization with < 10^{-7} M Photosens concentrations (Uzdensky et al., 2002).

PDT-Induced Death of Crayfish Neurons and Glial Cells

Neuron response to Photosensitization

The nuclei of the control neurons fluorochromed with Hoechst 33342 are round or oval with a granular inner structure (Figure 3A). They are bigger than the nuclei of numerous glial cells surrounding the neuron.

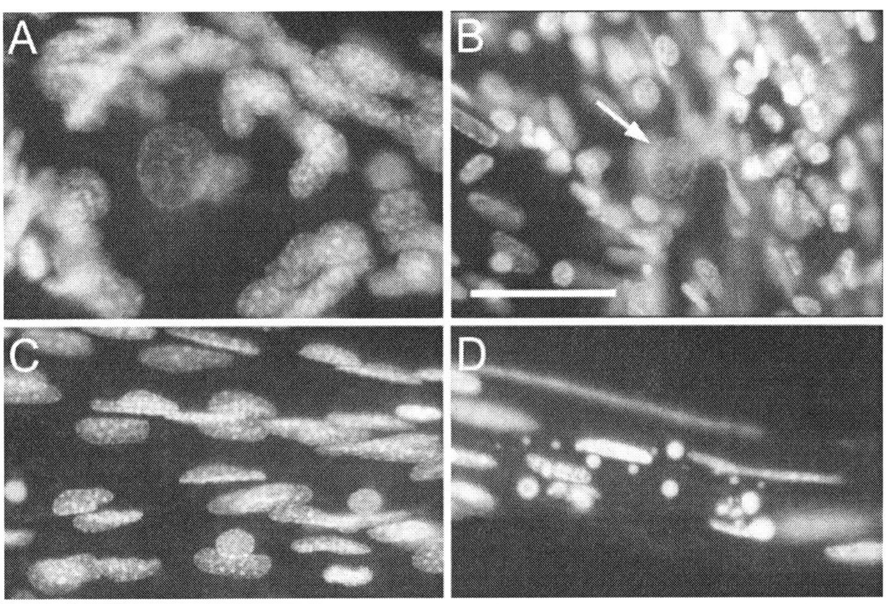

Uzdensky et al., 2005.

Figure 3. The nuclear morphology of the control (A and C) and photosensitized (B and D) stretch receptor neuron and surrounding glial cells. The preparation was double stained with propidium iodide fluorochroming only necrotic cells with injured plasma membrane and Hoechst 33342 fluorochroming all nuclear in 8 h after abolition of neuron activity and PDT finish. (A and C) – the neuron soma region, (B and D) – the axon region. Fragmented nuclei on (D) belong to apoptotic glial cells. Objective 40x. Scale bar 40 μm.

PDT induced shrinkage of the neuronal nuclei. The chromatin granules became larger and fluoresced brighter (Figure 3B) possibly due to denser chromatin packing (Uzdensky et al., 2002, 2005). Intense PDT (10^{-5} M Photosens) that accelerated neuronal firing caused red fluorescence of the neuronal nuclei just after PDT. This indicated the entry of propidium iodide into the neuronal nuclei through the compromised plasma membrane and necrosis that occurred during the treatment. At weaker PDT (10^{-7} M Photosens), red PI fluorescence appeared in the neuronal nuclei 1.7 h after PDT. Therefore, in these conditions neuronal necrosis was delayed (Uzdensky et al., 2002). This suggests that SRN necrosis induced by a relatively weak photodynamic treatment may be not a catastrophic but a controlled cell death mode.

Nucleus fragmentation or chromatin clumping into big dense masses characteristic for apoptosis (Arendt and Willie, 1991) did not occur either in PDT-treated SRN or in neurons treated with various proapoptotic chemical agents, whereas the apoptosis of glial cells was often observed (Figure 3D).

Glial Response to Photosensitization

Unlike SRN, surrounding glial cells died not only from PDT-induced necrosis but also from apoptosis. The level of glial necrosis in untreated crayfish stretch receptor preparations was about 10 %. These cells could be possibly damaged during stretch receptor isolation. PDT induced karyopyknosis (Figure 3B,D) and increased the fraction of necrotic glial cells with compromised plasma membranes to 30 % just after the treatment and to 50 % in the next 8 h (Uzdensky et al, 2005). After photosensitization some glial nuclei became fragmented (Figure 3D) that indicated apoptosis. The number of apoptotic glial nuclei was negligible in the first 4 h after PDT but significantly increased 4 h later. Therefore, 8-h post-treatment exposure was necessary for execution of all apoptosis stages resulting in nucleus destruction (Uzdensky et al., 2005).

Ultrastructural Changes in PDT-Treated Neurons and Surrounding Glial Cells

In order to study the dynamics of ultrastructural changes in SRN the following experimental groups were studied: (*a*) Control untreated neurons that fired regularly for 1 h; (*b*) Neurons fixed after 5-min photosensitization when changes in their activity by 2-4 Hz were higher comparing to the background fluctuations of firing frequency; (*c*) Neurons fixed just after PDT-induced firing abolition (approximately 30-min light exposure); (*d*) Neuron fixed 1 hr after 30-min PDT treatment (points *a, b, c, d* at Figure 4, respectively).

Control Preparations

Neurons

SRN morphology has been characterized in general earlier (Fisher et al., 1975; Tao-Cheng et al., 1981; Fedorenko et al., 1995; Fedorenko, Uzdensky, 2009a,b). Here we focus on the ultrastructural elements involved in bioenergetic and biosynthetic processes as well in the intraneuronal and neuroglial transport.

The SRN body is abundant with organelles and ribosomes (Figure 5) that indicates intense metabolism and high level of protein synthesis. Neurites, axon and dendrites, are filled with numerous microtubules (Figure 6B) involved in transport of proteins and other cell components to remote regions (Fedorenko, Uzdensky, 2009b). The microtubule bundles expanded from dendrites to the axon through the neuronal body separate the perikarion by 2-5 μm electron-dense areas called Nissl bodies, which are abundant with endoplasmic reticulum (ER), ribosomes and polysomes as protein-producing units, mitochondria as energy suppliers, and Golgi dictyosomes involved in protein processing, sorting and preparation for transport (Figure 5, 7,8A). Such spotted morphology of the perikarion at the optical microscopy level is called as "tigroid" (Fedorenko, Uzdensky, 2009a,b). Dictyosomes typically consist of 3-7 flattened cisterns and numerous vesicles. In Nissl bodies they are usually oriented so to be faced to microtubule bundles by their output trans-Golgi network (Figure 7). This shortens the transport pathways from the places of protein synthesis and processing to microtubular rails (Fedorenko, Uzdensky, 2009b). Mitochondria have mainly round or oval shape, moderately dense matrix and well-developed cristae that indicates their energized state (Figure 5,8). Mitochondrial aggregates consisting of several contacted mitochondria are distributed throughout the cytoplasm. Many mitochondria are located near the plasma membrane (Figure 5-8). The neuronal nuclei contain rare and small lumps of condensed chromatin located mainly at their periphery (Figure 5).

Glial Cells

The stretch receptor neuron is enwrapped by the multilayer envelope that consists of 10-30 glial layers separated by darker collagen layers (Figure 5-8). The number and width of glial layers vary in different SRN regions (Figure 7,8). Glial processes are the wide flattened sheets ending with finger-like protrusions, which intertwine with protrusions of other glial cells thus forming a lock that confine them around the neuron (Figure 5,6,8B).

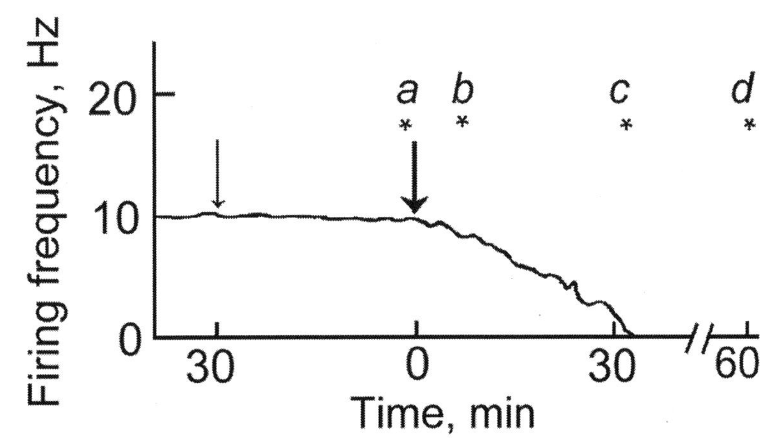

Fedorenko, Uzdensky, 2008.

Figure 4. The response of the stretch receptor neuron to PDT effect of 10^{-7} M Photosens. Thin arrow indicates the moment of Photosens application, thick arrow – start of 30-min light exposure. Asterisks *a, b, c* and *d* indicate the start of glutaraldehyde fixation: control, 5 min, 30 min, and 1 hr after PDT, respectively.

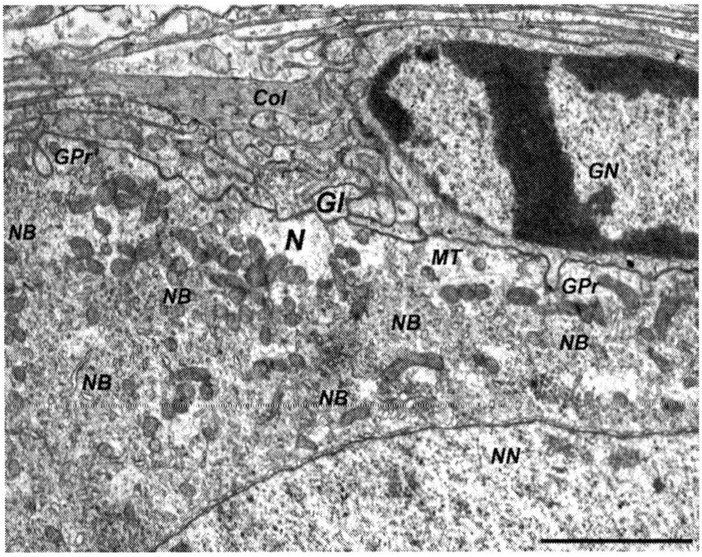

Fedorenko et al., 2011.

Figure 5. Control sample. The ultrastructure of the crayfish stretch receptor neuron (*N*) and its glial envelope (*Gl*). The neuron soma is separated by darker regions abundant with ribosomes, ER cisterns and mitochondria – Nissl bodies (*NB*). The electron-lucent regions between Nissl bodies contain microtubular bundles (*MT*); *Col* – collagen layers separating glial layers, *GPr* – glial protrusions, *GN* – glial nucleus, *NN* – neuronal nucleus. Scale bar – 2 μm.

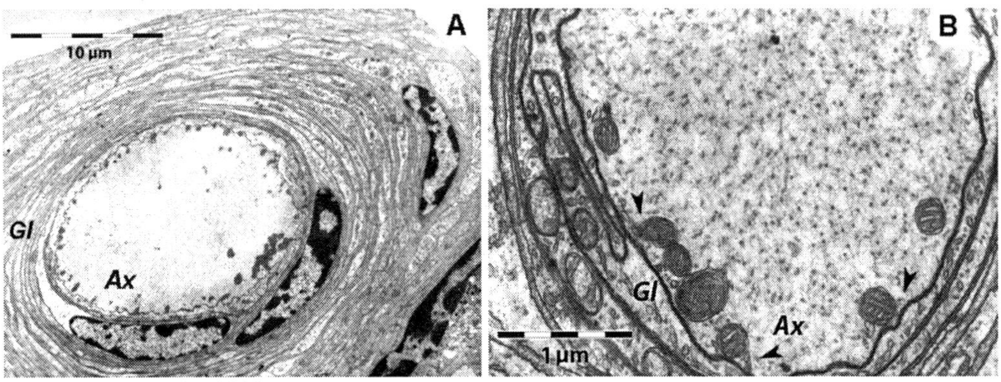

Fedorenko, Uzdensky, 2009a.

Figure 6. Control sample. The multilayer glial envelope around the crayfish stretch receptor neuron. The transversal section of the axon (*Ax*) and its glial envelope (*Gl*). Light layers – the cytoplasm of glial cells, darker sheets – collagen. Dark dots on (B) (rings at a higher magnification) are microtubules. The axon border is tortuous and that peripheral mitochondria communicate with the axonal membrane through membrane cisterns (arrowheads). *Gl* – glial cell, The scale bar on (A) – 10 μm, on (B) – 1 μm.

Together with the tortuosity of the neuroglial border (Figure 6) and glial protrusions into the neuronal cytoplasm (Figure 5,8A), this provides the robust attachment of the glial envelope to the neuron (Fedorenko, Uzdensky, 2009a). Flattened and elongated glial nuclei are lesser than the SRN nucleus (Figure 5). Unlike SRN nucleus that contains mostly the decondensed chromatin, which is involved in biosynthetic processes (Figure 5), in glial cells large dense masses of condensed chromatin are located at the nucleus periphery (Figure 5,

6A, 7). The glial cytoplasm contains much less organelles and ribosomes than the neuronal cytoplasm (Figure 7, 8). Therefore, the level of biosynthetic processes in glial cells is much lower than that in SRN.

The Structures Involved in Neuroglial Interactions

Glial cells and neurons are known to exchange metabolites, neuromediators, neurotrophins, proteins, mRNA and even ribosomes (Deitmer, 2001; Villegas et al., 2003). Small signaling molecules diffuse through the narrow intercellular space <10-15 nm width (Fedorenko, Uzdensky, 2009a). But how macromolecules are transferred between cells? The vesicular transport between the neuron perikarion and the plasma membrane and further to the neighboring glial cell is limited by the fibrillar microtubular layer surrounding the neuronal soma (Figure 7, 8) (Fedorenko, Uzdensky, 2009a). However, glial protrusions penetrate into the neuronal cytoplasm up to 0.5-1 μm (Figure 5, 8A). The tips of the corresponding invaginations of the neuronal membrane are sometimes captured and form double-wall vesicles (DWV) 0.2-0.5 μm in diameter (Figure 8B, C). Large masses of the glial cytoplasm that sometimes contain electron-dense inclusions, presumably fat aggregates, are thus transferred into the neuron. The flattened submembrane cisterns (SC) at the inner side of the neuronal membrane are involved in formation of neuronal invaginations. These cisterns remain often to be bound to the invagination tips and DWV surface (Figure 8) (Fedorenko, Uzdensky, 2009a, Fedorenko et al., 2011).

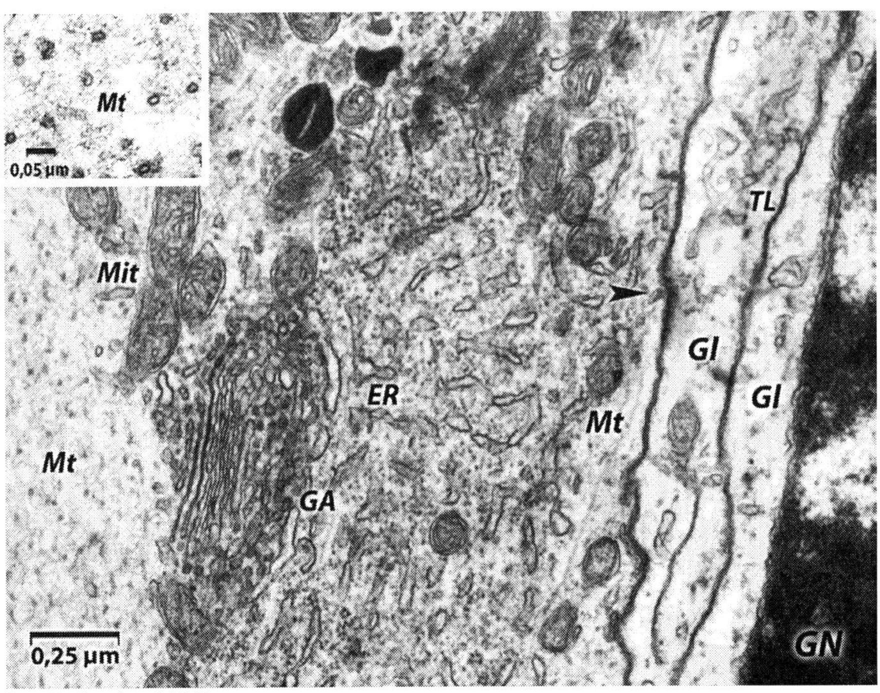

Fedorenko, Uzdensky, 2009b.

Figure 7. Control sample. The Nissl body at the periphery of the neuron perikarion is separated from the outer membrane by the microtubule bundle (*MT*). The *trans*–side of the Golgi dictyosome (*GA*) is faced to the inner microtubule bundle (*Mt*). *ER* – endoplasmic reticulum; *Gl* – glial cell, *GN* – glial nucleus; *Mit* – mitochondria; *TL* – tubular lattice.

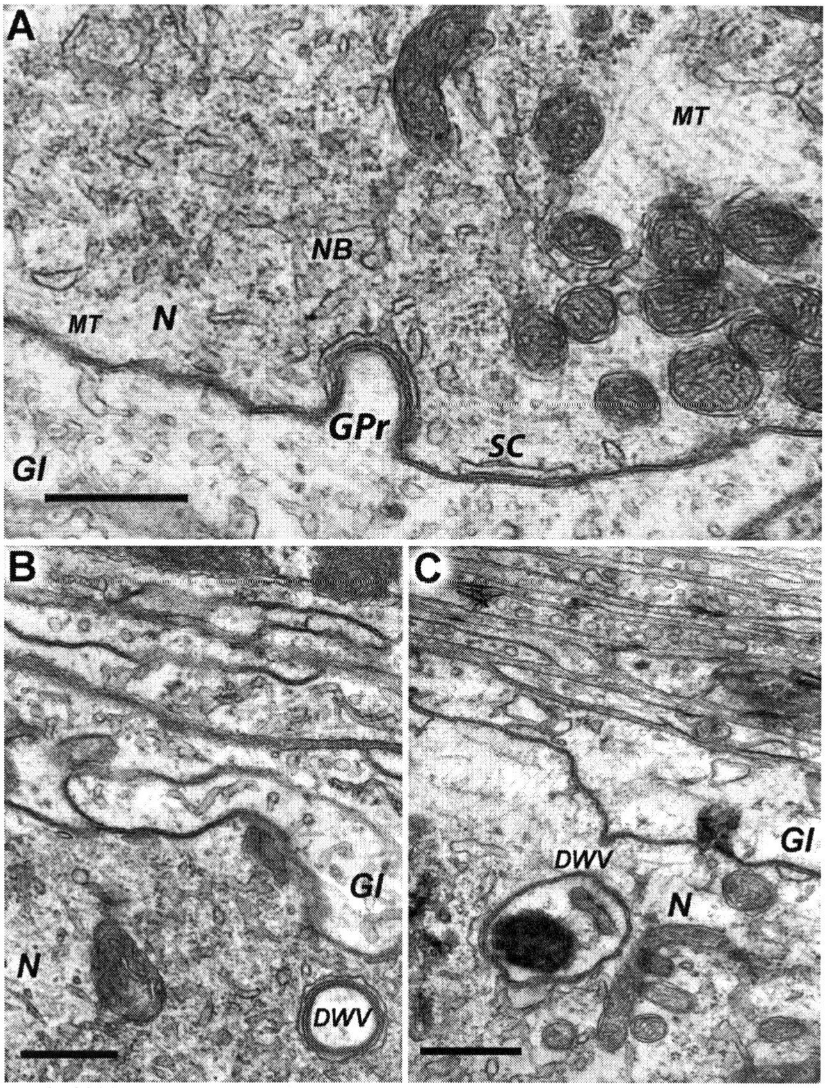

Fedorenko et al., 2011.

Figure 8. Control sample. Glial protrusion (A) and double-wall vesicles (B and C) near the neuroglial contact zones. *DWV* – double-wall vesicle, *Gl* – glial layers; *GPr* – glial protrusion, *MT* – microtubules, *N* – neuron, *NB* – Nissl body; *SC* – submembrane cistern. All scale bars – 0.5 μm.

Dynamics of Ultrastructural Changes in PDT-Treated Neurons and Glial Cells

5-Min PDT

At the initial stage of the neuron response to PDT treatment when its firing was shifted only slightly (2-4 Hz), noticeable changes of SRN ultrastructure were observed. The spotted tigroid morphology of the neuron body was saved (Figure 9A). Therefore the protein production and intraneuronal transport were not strongly impaired. However, large vacuoles (about 1 μm) with irregular, tortuous shape appeared within the cytoplasm (Figure 9A).

Numerous dictyosomes were observed in the neuron body (Figure 9A-D). Their structure was in general normal but some their cisterns in the middle and at the periphery moderately swelled (Figure 9B, C). Sometimes Golgi apparatus looked like a group of vesicles (Figure 9D,E) that could indicate its disassembly (Shorter, Warren, 2002). Numerous lysosomes were dispersed throughout the cytoplasm (Figure 9A-E).

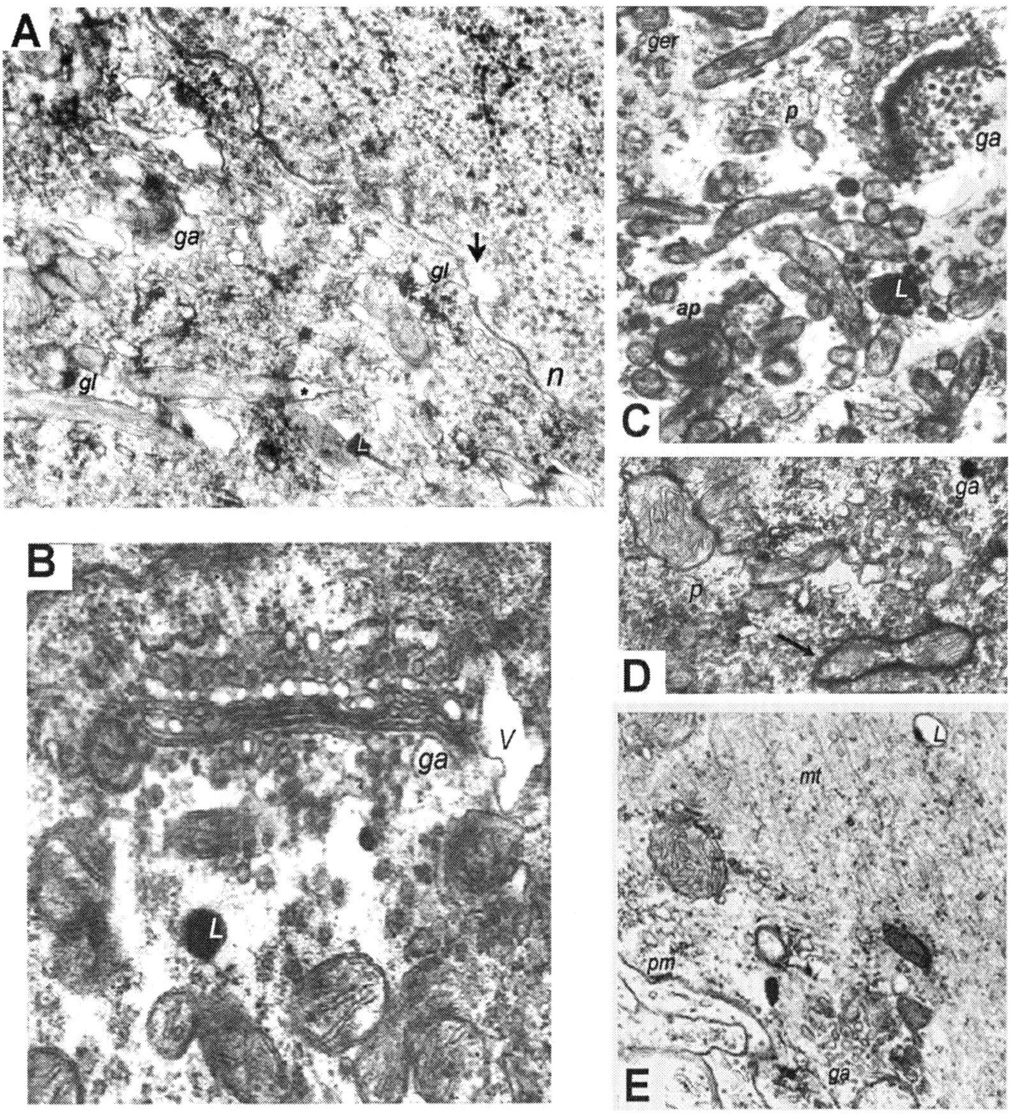

Fedorenko, Uzdensky, 2008.

Figure 9. Neuron ultrastructure just after 5-min photosensitization with 10^{-7} M Photosens. A – The nucleus and perikaryon, asterisk – the injured mitochondrial region, arrow – the electron-transparent nucleus region; x28000. B – the vacuole formation from the Golgi apparatus; *V* - the irregular-shape vacuole; x65000. C and D – the cytoplasm regions, x40000 and x50000, respectively. The arrow on D shows the membrane-coated mitochondria, from which an autophagosome (*ap* on C) might be formed. E – The peripheral region, x50000. *ga* – Golgi apparatus, *ger* – granular endoplasmic reticulum, *gl* – glycogen granules, *L* - lysosome, *mt* – microtubules (the normal section), *n* – nucleus, *p* – polysomes, *pm* – plasma membrane.

Sometimes mitochondria were enwrapped with the electron-dense envelope (Figure 9D, arrow) that could lead to formation of autophagosomes (Figure 9C). The ultrastructure of mitochondria was variable: some of them looked like the control ones, whereas others contained regions with electron-transparent matrix and destroyed cristae neighboring to the regions with normal ultrastructure (Figure 9A). Many mitochondria were elongated and formed aggregates (Figure 9A, C). Microtubules looked undamaged (Figure 9E). The nuclear chromatin was rarer than in the control samples. Condensed chromatin was practically absent and small electron-transparent regions (50-200 nm) appeared in the karyoplasm (Figure 9A).

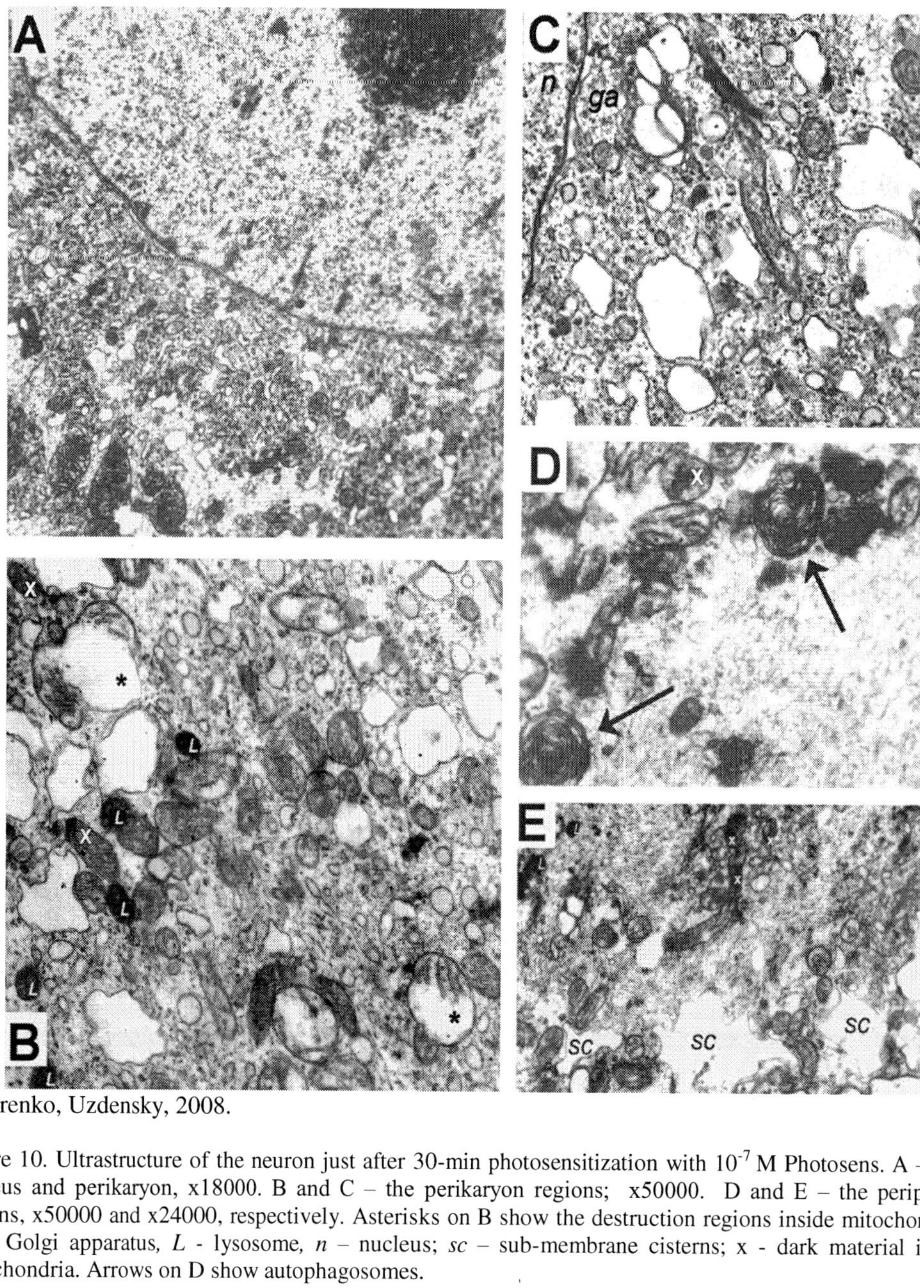

Fedorenko, Uzdensky, 2008.

Figure 10. Ultrastructure of the neuron just after 30-min photosensitization with 10^{-7} M Photosens. A – The nucleus and perikaryon, x18000. B and C – the perikaryon regions; x50000. D and E – the peripheral regions, x50000 and x24000, respectively. Asterisks on B show the destruction regions inside mitochondria. *ga* – Golgi apparatus, *L* - lysosome, *n* – nucleus; *sc* – sub-membrane cisterns; x - dark material inside mitochondria. Arrows on D show autophagosomes.

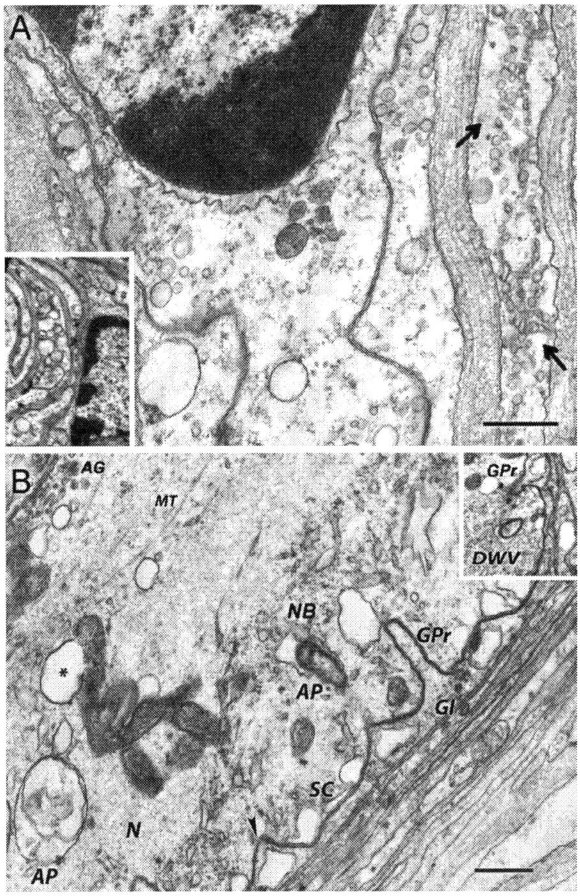

Fedorenko et al., 2011.

Figure 11. Ultrastructure of the stretch receptor just after 30-min photosensitization with 10^{-7} M Photosens. (A). Ultrastructure of glial cells. Cytoplasm vacuolization. Arrows indicate the chains of small vesicles, presumably the products of tubular lattice destruction. The fragment of swollen tubular lattice is shown in the inset. (B). The neuroglial border. Arrowhead indicates the rupture of the neuronal and glial membranes. *AG* – dictyosome, *AP* – autophagosome, *DWV* – double-wall vesicle, *Gl* – glial layers; *GPr* – glial protrusion, *MT* – microtubules, *N* – neuron, *NB* – Nissl body; *SC* – submembrane cistern. Asterisk – the swollen and destructed part of a mitochondrion. The scale bars are 0.5 μm.

30-Min PDT

Just after 30-min light exposure PDT that abolished neuronal activity the spotted tigroid morphology of the neuron soma looked in general saved (Figure 10A) and microtubules - undamaged (Figure 11B). Therefore, the intraneuronal transport pathways were not strongly impaired. However, the neuronal cytoplasm was abundant with numerous 0.2-1 μm irregular-shape vacuoles (Figure 10,11). They possibly originated from photodamaged mitochondria, Golgi apparatus, ER and sub-membrane cisterns. Mitochondria were very heterogeneous (Figure 10B, 11B). Some of them saved the normal ultrastructure, whereas others had completely destroyed cristae and lucid matrix. Curiously, in some mitochondria only one part swelled and lost matrix and cristae, whereas another part saved the normal structure. Some mitochondria contained the electron-dense material inside (Figure 10B,11B). These organelles could be converted into secondary lysosomes, abundant at this stage. Multilayer autophagosomes including large cytoplasm fragments were observed within the neuron soma

(Figure 10D,11B). The Golgi stacks were observed rarer than in control cells (Figure 10A-E). Some Golgi cisterns were swollen, whereas others – shrunk (Figure 10C). Granular ER and polysomes were not observed (Figure 10B-E,11B). Lysosomes were numerous. Significant changes in the neuronal nucleus were not detected (Figure 10A).

The general structure of the glial envelope was not significantly disturbed (Figure 11). Some but not all ER cisterns, dictyosomes and mitochondria swelled in the glial like in the neuron (Figure 11). The middle cisterns of dictyosomes only were especially swollen, whereas other cisterns and vesicles remained normal (Figure 11). Numerous vacuoles were observed in the glial cytoplasm (Figure 11). Unlike neurons, autophagosomes were not observed in glial cells. The spheres and tubules of tubular lattices involved in the inter-glial transport were expanded (Figure 11A, inset). Glial nuclei shrank. The nuclear chromatin was condensed and occupied the significant part of the nucleus (Figure 11A).

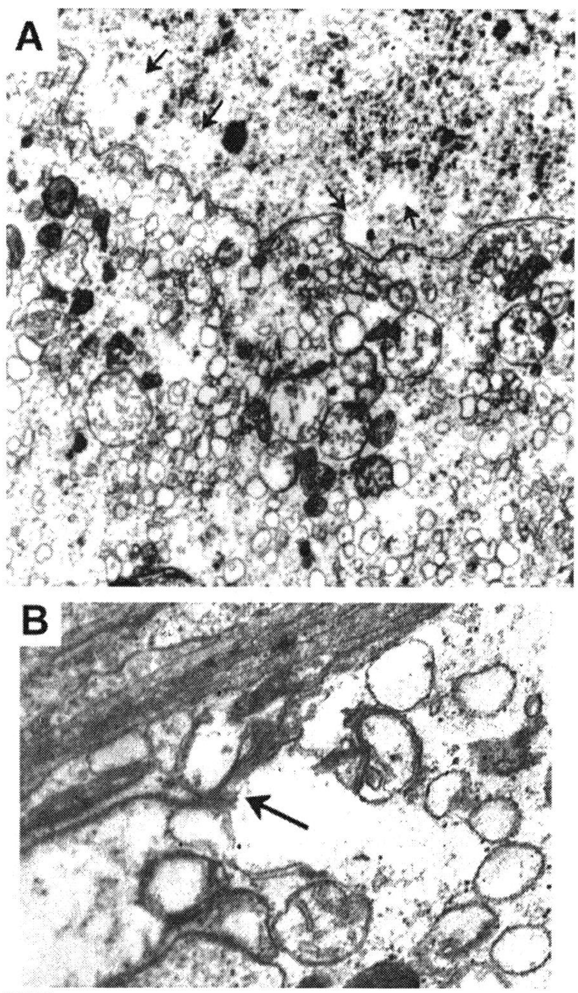

Fedorenko, Uzdensky, 2008.

Figure 12. Ultrastructure of the stretch receptor fixed 1 hr after 30-min photosensitization with 10^{-7} M Photosens. A – The nucleus and perikarion, x28000. B - Peripheral region; x50000. Arrows on A indicate electron-transparent regions in the nucleus; the arrow on B – the plasma membrane disruption.

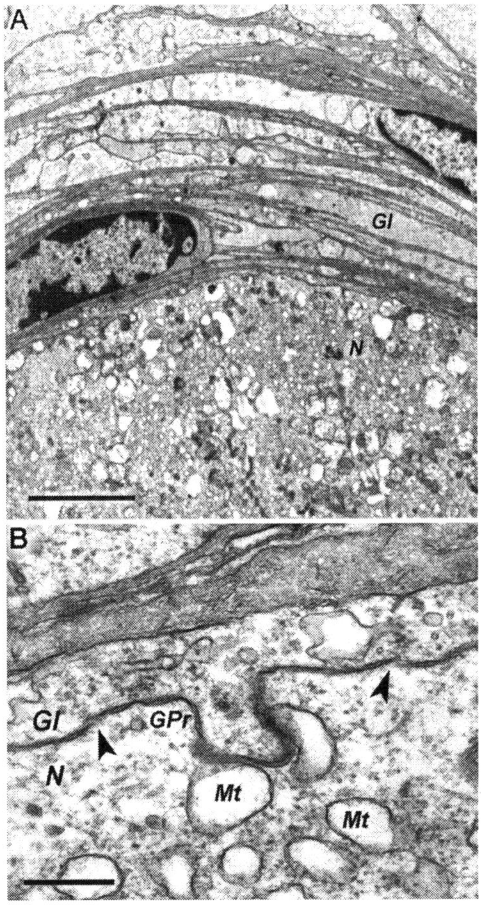

Fedorenko et al., 2011.

Figure 13. Ultrastructure of the stretch receptor fixed 1 hr after 30-min photosensitization with 10^{-7} M Photosens. Alterations of the glial envelope, neuronal cytoplasm and structures involved in neuroglial interactions. (A). Massive vacuolization of the neuronal cytoplasm and dilation of glial layers. (B) Electron-dense material is accumulated on tips of neuronal invaginations; associated submembrane vesicles and mitochondria (*Mt*) are swollen. Arrowheads – breaks in the glial and neuronal membranes. *Gl* – glial layers; *GPr* – glial protrusion, *N* – neuron, *SC* – submembrane cisterns. The scale bar on (A) – 2.5 μm, on (B) – 0.25 μm.

Glial protrusions at the neuroglial border and DWV, numerous in control samples, were seldom thus indicating impairment of the neuroglial exchange. Formation of glial protrusions was possibly prevented by swelling of submembrane cisterns. Creation of irregular vacuoles in the adaxonal glial layer curved the neuroglial border. Ruptures of neuronal and glial membranes in such regions opened the direct passage between neuron and glial cell (Figure 11B).

One Hour after PDT

PDT-induced ultrastructural changes were enhanced during subsequent 1-hour incubation in the darkness. At this moment the tigroid morphology of the neuronal soma was lost (Figure 12A,13A). This indicated the disorganization of the microtubular cytoskeleton and abolishment of the intraneuronal transport processes. Neuronal cytoplasm was strongly vacuolated and filled with numerous vacuoles formed from swollen cisterns of ER, dictyosomes and mitochondria. The alterations of organelles in glial cells were similar to that in the neuron (Figure 12-14).

The majority of mitochondria significantly swelled and rounded. Mitochondrial matrix was electron-transparent and cristae were disrupted, fragmented or absent. Almost all ER cisterns were swollen. Dictyosomes were not observed (Figure 12-14). The cytoplasm was abundant with small electron-dense structures (<0.5 μm) that were possibly the degraded mitochondria or secondary lysosomes. Small 50-200 nm electron-dense granules as well as electron-transparent regions were observed inside the neuronal nucleus (Figure 12A).

Glial layers, especially in the middle of the envelope, dilated. The cytoplasm of remote layers contained less vesicles and organelles and was more lucid than that in 3-4 layers adjacent to the neuron. Like in SRN, almost all mitochondria and vesicles in glial layers swelled and large irregular vacuoles appeared (Figure 14). Golgi apparatus was practically absent. The glial nuclei became pyknotic. In some of them the condensed chromatin occupied the major part of their area. Some nuclei became fragmented (Figure 15) that was characteristic for apoptosis. Although more profound fragmentation of glial nuclei was observed later, at 6-8 hours after PDT (Uzdensky et al., 2005), these data indicate the earlier start of the apoptotic process in the photosensitized glial cells.

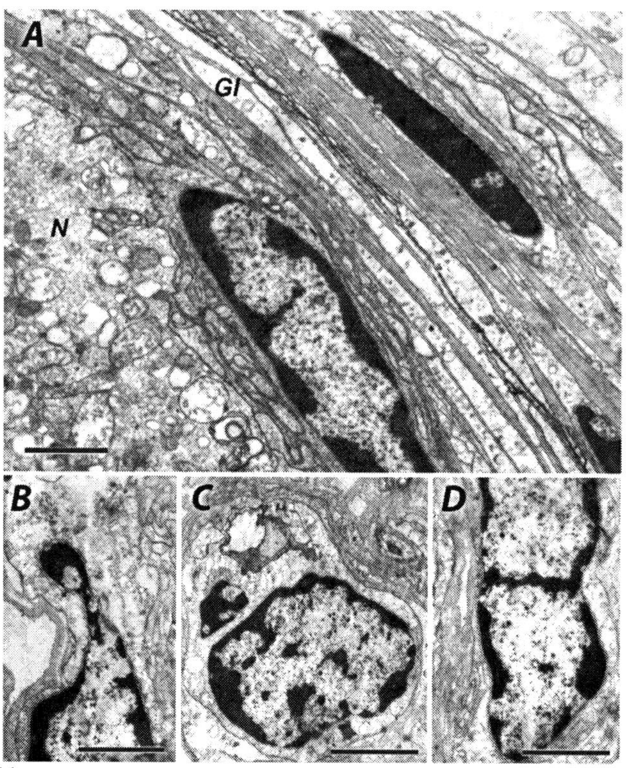

Fedorenko et al., 2011.

Figure 14. Ultrastructure of the stretch receptor fixed 1 hr after 30-min photosensitization with 10^{-7} M Photosens. Alterations on the glial nuclei. (A) Karyopyknosis. (B-D) Different stages of the nucleus fragmentation. All scale bars – 2 μm.

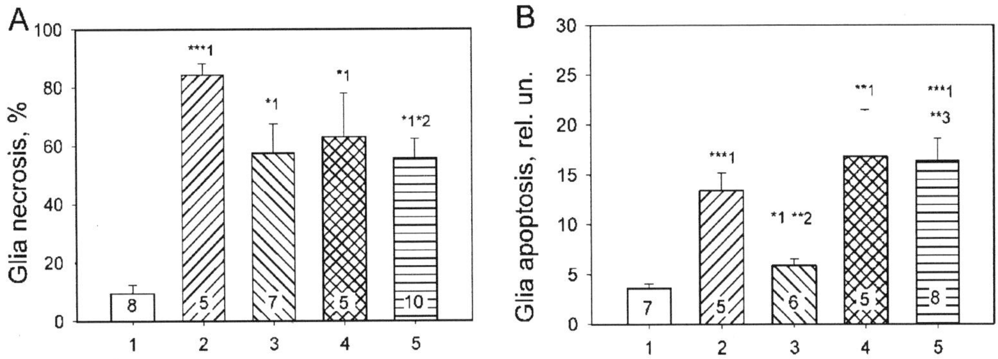

According to: Lobanov, Uzdensky, 2005.

Figure 15. PDT effect of 10^{-7} M Photosens on satellite glial cells after collagenase treatment (0.02 %, 60 min). A. Percent of necrotic glial cells observed 6-7 h after PDT-induced abolition of firing; B. The mean number of apoptotic glial cells with fragmented nuclei per 2 mm of axon length. *** p<0.001.

The structures involved in neuroglial interactions were significantly impaired. Glial protrusions and corresponding neuronal invaginations were more seldom comparing to control samples. Double-wall cisterns were absent (Figure 12-14). The formation of invaginations of the neuronal membrane and DWV was prevented by deterioration of the submembrane cisterns, which were significantly swollen but still attached to the neuronal membrane (Figure 12-14). As a whole, the ultrastructural alterations in photosensitized neurons observed at this stage were more uniform and typical for a dying necrotic cell.

The intracellular organelles: mitochondria, Golgi complex and ER as well as structures involved in intraneuronal transport and neuroglial interactions were especially sensitive to photodynamic treatment.

Neuroglial Interactions in the Photosensitized Crayfish Stretch Receptor

Intercellular neuroglial interactions are very significant for resistance of the nervous tissue to harmful impacts. Their role in the mutual maintaining of cell survival have been demonstrated in the cases of trauma and axotomy (Kopp et al., 1997), hypoxia/ischemia (Largo et al., 1996), deprivation of trophic factors (Grinspan et al., 1996). Neuron injury induces death of surrounding Schwann cells (Kopp et al., 1997; Grinspan et al., 1996). In turn, selective glia destruction suppresses neuronal functions and induces death of neurons (Largo et al., 1996). All types of glial cells (astroglia, oligodendroglia, microglia and Schwann cells) participate in neuron responses to harmful effects. Nerve and glial cells exchange with growth factors and neuromediators thus supporting mutual survival (Barres and Barde, 2000; Kopp et al., 1997). Neurons are protected by growth factors secreted by glial cells (Kopp et al., 1997; Rossner et al., 1997; Barres and Barde, 2000). Selective destruction of glial cells by gliotoxin fluorocytrate inhibits neuronal functions and induces neuron death (Largo et al., 1996). On the other hand, neurons release neuregulins that rescue glial cells and control their proliferation (Grinspan et al., 1996; Kopp et al., 1997; Barres and Barde, 2000).

In order to study the possible role of neuroglial interactions in survival and death of photosensitized neurons and glial cells, the crayfish stretch receptors were pretreated with proteolytic enzymes pronase or collagenase that destruct proteins in the intercellular medium and partly on the cell surface. Then the preparation was photosensitized with 10^{-7} M Photosens. Such proteolytic pretreatment did not influence PDT-induced changes in the neuron firing and necrosis as well as necrosis of glial cells. However the level of PDT-induced apoptosis was significantly reduced (Figure 15). Therefore, some intercellular or surface proteins are involved in PDT-induced apoptosis of crayfish glial cells (Lobanov, Uzdensky, 2005).

In another experiment we first inactivated SRN by the intense laser beam focused onto its body (633 nm; 200 W/cm^2, 90 μm diameter) after 30 min pre-incubation with 10^{-7} M Photosens, and then the whole stretch receptor preparation was irradiated with a lower light intensity (0.3 W/cm^2) as in other PDT experiments. The intense irradiation rapidly stimulated neuronal firing up to 40-60 Hz for about 3 min. Then neuronal activity was abruptly and irreversibly blocked, and necrosis of neurons and glial cells in the irradiation zone occurred that was revealed by appearance of red propidium iodide fluorescence in nuclei of these cells. The localized neuron injury did not influence necrosis of glial cells outside the irradiated region caused by following total photosensitization of the whole preparation. However, it significantly increased (about 8 times) the number of apoptotic glial cells enwrapping the initial part of the axon that was not subjected to the previous intense irradiation (Table 2) (Kolosov, Uzdensky, 2006). Therefore, the neuronal body protects glial cells in the non-irradiated zone from PDT-induced apoptosis but not necrosis.

According to ultrastructural data, crayfish glial cells supply SRN with the big cytoplasm fragments that are captured from glial protrusions. The present data demonstrated that in the crayfish stretch receptor neuroglial interactions are directed not only from glia to SRN but also from SRN to glia. How the neuronal soma controls apoptosis of remote glial cells located along an axon? It is tempted to suggest that SRN secrets some signaling molecules, messengers that influence glial cells. What is the molecular nature of such "neuron-to-glia signals"?

Neurotrophic factors and neuregulins initiating the pro-survival cascades in the crayfish neurons and glia could be involved in these processes (Grinspan et al., 1996; Kopp et al., 1997; Chao et al., 2006; Reichardt, 2006). These extracellular molecular signals are recognized by receptor tyrosine kinases and initiate various intracellular signaling pathways. Therefore, we investigated the possible involvement of different signal transduction pathways in responses of glial cells to PDT.

Table 2. Response of glial cells to total PDT treatment after local injury of the neuron body by a focused laser beam

Cell death type	Total PDT	Local PDT + total PDT
Necrosis, %	77 ± 5	80 ± 12
Apoptosis, rel. units	0,5 ± 0,4	3,9 ± 1,4*

* p<0,05. (According to: Kolosov, Uzdensky, 2006).

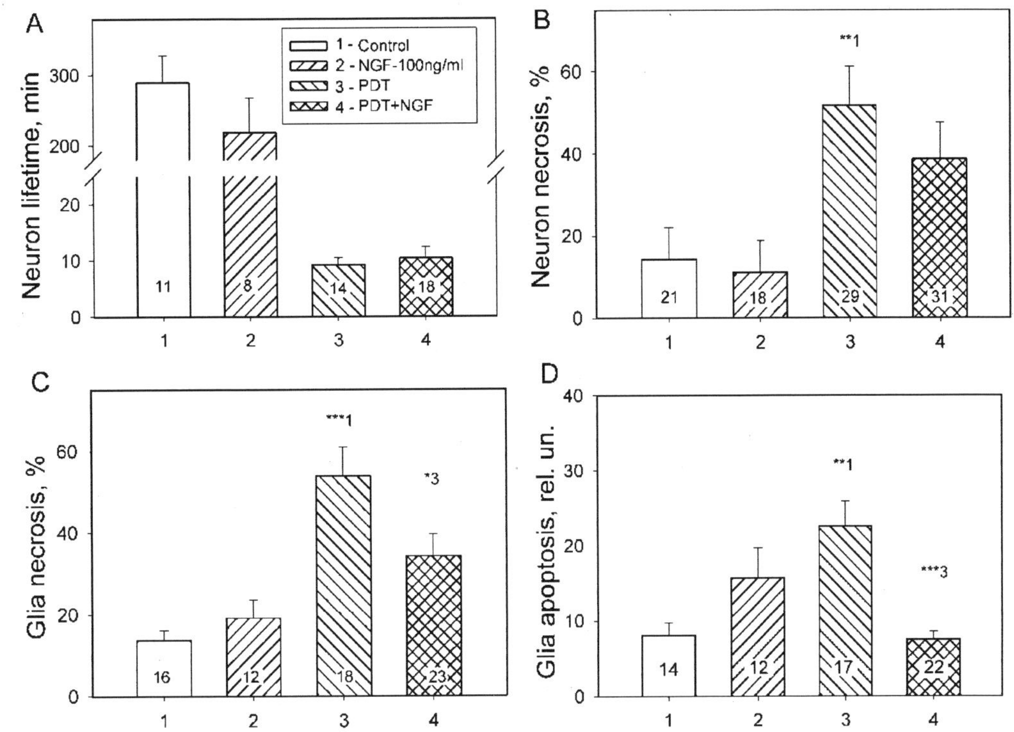

Lobanov, Uzdensky, 2009.

Figure 16. Photodynamic effect of 10^{-7} M Photosens on SRN and glial cells in the presence of 100 ng/ml NGF. A. Neuron lifetime, min; B. Percent of necrotic neurons in 8 h after PDT; C. Percent of necrotic glial cells in 8 h after PDT; D. Number of fragmented nuclei of apoptotic glial cells per 2 mm of proximal axon fragment in 8 h after PDT. The number of the experiments is shown inside bars. * $p<0.05$; ** $p<0.01$; *** $p<0.001$.

Neurotrophic Protection of Crayfish Glial Cells from PDT-Induced Death

Neurotrophic factors are widespreaed in the vertebrate nervous system. They maintain cell survival and play a key role in the organization of neural networks. There are several neurotrophins found in the brain: nerve growth factor NGF, brain-derived neurotrophic factor BDNF, neurotrophins NT-3 and NT-4/5. These are recognized by receptor tyrosine kinases TrkA, TrkB and TrkC, which initiate pro-survival intracellular signaling pathways. Additionally, the pro-apoptotic neurotrophin receptor $p75^{NTR}$ that recognizes all these neurotrophins and induces apoptosis is found in diverse neurons (Chao et al., 2006; Reichardt, 2006). Glia-derived neurotrophic factor GDNF recognized by receptor protein GFRα1 associated with transmembrane tyrosine kinase Ret also exerts a potent pro-survival effect (Airaksinen and Sarma, 2002).

We have studied the possible involvement of intercellular neurotrophic signaling in death and survival of PDT-treated crayfish neurons and glial cells. However, the problem is that neurotrophic factors and their receptors are still not found in the majority of invertebrates including crayfish (McKay et al., 1999, Jarro et al., 2001). For that reason, we explored the effect of some available mammalian neurotrophic factors (NGF, BDNF, and GDNF) on PDT-induced inactivation, necrosis or apoptosis of crayfish neurons and glial cells. The effect of epidermal growth factor (EGF) was studied as a control of non-specific interactions (Lobanov, Uzdensky, 2009).

None these neurotrophic factors influenced significantly PDT-induced firing abolition and necrosis of neurons. However, murine NGF protected crayfish glial cells from PDT-induced necrosis and apoptosis (Figure 16C and D). Similar effects were observed with GDNF (Uzdensky et al., 2008a). Unlike, BDNF and EGF did not influence PDT-induced necrosis and apoptosis of glial cells (Lobanov, Uzdensky, 2009).

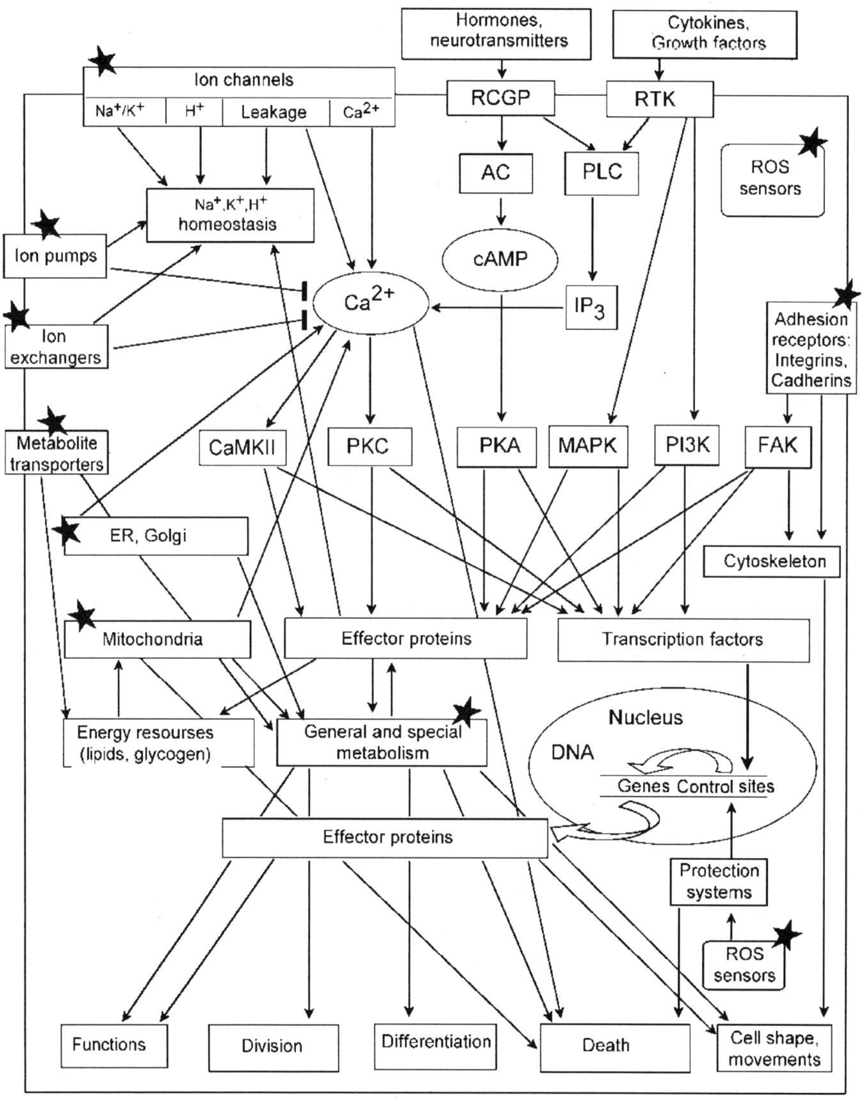

Figure 17. Main cell signaling pathways and cell functions they influence. The potential PDT targets are shown by astrerisks. AC – adenylate cyclase, CaMKII – calmodulin-dependent protein kinase II, ER-endoplasmic reticulum, FAK – focal adhesion kinase, IP_3 – inositol 1,4,5-trisphosphate, MAPK mitogen-activated protein kinase, PI3K – phosphatidylinositol 3-kinase, PKC – protein kinase C, PKA – protein kinase A, PLC – phospholipase C, RCGP – receptors coupled to G-proteins, RTK – receptor tyrosine kinase.

These data indicate the possible presence of receptors capable of recognizing recombinant murine NGF and human GDNF in crayfish glial cells. In stress conditions crayfish analogs of these neurotrophic factors could be involved in the neuron-to-glia signaling. Photodynamic impairment of their production and/or secretion could decrease the cell protection potential. In this situation application of exogenous NGF or GDNF could maintain survival of glial cells.

These results are of interest from the evolutionary point of view. The emergence of neurotrophic signaling has been suggested to be a key event in the significant increase of the number of neurons and formation of the complex head brain in higher organisms (McKay et al., 1999; Jarro et al., 2001, Jarro and Fainzilber, 2006). The evolution stage, in which neurotrophic factors have emerged, is unknown. It is also unknown, whether invertebrates, in particular crayfish, have the neurotrophic signaling. It was not found in the most of invertebrates. For example, the careful genome inspection showed that fly *Drosophila* and nematode *C. elegans* lack neurotrophins and their receptors (Hidalgo et al., 2006). On the other hand, analogs of neurotrophic factors or their receptors have been discovered in earthworm *Esenia foetida* (Davoli et al., 2002) and mollusks *Lymnaea stagnalis* (Fainzilber et al., 1996; Beck et al., 2004) and *Aplysia californica* (Ormond et al., 2004). The identification of neurotrophic signaling in diverse invertebrates is a very complicated task. The indirect indications on the presence of neurotrophic signaling in various invertebrate species may help in this problem. Since the commercially available invertebrate neurotrophins and antibodies to them and to their receptors are absent, the data on the effects of vertebrate neurotrophic factors on invertebrate neuronal and glial cells are of importance. According to the present data, crayfish *Astacus leptodactylus* as well as snail *Lymnaea stagnalis* (Ridgway et al., 1991; Wildering et al., 1995); squid *Loligo peallei* (Moreno et al., 1998) and silk moth *Bombyx mori* (Kim et al., 2005) may represent invertebrates that possess neurotrophic signaling.

The Role of Intracellular Signaling Processes in PDT-Induced Injury of Crayfish Neurons and Glial Cells

The Principal Signaling Pathways

Main signaling pathways and presumable PDT targets are schematically shown at Figure 17 (Uzdensky, 2008; 2010). Extracellular signaling molecules (hormones, neuromediators, or growth factors) are recognized by receptors coupled to G-proteins (RCGP), or receptor tyrosine kinases (RTK). Several intracellular signaling pathways lead from these receptors into the cytoplasm and the nucleus.

The calcium pathway is initiated by Ca^{2+} influx through the plasma membrane, or by Ca^{2+} release from ER and mitochondria. Their damage leads to accumulation of cytosolic Ca^{2+} and activation of Ca^{2+}-dependent proteins such as calmodulin, calmodulin-dependent kinase II, and protein kinase C. In the cAMP signaling pathway the binding of a ligand to RCGP activates G-protein, which stimulates adenylate cyclase to produce cAMP. cAMP-dependent protein kinase A activates then a variety of cellular reactions. Receptor tyrosine kinases initiate signaling pathways mediated by: (i) phospholipase C and Ca^{2+}; (ii) phosphatydilinositol 3-kinase and protein kinase B/Akt; (iii) mitogen-activated protein kinases (MAPK). The MAPK family includes extracellularly regulated kinase (ERK), c-Jun terminal kinase (JNK), and protein kinase p38. ERK is stimulated by growth factors, cytokines or chemical signals. It regulates cell proliferation and survival. JNK and p38 are involved in cell reactions to environmental stress that include survival, proliferation, or apoptosis. Integrins, cell adhesion receptors, activate focal adhesion kinase that regulates cytoskeleton remodeling, cell shape changes and motility.

All these protein kinases regulate numerous effector proteins and transcription factors, which determine cell responses. Each cell type has an individual set of such proteins that provide specific reactions to different impacts. Understanding the mechanism of action of various injurious factors requires comprehensive investigation of all cellular systems.

Using specific inhibitors of different signaling enzymes, we estimated the possible involvement of these pathways in PDT-induced necrosis of neurons and glia, and in apoptosis of glial cells. These data have been presented in our papers as diagrams like that shown at Figure 16, in which the data on the duration of neuron firing ("neuron lifetime"), the percentage of necrotic neurons and glial cells and the level of glial apoptosis in four experimental groups: (1) control, (2) the effect of an inhibitor in the darkness, (3) the PDT effect and (4) the combined effect of PDT and an inhibitor are shown. In order to avoid the presentation of numerous diagrams and tables, we display here the compressed summary of these data. The Table 3 demonstrates only significant ($p<0.05$) PDT-induced changes (increase or decrease) in the neuron activity, in the levels of necrosis of neurons and glial cells, and in the level of glial apoptosis.

Tyrosine Kinases and Phosphatases

In response to binding the extracellular messengers such as growth factors, receptor tyrosine kinases initiate several intracellular signaling pathways associated with phospholipase Cγ and calcium mobilization, MAPK and PI3K, which maintain the integrity of complex tissues and control the organism development and survival.

In photosensitized crayfish stretch receptors genistein, an inhibitor of various tyrosine kinases, reduced the mean neuron lifetime and decreased the percent of necrotic neurons (Table 3). In contrast, sodium orthovanadate, the inhibitor of tyrosine phosphatases, which removes phosphate groups from tyrosines and thus stops phosphotyrosine signaling, delayed PDT-induced inactivation of SRN. It, however, did not influence significantly the photoinduced necrosis of neurons but reduced necrosis of glial cells (Table 3). This indicates involvement of tyrosine phosphorylation of cellular proteins in the maintaining of neuronal activity and resistance of the plasma membrane to photodynamic injury. Neither genistein, nor sodium orthovanadate influenced PDT-induced apoptosis of glial cells (Uzdensky et al., 2005). Thus, phosphotyrosine-mediated signaling pathways were involved in the neuron photoinactivation and in the PDT-induced necrosis of neurons and glial cells but not in glial apoptosis.

However, genistein eliminated antiapoptotic effect of nerve growth factor on photosensitized glial cells (Lobanov, Uzdensky, 2009). Therefore, tyrosine phosphorylation was involved in the antiapoptotic activity of NGF. One can suggest that the intercellular signaling of NGF-like neurotrophin, involved in the protection of crayfish glial cells from PDT-induced apoptosis, could be mediated by receptor tyrosine kinases. Further we studied the possible role of three main signaling pathways initiated by receptor tyrosine kinases in photodynamic injury of crayfish neurons and glia.

Table 3. The summary of results of the inhibitory analysis of involvement of diverse intracellular signaling enzymes in PDT-induced inactivation and necrosis of the crayfish stretch receptor neurons and necrosis or apoptosis of surrounding glial cells

Enzyme	Inhibitor, concentration	Neuron lifetime	Neuron necrosis	Glial necrosis	Glial apoptosis	References
PI-Phospholipase C	ET-18, 20 μM	↑	—	↓	↓	Uzdensky et al., 2008
PC-phospholipase C	D-609, 7.5 μM	—	—	—	—	Uzdensky et al., 2007
Calmodulin	Fluphenasine, 1 μM	—	↓	↓	—	Uzdensky et al., 2007
CaMKII	KN-93 2 μM	—	↓	↓	—	Uzdensky et al., 2007
Protein kinase C	Staurosporine 1nM	↑	—	↓	↑	Uzdensky et al., 2007
Adenylate cyclase	Inhibitor MDL-12330A, 50 nM	↓	↓	↓	—	Uzdensky et al., 2005
	Activator forskolin, 10 μM	↑	—	—	↓	Uzdensky et al., 2005
Protein kinase A	H-89 1 μM	↓	↑	—	↑	Uzdensky et al., 2007
Tyrosine kinase	Genistein, 50 μM	↓	↓	—	—	Uzdensky et al., 2005
Tyrosin phosphatase	Orthovanadate Na, 500 μM	↑	—	↓	—	Uzdensky et al., 2005
MEK1/2	U0126, 2 μM	—	—	—	—	Uzdensky et al., unpublishe d data
JNK/SAPK	SP-600125, 10 μM	—	—	—	—	Lobanov, Uzdensky, 2009
P38	SB 202190, 5 μM	—	—	↓	—	Lobanov, Uzdensky, 2009

Table 3. (Continued)

Enzyme	Inhibitor, concentration	Neuron lifetime	Neuron necrosis	Glial necrosis	Glial apoptosis	References
PI 3-kinase	Wortmannin, 120 nM	—	—	—	—	Komandirov et al., 2011
	LY294002, 10 μM	—	—	—	—	Komandirov et al., 2011
Protein kinase B/Akt	Protein kinase B/Akt inhibitor IV, 1 μM	—	↓	↓	—	Komandirov et al., 2011
Glycogen synthase kinase 3β	TDZD-8, 2 μM	—	↓	—	↓	Komandirov et al., 2011
NGF	100 ng/ml	—	—	↓	↓	Lobanov, Uzdensky,2009
GDNF	10 ng/ml	—	—	↓	↓	Uzdensky et al., 2008b
BDNF	1 ng/ml	—	—	—	—	Lobanov, Uzdensky,2009
EGF	100 ng/ml	—	—	—	—	Lobanov, Uzdensky,2009

The significant changes ($p<0.05$) are shown with ↑ – increase or ↓ - decrease; — no changes.

MAP Kinases

Upon binding of a trophic factor, receptor tyrosine kinases dimerize and mutually phosphorylate tyrosines in the intracellular domains of each other. Recognitions of intracellular phosphotyrosines by the protein Grb/SOS initiates the enzyme cascade Ras-Raf-MEK1/2-ERK1/2, which activates a set of executive proteins, specific for different cell lines, which change the functional activity and execute various cell programs, e.g. protect the cell from apoptosis. Two other pathways activated by environmental stress are mediated by MAP kinases p38 or JNK (Robinson, Cobb, 1997).

Application of U0126 (2 μM), an inhibitor of MEK1/2, did not influence PDT-induced SRN inactivation and necrosis as well as necrosis and apoptosis of glial cells. It also did not influence significantly the protective effects of NGF and GDNF on PDT-induced death of neurons and glia. Therefore, we did not get evidence of involvement of the MEK1/2-ERK1/2 pathway in responses of SRN and glial cells to photodynamic treatment and in anti-necrotic and anti-apoptotic effects of NGF and GDNF on photosensitized glial cells.

The inhibition of MAP kinase p38 with SB202190 did not influence the SRN responses to PDT and PDT-induced apoptosis of glial cells. However, it reduced glial necrosis (Table 3). Therefore, p38 was involved in necrotic damage to photosensitized glial cells. The application of SP600125, the inhibitor of JNK did not influence photodynamic injury of neurons and glial cells (Figure 18; Table 3). However, SP600125 but not SB202190 eliminated the antiapoptotic effect of NGF on photosensitized glial cells (Figure 18). Therefore, JNK but not p38 was involved in the antiapoptotic activity of NGF (Lobanov, Uzdensky, 2009).

PI3K/Akt/GSK3β Pathway

The signaling pathway involving phosphatidylinositol 3-kinase (PI3K), protein kinase B/Akt (Akt) and glycogen synthase kinase-3β (GSK-3β) regulates cellular responses to external impacts and plays usually the protective, anti-apoptotic role (Krasilnikov, 2000; Duronio, 2008).

In photosensitized crayfish stretch receptor inhibition of phosphatidylinositol 3-kinase with wortmannin or LY294002 did not influence PDT-induced necrosis and apoptosis of neurons and glial cells (Table 3). This indicates non-participation of PI3K in photoinduced death of these cells. Specific inhibition of protein kinase B/Akt with protein kinase B/Akt inhibitor IV did not influence photoinduced abolition of neuronal activity and glial apoptosis. However, it significantly decreased the level of PDT-induced necrosis of neurons and glial cells (Table 3). Therefore, Akt was involved in photoinduced necrosis but not in apoptosis of neurons and glia. Glycogen synthase kinase-3β was shown to participate in photoinduced apoptosis of glial cells and in necrosis of neurons (Table 3). One can conclude that, the PI3K/Akt/GSK-3β pathway was not involved as a whole in photodynamic injury or protection of crayfish neurons and glia but its components, Akt and GSK-3β independently and cell-specifically regulated death of neurons and glial cells (Komandirov et al., 2011).

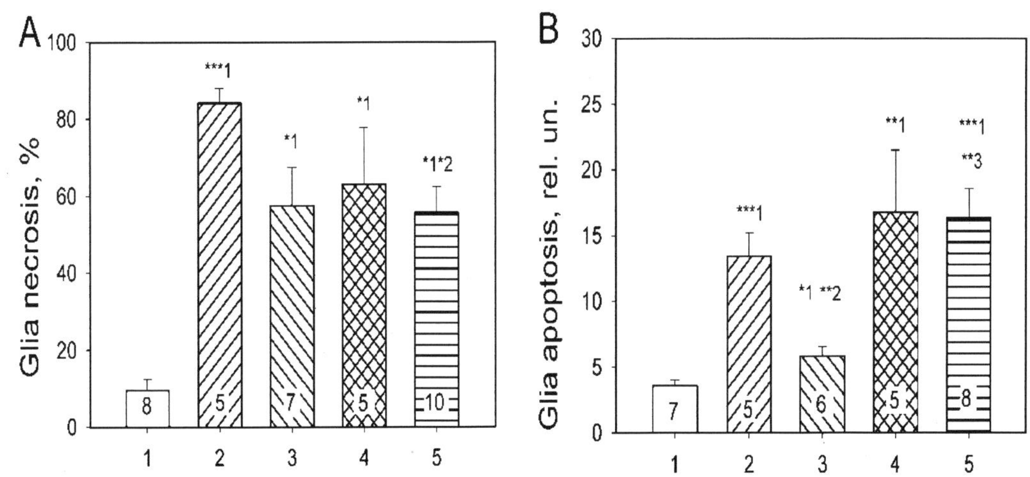

Lobanov, Uzdensky, 2009.

Figure 18. Modulation of the PDT effect of 10^{-7} M Photosens on crayfish glial cells by NGF (100 ng/ml) and JNK inhibitor SP600125 (10 μM). A. Percent of necrotic glial cells in 8 hours after PDT; B. Number of fragmented nuclei of apoptotic glial cells per 2 mm of proximal axon fragment in 8 hours after PDT. The number of the experiments is indicated inside each bar. Bars: 1 – control; 2 – PDT treatment; 3 – PDT+NGF; 4 – PDT+SP600125; 5 – PDT+NGF+SP600125. * $p<0.05$; ** $p<0.01$; *** $p<0.001$.

cAMP-Dependent Pathway

The cAMP-dependent signaling pathway is triggered by receptors coupled to G-proteins. The binding of extracellular messengers (hormones, neuromediators) to RCGP stimulates adenylate cyclase to produce cyclic adenosine monophosphate (cAMP), which then activates protein kinase A (PKA) and numerous downstream processes in the cell. This pathway is generally involved in cell protection (Michel, Agid, 1996).

The inhibition of adenylate cyclase by MDL-12330A or inhibition of downstream protein kinase A by H89 shortened the neuronal response to photodynamic impact, whereas the activation of adenylate cyclase by forskolin prolonged neuronal activity (Table 3). This suggests the involvement of adenylate cyclase and PKA in the maintaining of neuronal activity (Uzdensky et al., 2005, 2007).

The inhibition of adenylate cyclase by MDL-12330A also significantly reduced PDT-induced necrosis of neurons and glial cells (Table 3). Therefore, cAMP produced by this enzyme was involved in photoinduced death of these cells. However, this effect was not mediated by protein kinase A those inhibitor H89, oppositely, enhanced SRN necrosis (Table 3) that indicated the protective role of PKA. We did not find any significant influence of H89 on PDT-induced necrosis of glial cells as well as any effects of forskolin, the activator of adenylate cyclase, on necrosis of neurons and glial cells (Uzdensky et al., 2005, 2007). PDT-induced apoptosis of glial cells was reduced under adenylate cyclase activation by forskolin. PKA inhibition by H89, oppositely, enhanced glial apoptosis (Table 3). Therefore, both adenylate cyclase and protein kinase A were consecutively involved in protection of glial cells from PDT-induced apoptosis (Uzdensky et al., 2005, 2007).

Ca^{2+}-Dependent Pathways

Ca^{2+} is the key regulator of a variety of cellular processes. Its concentration in living cells is maintained at a very low level of $10^{-7} - 10^{-8}$ M. External stimuli like nervous stimulation of glands or muscles induce Ca^{2+} influx through the plasma membrane or release it from intracellular stores (mitochondria or ER) so that the cytosolic Ca^{2+} level increases up to 10^{-6} M. This induces various physiological reactions, e.g. hormone secretion or muscle contraction. Stressful factors damaging cellular membranes may increase the intracellular Ca^{2+} level to $10^{-4} - 10^{-3}$ M and thus induce cell death, necrosis or apoptosis (Trump, Berezesky, 1992). Ca^{2+} activates numerous enzymes and initiates various signaling pathways (Figure 17). Photodynamic treatment rapidly increases the cytosolic Ca^{2+} level (see reviews Castano et al., 2005; Buytaert et al., 2007, Uzdensky, 2008, 2010).

As has been shown earlier, cytosolic Ca^{2+}, independently on its source, was involved in PDT-induced inhibition of SRN activity and following firing abolition. Actually, PDT shortened neuron lifetime in the presence of agents elevating the intracellular Ca^{2+} level: 3-fold extracellular Ca^{2+} concentration, Ca^{2+} ionophore ionomycin, thapsigargin (inhibitor of Ca^{2+}-ATPase of endoplasmic reticulum), or caffeine that induces Ca^{2+} release from ER. In contrast, agents decreasing cytosolic Ca^{2+} concentration: 1/3-fold extracellular Ca^{2+} concentration, or $CdCl_2$, a blocker of Ca^{2+}-channels, or dantrolene, an inhibitor of calcium release from ER, reduced PDT-induced firing inhibition and increased neuron lifetime (Uzdensky et al., 2000, 2002).

Calcium ions can penetrate into the photosensitized cell through the compromised plasma membrane, or be released from photodamaged mitochondria and ER. However, in the complex tissue such as the neuroglial preparation, a significant amount of Ca^{2+} may be also released from ER upon binding the extracellular signaling molecules. In fact, both RCGP and RTK can activate phosphatidylinositol-dependent phospholipase Cγ (PI-PLC), which releases inositol 1,4,5-trisphosphate (IP_3) that in turn induces calcium liberation from ER. In our experiments the inhibition of PI-PLC with ET-18 slowed down PDT-induced firing inhibition and thus increased neuron lifetime. It also significantly reduced PDT-induced necrosis and apoptosis of glial cells (Table 3). Therefore, PI-PLC was involved in PDT-induced firing inhibition and in necrosis and apoptosis of glial cells. These effects could be a result of IP_3-mediated calcium release from ER. Unlike, another phospholipase Cγ that splits phosphatidylcholine-containing phospholipids (PC-PLC) did not influence PDT effects on neurons and glial cells. Therefore, these effects where IP_3^- and Ca^{2+}-specific (Uzdensky et al., 2007, 2008b).

We have also studied the involvement of two central Ca^{2+}-dependent signaling pathways, calmodulin- and protein kinase C-dependent, in PDT-induced death of crayfish neurons and glia. Being activated by Ca^{2+}, calmodulin (CaM) controls diverse cellular processes including various protein kinases, cytoskeleton, ion transport, neuronal activity and survival. Calmodulin-dependent kinase II (CaMKII) is especially expressed in the nervous system. It is activated by calmodulin in response to increased Ca^{2+} level. CaMKII regulates metabolism and various cell functions. It is considered as a "decoder" of calcium information in the cell (Soderling, 1999; Colbran, 2004).

The inhibition of CaM with fluphenasine or the inhibition of CaMKII with KN-93 in our experiments did not influence PDT-induced SRN inactivation and apoptosis of glial cells. However, both inhibitors significantly reduced the percent of necrotic neurons and glial cells observed at 8 h after PDT (Table 3) that indicated the consecutive involvement of CaM and CaMKII in PDT-induced deterioration of the plasma membrane and subsequent necrosis of SRN and glial cells (Uzdensky et al., 2007).

Protein kinase C (PKC) controls gene expression, enzyme activity, ionic currents, cytoskeleton functions, and apoptosis (Newton, Johnson, 1998). Its activation with 12-0-tetradecanoylphorbol 13-acetate (TPA) enhanced PDT-induced firing inhibition and decreased the lifetime of photosensitized neurons. In contrast, PKC inhibitors staurosporine, chelerythrine or hypericin slowed down PDT-induced firing inhibition and increased neuron lifetime thus protecting SRN against photodynamic inactivation (Bragin et al.,2003). Inhibition of PKC with staurosporine significantly increased neuron lifetime and the level of PDT-induced glial apoptosis. It also decreased the percent of necrotic glial cells in the photosensitized preparations (Table 3). Therefore, PKC was involved in PDT-induced inhibition of neuronal activity and in necrosis of glial cells. This could be related to PKC-mediated activation of calcium channels in the plasma membrane and following Ca^{2+} influx. At the same time, PKC protected glial cells from PDT-induced apoptosis (Uzdensky et al., 2007).

On the Mechanisms of Neuron and Glial Cell Responses to PDT

Structural Aspects of PDT-Induced Death of Neurons and Glia

The parallel study of dynamics of changes in bioelectric activity and ultrastructure of the stretch receptor neuron may answer the important questions: (*i*) What are the primary PDT targets in the cell? (*ii*) How the primary changes are generalized onto the whole cell? (*iii*) How cell damage develops after the treatment? (*iv*) How functional and structural changes correlate? (*v*) What cell type, necrosis or apoptosis, was induced by PDT in neurons and glial cells? (*vi*) Which biochemical processes lead to cell death in neurons and glial cells?

Singlet oxygen, the main reactive oxygen species in photosensitized cells (Dougherty et al., 1998; Castano et al., 2004), is known to diffuse not more than 10-20 nm during its short lifetime and oxidize only the nearest cell structures (Moan, 1990). Therefore, the mechanism of cell damage is determined by the intracellular localization of a photosensitizer. Photosens is a mixture of sulphonated aluminum phthalocyanines $AlPcS_2$, $AlPcS_3$ and $AlPcS_4$, which differ in hydrophobicity and, therefore, in the intracellular localization. More lipophilic $AlPcS_2$ and $AlPcS_3$ are preferentially localized to cellular membranes including the plasma membrane and the membranes of intracellular organelles. Hydrophylic $AlPcS_4$ does not cross the plasma membrane but may be adsorbed on its surface and photosensitize it from the outer side.

In the complex electrophysiological and electron-microscopic study functional changes correlated with structural alterations. The first changes in the neuronal activity indicating the alteration of the neuronal membrane occurred simultaneously with the earliest ultrastructural changes. 5-min light exposure that changed SRN firing only slightly injured some mitochondria and induced slight swelling of some ER and Golgi cisterns. Longer irradiation (30 min) that completely abolished neuronal activity caused strong damage to mitochondria, Golgi apparatus and ER. Therefore, cellular membranous structures are the primary PDT targets possibly due to higher accumulation of Photosens. Other researchers also showed that mitochondria, ER and Golgi stacks are the most vulnerable structures in the photosensitized brain (Yoshida et al., 1992a,b).

If mitochondria were traditionally considered to be the main PDT target, our data showed also significant alterations in the Golgi apparatus. This is in agreement with the accumulating literature data that different hydrophobic photosensitizers such as zinc phthalocyanine (Fabris et al., 2001), pheophorbide-a methylester (Matroule et al., 2001), chlorine p6 derivative (Feofanov et al., 2002); Photofrin (Hsien et al., 2003), Foscan (Teiten et al., 2003), or 2,4,5,7-tetrabromrodamine 123 bromide (Ogata et al., 2003) target the Golgi apparatus. PDT-induced Golgi dysfunction may lead to apoptosis (Matroule et al., 2001; Ogata et al., 2003). Activation of the Golgi apparatus at the early response stage could be related to formation of numerous lysosomes directed to dye excretion and compensation of the primary damages. The swelling of Golgi stacks led also to cytoplasm vacuolization. It is of interest that vacuoles formed at the periphery of Golgi cisterns were not round but had a complex irregular shape that implies the presence of some inner or outer skeleton. (Fedorenko, Uzdensky, 2008).

Prolonged incubation of the stretch during 1 h after PDT treatment induced not only cistern swelling but also fragmentation and disassembly of the Golgi apparatus. The loss of dictyosomes was a sign of impairment of the intraneuronal transport that play a key role in the integration of structure and function of such large and complex cell as a neuron. The loss of tigroid morphology was another essential sign that the supply of remote neuron regions with proteins and other necessary components was stopped. Specific ultrastructural elements involved in the intercellular transport between SRN and adjacent glial cells: submembrane cisterns, glial protrusions into the neuronal cytoplasm, double-wall vesicles and tubular lattices were among the structures, the most sensitive to photodynamic injury (Fedorenko, Uzdensky, 2011). At this time almost all these structures were lost. Mitochondria significantly swelled and their inner structure was destroyed. This demonstrated the deterioration of the cellular energetics. Altogether, these ultrastructural alterations indicated the development of necrotic processes in the neuron within the first hour after photodynamic treatment. Fluorescence-microscopic study showed that the neuronal membrane was compromised and red fluorescence of propidium iodide appeared in the nucleus at 1.7±0.3 h after PDT-induced firing abolition. This was the additional evaluation of the time interval necessary for necrosis accomplishment in the stretch receptor neuron. Glial cells were more sensitive to PDT, and necrosis of 30 % cells was observed just after 30-min photodynamic treatment comparing to 10 % in control samples.

There are three main cell death modalities known: apoptosis, autophagy and necrosis (Golstein, Kroemer, 2006; Galluzzi et al., 2007). Apoptosis of glial cells with nucleus disintegration by small fragments was often observed but apoptotic fragmentation of the nucleus or chromatin clumping into big masses was not observed in SRN either after photodynamic treatment or action of different chemical agents. One can suggest that the apoptotic program is intrinsically blocked in this unique neuron, which is essential for control of animal movements and, therefore, for crayfish capabilities of hunting and avoiding a danger. The functions of this neuron are not doubled (Uzdensky et al., 2002, 2005). The apoptosis block is characteristic for neurons of adult animals. The intrinsic antiapoptotic program may be blocked in the postembryonic period by different mechanisms from inhibition of diverse apoptosis stages to prevention of the transfer of proapoptotic signals from Fas receptors inside the cell (Benn and Woolf, 2004).

The earlier appearance of autophagosomes in photosensitized SRN indicated the possible protective role of autophagic processes in agreement with the literature (Kessel et al., 2006). Autophagosomes were practically absent in photosensitized glial cells. Therefore, autophagy was not a reason of photodynamic killing of glial and neuronal cells. In SRN autophagic processes were overcome by more potent necrotic alterations associated with disruption of the plasma membrane and loss of the energy sources. Thus, we classified photodynamic SRN death as a delayed necrosis (Uzdensky et al., 2002, Fedorenko, Uzdensky,2008, Uzdensky, 2010).

Although apoptosis was not characteristic for SRN death, satellite glial cells surrounding this neuron dyed from apoptosis rather often. Therefore, unlike SRN, the apoptotic program was not blocked in the crayfish glia.

Biochemical Regulation of PDT-Induced Death of Neurons and Glia

PDT-induced swelling and destruction of mitochondria, Golgi apparatuses and ER could release Ca^{2+} from these organelles into cytosol. We hypothesize that free cytosolic Ca^{2+}

played a central role in PDT-induced neuron inactivation and death. The following arguments support this suggestion:

- Photodynamic treatment rapidly, with the first minute increases the cytosolic Ca^{2+} level in various cells lines (Ben-Hur, Dubbelman, 1991; Humber et al., 1996; Ruck et al., 2000). Although we did not measure the cytosolic calcium concentration, the universality of this effect suggests that it could take place in the photosensitized SRN.
- According to the ultrastructural data, the primary alterations in photosensitized neurons and glial cells occur in Ca^{2+}-storing organelles, mitochondria, ER and Golgi apparatus. Their photodynamic damage should liberate the stored Ca^{2+}.
- Diverse pharmacological agents increasing $[Ca^{2+}]_i$ shortened the lifetime of photosensitized neurons in our experiments, whereas those reducing $[Ca^{2+}]_i$ prolonged SRN firing independently on the pathways leading to increase or decrease in $[Ca^{2+}]_i$.
- The inability of Ca^{2+}-ATPases, which are very vulnerable to PDT (Kondo, Kasai, 1974), to pump out Ca^{2+} or to accumulate it in the intracellular stores might be also involved in the elevation of the intracellular calcium level.
- Ca^{2+} is responsible for firing inhibition phases in the bioelectric response of SRN to PDT. It may activate Ca^{2+}-dependent K^+-channels in the neuronal membrane. The outward potassium current may hyperpolarize the neuron and inhibit its firing (Stocker et al., 2004).
- The excessive amount of free Ca^{2+} in the cytosol activates Ca^{2+}-dependent proteolytic enzymes and induces necrosis or apoptosis (Trump, Berezesky, 1992; Orrenius et al., 1994).

The signal transduction processes play a very important role in cell response to photodynamic treatment. Our experiments showed the involvement of Ca^{2+}/CaM/CaMKII; Ca^{2+}/PKC; adenylate cyclase/cAMP/PKA; tyrosine kinase/tyrosine phosphatase, MAPK p38 and JNK; protein kinase B/Akt- and GSK-3β-dependent signaling pathways as well as neurotrophic factors NGF and GDNF in the regulation of neuron and glia responses to PDT (Figure 19, 20). All of them can interact with each other and are united in the signaling network.

It is of importance that at relatively weak PDT (10^{-7} M Photosens) the necrotic processes in SRN developed rather slowly, during 1-2 hours after the end of photosensitization. As shown in the experiments with chemical modification, not only apoptosis but also necrosis was controlled by different signaling enzymes. Therefore, necrosis is not a catastrophic and passive cell damage but it may be regulated by the cellular signal transduction system (Uzdensky, 2010). These data are in consistence with the data of other authors that are accumulated in last years (Syntichaki, Tavernarakis, 2002; Golstein, Kroemer, 2006).

The problem of signaling control of neuronal death, especially THE control of necrosis is highly important for neuropathology. At stroke, necrosis is the main mechanism of cell death in the focal insult zone (Dirnagl et al., 1999). Inhibition of the early necrosis stages could limit the brain damage. Although necrosis has been known much earlier than apoptosis, the mechanisms controlling this cell death modality are much less studied. The data on signaling regulation of necrosis are practically absent in the literature (Syntichaki, Tavernarakis, 2003; Golstein, Kroemer, 2006). Therefore, the experimental data on the pharmacological regulation of oxidative stress and following necrosis of glial and neuronal cells may be used in a search of potential drugs for treatment of various neurodegenerative diseases.

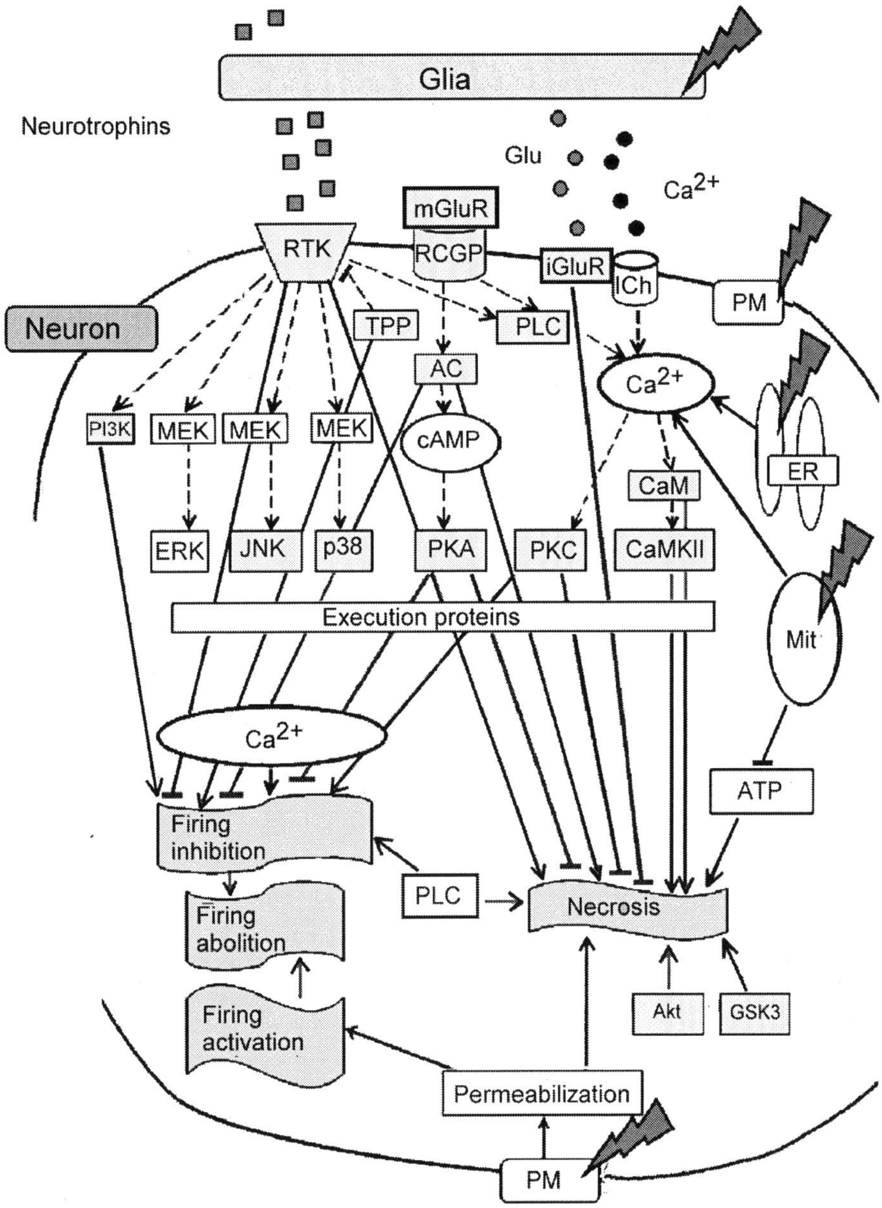

Abbreviations: RTK – receptor tyrosine kinase; RCGP – receptor coupled with G-protein; ICh – ionic channels; PM – plasma membrane; TPP – tyrosine protein phosphatase; AC – adenylate cyclase; PDE – phosphodiesterase; PI3K – PI 3-kinase; MEK – MAP kinase kinase; cAMP – cyclic adenosine mononucleotide; ERK, JNK and p38 – MAP kinases; PKA – protein kinase A; PKC – protein kinase C; CAMKII – calmodulin-dependent kinase II; ER – endoplasmic reticulum; Mit – mitochondria.

Figure 19. Neuron-to-glia signaling in the photosensitized crayfish stretch receptor.

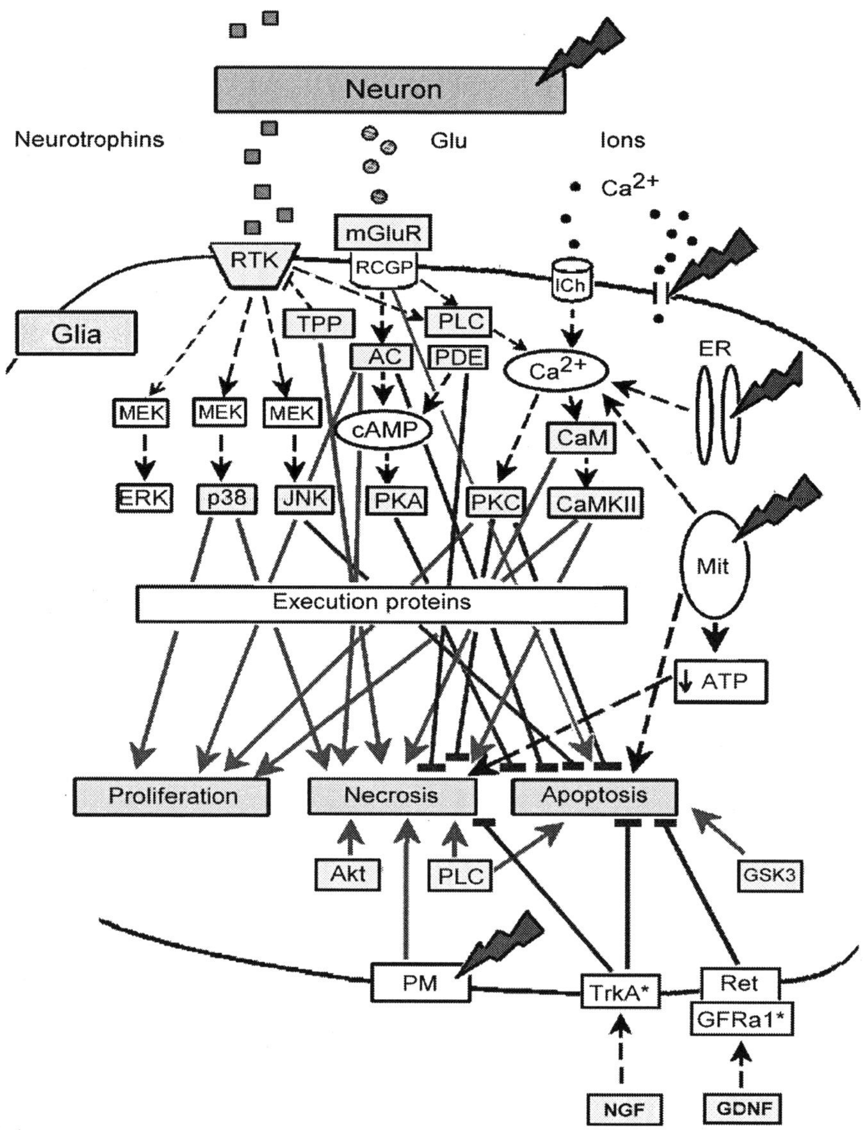

Abbreviations: RTK – receptor tyrosine kinase; RCGP – receptor coupled with G-protein; ICh – ionic channels; PM – plasma membrane; TPP – tyrosine protein phosphatase; AC – adenylate cyclase; PDE – phosphodiesterase; PI3K – PI 3-kinase; MEK – MAP kinase kinase; cAMP – cyclic adenosine mononucleotide; ERK, JNK and p38 – MAP kinases; PKA – protein kinase A; PKC – protein kinase C; CAMKII – calmodulin-dependent kinase II; ER – endoplasmic reticulum; Mit – mitochondria.

Figure 20. Glia-to-neuron signaling in the photosensitized crayfish stretch receptor.

Another important point of the present study is the difference of signaling mechanisms that control cell death in neurons and glia (Figure 19, 20). This may be due to different molecular messengers and their receptors involved in the intercellular signaling as well as the different intracellular signaling systems in these cells. Use of such cell-specific features may provide different damage to different cells in the complex tissue. It would be very desirable to design pharmacological tools, which increase selectively the photosensitivity of glioma and simultaneously protect surrounding normal neurons and glial cells.

Acknowledgment

The work was supported by Russian Foundation for Basic Researches; grants 05-04-48440, 08-04-01322 and Minobrnauki, grant 2.1.1/6185.

References

Airaksinen, MS; Sarma, M. The GDNF family: signaling, biological functions and therapeutical value. *Nat. Rev. Neurosci.* 2002, 3, 383-394.

Arends, MJ; Willie, AH. Apoptosis: mechanisms and role in pathology. *Int. Rev. Exp. Pathol.* 1991, 32, 223-254.

Barres, BA; Barde, YA. Neuronal and glial cell biology. *Curr. Opin. Neurobiol.* 2000, 10, 642-648.

Beck, G; Munno, DW; Levy, Z; Dissel, HM; Van-Minnen, J; Syed, NI; Fainzilber, M. Neurotrophic activities of Trk receptors conserved over 600 million years of evolution. *J. Neurobiol.* 2004, 60,12-20.

Benn, SC; Woolf, CJ. Adult neuron survival strategies-slamming on the brakes. *Nat. Rev. Neurosci.* 2004, 5, 686-700.

Ben-Hur, E; Dubbelmann, TMAR; van Steveninck, J. Phthalocyanine-induced photodynamic changes of cytoplasmic free calcium in Chinese hamster cells. *Photochem. Photobiol.* 1991, 54, 163-166.

Bragin, DE; Kolosov, MS; Uzdensky, AB. Photodynamic inactivation of isolated crayfish neuron requires protein kinase C, PI 3-kinase and Ca^{2+}. *J. Photochem. Photobiol. B: Biol.*, 2003, 70, 99-105.

Brown, SB; Brown, EA; Walker, I. The present and future role of photodynamic therapy in cancer treatment. *Lancet Oncology*. 2004. 5: 497-508.

Buytaert, E; Dewaele, M; Agostinis, P. Molecular effectors of multiple cell death pathways initiated by photodynamic therapy. *Biochim. Biophys. Acta*. 2007, 1776, 86-107.

Castano, AP; Demidova, TN; Hamblin MR. Mechanisms in photodynamic therapy: part one – photosensitizers, photochemistry and cellular localization. *Photodiagn. Photodyn. Ther.* 2004. 1, 279-293.

Castano, AP; Demidova, TN; Hamblin, MR. Mechanisms in photodynamic therapy: part two – cellular signaling, cell metabolism and modes of cell death. *Photodiagn. Photodyn. Ther.* 2005, 2, 1-23.

Chao, MV; Rajagopal, R; Lee FS. Neurotrophin signalling in health and disease. *Clin. Sci.* (Lond). 2006, 110, 167-173.

Colbran, RJ. Targeting of calcium/calmodulin-dependent protein kinase II. *Biochem. J.* 2004, 378 (Pt 1), 1-16.

Davoli, C; Marconi, A; Serafino, A; Iannoni, C; Marcheggiano, A; Ravagnan, G. Expression of nerve growth factor-like polipeptides and immunoreactivity related to the two types of neurotrophin receptors in earthworm tissues. *Cell. Mol. Life Sci.* 2002, 59, 527-539.

Deitmer, JW. Strategies for metabolic exchange between glial cells and neurons. *Respir. Physiol.* 2001, 129, 71-81.

Detty, MR; Gibson, SL; Wagner, SJ. Current clinical and preclinical photosensitizers for use in photodynamic therapy. *J. Med. Chem.* 2004, 47, 3897-3915.

Dirnagl, U; Iadecola, C; Moskowitz, MA. Pathology of ischaemic stroke: an integrated view. *Trends Neurosci.* 1999, 22, 391-397.

Dougherty, T; Gomer, CJ; Henderson, BW; Jori, G; Kessel, D; Korbelik, M; Moan, J; Peng, Q. Photodynamic therapy. *J. Natl. Cancer Inst.* 1998, 90, 889-903.

Dubinina, EE. Products of oxygen metabolism in functional activity of cells. Life and death. Creation and destruction. Sankt-Petersburg: Medical Press; 2006. (in Russian).

Duronio, V. The life of a cell: apoptosis regulation by the PI3K/PKB pathway. *Biochem. J.* 2008, 415, 333-344.

Elahi, MM; Kong, YX; Matata, BM. Oxidative stress as a mediator of cardiovascular disease. *Oxid. Med. Cell Longev.* 2009, 2, 259-269.

Eljamel, MS. Brain PDD and PDT unlocking the mystery of malignant gliomas. *Photodiagn. Photodyn. Ther.* 2004, 1, 303—310.

Fabris, C; Valduga, G; Miotto, G; Borsetto, L; Jori, G; Garbisa, S; Reddi, E. Photosensitization with zinc (II) phthalocyanine as a switch in the decision between apoptosis and necrosis. *Cancer Res.* 2001, 61, 7495-7500.

Fainzilber, M; Smit, AB; Syed, NI; Wildering, WC; Hermann; van der Schors, RC; Jiménez, C; Li KW; van Minnen, J; Bulloch, AG; Ibáñez, CF; Geraerts, WP. CRNF, a molluscan neurotrophic factor that interacts with the p75 neurotrophin receptor. *Science*, 1996, 274, 1540-1543.

Fatokun, AA; Stone, TW; Smith, RA. Oxidative stress in neurodegeneration and available means of protection. *Front Biosci.* 2008, 13, 3288-3311.

Fedoreno, GM; Gusatinsky, VN; Kaminsky, II; Kondratyeva, LA; Korzak, VM. Changing of an isolated neurone ultrastructure during a prolonged impact of mediator. *Neuroreport.* 1995, 6, 2325-2332.

Fedorenko, GM; Uzdensky, AB. Dynamics of ultrastructural changes in the isolated crayfish mechanoreceptor neuron under photodynamic impact. *J. Neurosci. Res.* 2008, 86, 1409 – 1416.

Fedorenko, GM; Uzdensky, AB. Ultrastructure of neuroglial contacts in crayfish stretch receptor. *Cell Tissue Res.* 2009a, 337, 477-490.

Fedorenko, GM; Uzdensky, AB. Cellular structures involved in the transport processes and neuroglial interactions in the crayfish stretch receptor. *J. Integr. Neurosci.* 2009b, 8, 433-440.

Fedorenko, G.M; Fedorenko, Y.P; Fedorenko, A.G; Uzdensky, A.B. Dynamics of Ultrastructural Alterations in Photosensitized Crayfish Glial and Neuronal Cells: Structures Involved in Transport Processes and Neuroglial Interactions. *J. Neurosci. Res.* 2011, 89, 341–351.

Feofanov, A; Grichine, A; Karmakova, T; Pljutinskaya, A; Lebedeva, V; Filyasova, A; Yakubovskaya, R; Mironov, A; Egret-Charlier, M; Vigny, P. Near-infrared photosensitizer based on a cycloimide derivative of chlorin p6: 13,15-N-(3'-hydroxypropyl)cycloimide chlorin p6. *Photochem. Photobiol.* 2002, 75, 633-643.

Fischer, W; Fischer, H; Uerlings, I; David, H. Light and electron microscopic studies of slowly adapting abdominal stretch receptors of the American river crayfish *Orconectes limosus* (RAF). *Z. Mikrosk. Anat. Forsch.* 1975, 89, 340-366.

Florey, E; Florey, E. Microanatomy of the abdominal stretch receptors of the crayfiish (Astacus fluviatilis, leptodactylis). *J.Gen.Physiol.*, 1955, 39, 69-85.

Galluzzi, L; Maiuri, MC; Vitale, I; Zischka, H; Castedo, M; Zitvogel, L; Kroemer, G. Cell Death Modalities: Classification and Pathophysiological Implications. *Cell Death Differ.*, 2007, 14, 1237–1243.

Golstein, P; Kroemer, G. Cell Death by Necrosis: Towards a Molecular Definition, TIBS, 2006, 32, 37–43.

Grinspan, JB; Marchionni, MA; Reeves, M; Coulaloglou, M; Scherer, SS. Axonal interactions regulate Schwann cell apoptosis in developing peripheral nerve: neuregulin receptors and the role of neuregulins. *J. Neurosci.* 1996, 16, 6107-6118.

Hidalgo, A; Learte, AR; McQuilton, P; Pennack, J; Zhu, B. Neurotrophic and gliatrophic contexts in Drosophila. Brain, *Behav., Evol.* 2006, 68, 173-180.

Hsien, YJ; Wu, CC; Chang, CJ; Yu, JS. Subcellular localization of Photofrin determines the death phenotype of human epidermoid carcinoma A431 cells triggered by photodynamic therapy: when plasma membranes are the main targets. *J. Cell Physiol.* 2003,194, 363-375.

Hubmer, A; Hermann, K; Uberriegler, Krammer, B. Role of calcium in photodynamically induced cell damage of human fibroblasts. *Photochem. Photobiol.* 1996, 64, 211 –215.

Jarro, H; Beck, G; Conticello, SG; Fainzilber, M. Evolving better brains: a need for neurotrophins? *TiNS*, 2001, 24 79-85.

Jarro, H; Fainzilber, M. Building complex brains – missing pieces in an evolutionary puzzle. *Brain Behav. Evol* . 2006, 68, 191-195.

Karageuzyan, KG. Oxidative stress in the molecular mechanism of pathogenesis at different diseased states of organism in clinics and experiment. Curr. Drug Targets Inflamm. *Allergy*. 2005, 4, 85-98.

Kessel, D; Vicente MGH; Reiners JJ. Initiation of apoptosis and autophagy by photodynamiv therapy. *Lasers Surg. Med.* 2006, 38, 482-486.

Kim, JH; Sung, DK; Park, CW; Park, HH; Park, C; Jeon, SH; Kang, PD; Kwon, OY; Lee, BH. Brain-derived neurotrophic factor promotes neurite growth and survival of antennal lobe neurons in brain from the silk moth, Bombyx mori in vitro. *Zoolog. Sci.* 2005, 22, 333-342.

Klaunig, JE; Kamendulis LM; Hocevar BA. Oxidative stress and oxidative damage in carcinogenesis. *Toxicol. Pathol.* 2010, 38, 96-109.

Kolosov, M; Uzdensky, A. Crayfish mechanoreceptor neuron prevents photoinduced apoptosis of satellite glial cells. *Brain Res. Bull.* 2006, 69, 495-500.

Komandirov, MA; Knyazeva, EA; Fedorenko, YP; Rudkovskii, MV; Stetsurin, DA; Uzdensky, AB. On the Role of Phosphatidylinositol 3-Kinase, Protein Kinase B/Akt, and Glycogen Synthase Kinase-3β in Photodynamic Injury of Crayfish Neurons and Glial Cells. *J Mol. Neurosci.* 2011 Feb 12. [Epub ahead of print].

Kondo, M; Kasai, M. Photodynamic inactivation of sarcoplasmic reticulum vesicule membranes by xanthene dyes. *Photochem. Photobiol.* 1974, 19, 35-41.

Kopp, DM; Trachtenberg, JT; Thompson, WJ. Glial growth factor rescues Schwann cells of mechanoreceptors from denervation-induced apoptosis. *J. Neurosci*, 1997, 17, 6697-6706.

Kostron, H. Photodynamic diagnosis and therapy and the brain. In: Gomer, C.J. (Ed.) Photodynamic therapy. Methods and Protocols. Methods in Molecular Biology. Springer Science + Business media. 2010, 635, 261-280.

Krasilnikov, MA. Phosphatidylinositol-3 kinase dependent pathways: the role in control of cell growth, survival, and malignant transformation. *Biochemistry* (Moscow). 2000, 65, 59-67.

Largo, C; Cuevas, P; Herreras, O. Is glia disfunction the initial cause of neuronal death in ischemic penumbra? *Neurol. Res.*, 1996, 18, 445-448.

Lobanov, A.V; Uzdensky, A.B. PDT-induced death of sensory neurons and glial cells in the isolated crayfish stretch receptor after proteolytic treatment. *J. Neurosci. Res.* 2005, 82, 866-874.

Lobanov, AV; Uzdensky, AB. Protection of Crayfish Glial Cells but not Neurons from Photodynamic Injury by Nerve Growth Factor. *J. Mol. Neurosci.* 2009, 39, 308-319.

Matroule, JY; Carhy, CM; Granville, DJ; Jolois, O; Hunt, DWC; Piette, J. Mechanism of colon cancer cell apoptosis mediated by pheophorbide-a methylester photosensiti-zation. *Oncogene.* 2001, 20, 4070-4084.

McKay, S; Purcell, AL; Carew, TJ. Regulation of synaptic function by neurotrophic factors in vertebrates and invertebrates: implications for development and learning. *Learning and Memory*, 1999, 6, 193–215.

Michel, PP; Agid, Y. Chronic activation of the cyclic AMP signaling pathway promotes development and long-term survival of mesencephalic dopaminergic neurons. *J. Neurochem.* 1996, 67, 1633-1642.

Moan, J. On the diffusion length of singlet oxygen in cells and tissues. *J. Photochem. Photobiol. B. Biol.* 1990, 6, 343-347.

Moreno, H; Nadal, M; Leznik, E; Sugimori, M; Lax, I; Schlessinger, J; Llinás, R. Nerve growth factor acutely reduces chemical transmission by means of postsynaptic TrkA-like receptors in squid giant synapse. *Proc. Natl. Acad. Sci. USA* 1998, 95, 14997-15002.

Newton, AC; Johnson, JE. Protein kinase C: a paradigm for regulation of protein function by two membrane-targeting modules. *Biochim. Biophys. Acta.* 1376, 1998, 155-172.

Niizuma, K; Endo, H; Chan, PH. Oxidative stress and mitochondrial dysfunction as determinants of ischemic neuronal death and survival. *J. Neurochem.* 2009, 109, Suppl 1, 133-138.

Ogata, M; Inanami, O; Nakajima, M; Nakajima, T; Hiraoka, W; Kuwabara, M. Ca(2+)-dependent and caspase-3-independent apoptosis caused by damage in Golgi apparatus due to 2,4,5,7-tetrabromorhodamine 123 bromide-induced photodynamic effects. *Photochem. Photobiol.* 2003, 78, 241-247.

Ormond, J; Hislop, J; Zhao, Y; Webb, N; Vaillaincourt, F; Dyer, J.R; Ferraro, G; Barker, P; Martin, K.C; Sossin, W.S. ApTrkl, a Trk-like receptor, mediates serotonin- dependent ERK activation and long-term facilitation in Aplysia sensory neurons. *Neuron.* 2004, 44, 715-728.

Orrenius, S; Zhivotovsky, B; Nicotera, P. Regulation of cell death: the calcium-apoptosis link. *Nat. Rev. Mol. Cell Biol.* 2003, 4, 552-65.

Reichardt, LF. Neurotrophin-regulated signalling pathways. *Philos. Trans. R. Soc. B. Biol. Sci.* 2006, 361, 1545-1564.

Ridgway, RL; Syed, NI; Lukowiak, K; Bulloch, AG. Nerve growth factor (NGF) induces sprouting of specific neurons of the snail, Lymnaea stagnalis. *J. Neurobiol.* 1991, 22, 377-390.

Rossner, S; Schliebs, R; Hartig, W; Perez-Polo, JR; Bigl, V. Selective induction of c-Jun and NGF in reactive astrocytes after cholinergic degenerations in rat basal forebrain. *Neuroreport,* 1997, 8, 2199-2202.

Ruck, A; Heckelsmiller, K; Kaufmann, R; Grossman, N; Haseroth, E; Akgun, N. Light-induced apoptosis involves a defined sequence of cytoplasmic and nuclear calcium release in AlPcS4-photosensitized rat bladder RR 1022 epithelial cells. *Photochem. Photobiol.* 2000, 72, ,210-216.

Shorter, J; Warren, G. Golgi architecture and inheritance. *Annu. Rev. Cell Dev. Biol.* 2002, 18, 379-420.

Soderling, T. R. The Ca^{2+}–calmodulin-dependent protein kinase cascade. *Trends Biochem. Sci.* 1999, 24, 232-236.

Stocker, M; Hirzel, K; D'hoedt, D; Pedarzani, P. Matching molecules to function: neuronal Ca^{2+}-activated K^+ channels and afterhyperpolarizations. *Toxicon.* 2004, 43, 933-949.

Stylli, SS; Kaye, AH. Photodynamic therapy of cerebral glioma - a review. Part I. A biological basis. *J. Clin. Neurosci.* 2006, 13, 615-625.

Tao-Cheng, JH; Hirosawa, K; Nakajima, Y. Ultrastructure of the crayfish stretch receptor in relation to its function. *J. Comp. Neurol.* 1981, 200, 1-21.

Teiten, MH; Marchal, S; D'Hallewin, MA; Guillemin, F; Bezdetnaya, L. Primary photodamage sites and mitochondrial events after Foscan photosensitization of MCF-7 human breast cancer cells. *Photochem. Photobiol.* 2003, 78, 9-14.

Trump, BF; Berezesky, IK. The role of cytosolic calcium in cell injury, necrosis and apoptosis. *Curr. Opin. Cell Biol.* 1992, 4, 227-232

Uzdensky, AB. Signal Transduction and Photodynamic Therapy. *Curr. Signal Transduct. Ther.* 2008, 3, 55-74.

Uzdensky, AB. Cellular and molecular mechanisms of photodynamic therapy. Sankt-Petersburg: Nauka. 2010. (in Russian).

Uzdensky, AB; Bragin, DE; Kolosov, MS; Zhavoronkova, AA. PDT effect of different photosensitizers on a single nerve cell: Electrophysiological and Pharmacological Study. *IEEE J. of Selected Topics in Quantum Electronics*, 2001, 7, 989-995.

Uzdensky, AB; Bragin, DE; Kolosov, MS; Dergacheva, OY; Fedorenko, GM; Zhavoronkova, AA. Photodynamic inactivation of isolated crayfish mechanoreceptor neuron: different death modes under different photosensitizer concentrations. *Photochem. Photobiol.* 2002, 76, 431-437.

Uzdensky, A; Kolosov, M; Bragin, D; Dergacheva, O; Vanzha, O; Oparina, L. Involvement of adenylate cyclase and tyrosine kinase signaling pathways in response of crayfish stretch receptor neuron and satellite glia cell to photodynamic treatment. *Glia*, 2005, 49, 339-348.

Uzdensky, AB; Kolosov, MS; Lobanov, AV. Neuron and gliocyte death induced by photodynamic treatment: Signal processes and neuroglial interactions. *Neurosci. Behav. Physiol.* 2008b, 38, 727-735.

Uzdensky, A; Lobanov, A; Bibov, M; Petin, Y. Involvement of Ca^{2+}- and cyclic adenosine monophosphate-mediated signaling pathways in photodynamic injury of isolated crayfish neuron and satellite glial cells. *J. Neurosci. .Res.* 2007, 85, 860-870.

Uzdensky, AB; Lobanov, AV; Romanenko, NP. Neuron-to-glia signaling controls death of photosensitized crayfish glial cells. Cold Spring Harbor Laboratory Meetings. *Glia in Health and Diseases.* July 17 –21, New York. 2008a, 118-118.

Uzdensky, AB; Mironov, AF. Photodynamic inactivation of the single crayfish nerve cell: dynamics of electrophysiological responses and comparison of photosensitizers. *Lasers Med. Sci.* 1999, 14, 185-195.

Uzdensky, AB; Zhavoronkova, AA; Dergacheva, OY. Firing inhibition processes in the response dynamics of isolated crayfish nerve cell to the photodynamic effect of sulphonated aluminum phthalocyanine: participation of free radicals and Ca^{2+}. *Lasers Med. Sci.* 2000, 15, 123-130.

Villegas, SRN; Poletta, FA; Carri, NG. Glia: A reassessment based on novel data on the developing and mature of the central nervous system. *Cell Biol. Int.* 2003, 27, 599-609.

Yoshida, Y; Dereski, MO; Garcia, JH; Hetzel, FW; Chopp, M. Neuronal injury after photoactivation of photofrin II. *Am. J. Pathol.* 1992a, 141, 989-997.

Yoshida, Y; Dereski, MO; Garcia, JH; Hetzel, FW; Chopp, M. Photoactivated Photofrin II: astrocytic swelling precedes endothelial injury in rat brain. *J. Neuropathol. Exp. Neurol.* 1992b, 51, 91-100.

Wildering, WC; Lodder, JC; Kits, KS; Bulloch, AG. Nerve growth factor (NGF) acutely enhances high-voltage-activated calcium currents in molluscan neurons. *J. Neurophysiol.* 1995, 74, 2778-2781

Zenkov, NK; Lankin, VZ; Menschikova, EB. Oxidative stress. Moscow: MAIK Nauka. 2001. (in Russian).

In: Neuronal Cell Apoptosis
Editors: C. J. Schmid and J. L. Wolfe
ISBN: 978-1-61324-633-7

Chapter IV

Cell Death During Cerebellar Development

Laura Lossi, Graziana Gambino, Silvia Alasia, and Chiara Salio
Department of Veterinary Morphophisiology,
University of Turin, Grugliasco (TO), Italy

Abstract

Development of the nervous system not only involves the generation of new neurons and glia, but also the death of cells that are no longer required and/or are produced in excess. Because the process of cell death follows a complex framework of regulated events during normal development, it is usually referred to as programmed cell death.

Programmed cell death is a gene-regulated process that plays important roles in several normal and pathological conditions. It takes several different forms, but, most commonly, eukaryotic cells die following a conserved series of molecular and cellular modifications usually referred to as apoptosis, a form of "cell suicide" most often found during embryonic development, but also in normal cell and tissue turnover during adulthood and senescence. Another form of cell death is autophagy, a catabolic process involving the degradation of cellular components through the lysosomal machinery.

There is no conclusive evidence as regarding the involvement of autophagy in apoptosis, and a causative relationship between autophagy and cell death has not yet been established in full. Nevertheless, observations of cells with autophagic features undergoing programmed cell death have led to define an autophagic type of death (also known as cytoplasmic cell death or type II cell death), with morphological features somehow different from those of apoptosis. Proliferative and/or apoptotic events have been extensively characterized not only during embryonic development but also in several areas of the post-natal and adult brain such as the cerebellar cortex. The post-natal cerebellar cortex undergoes significant structural remodeling mainly because of the generation, differentiation, migration, and apoptosis of the granule cells. Therefore, it has been widely investigated as a general model for studying neuronal cell apoptosis. The existence of two waves of apoptotic cell death has been demonstrated in cerebellum and other areas of the central nervous system, affecting different types of neurons in normal development and in certain pathological conditions. The first wave regards neuronal progenitors and pre-migratory neuroblasts, the second post-migratory neuroblasts and mature neurons. In parallel, autophagic phenomena have also been observed to mainly affect the granule cells and the Purkinje neurons. These may be of particular relevance in the granule cells, for the regulation of the cellular levels of B cell lymphoma-2, one of the most important anti-apoptotic proteins in mammals.

Introduction

Programmed cell death (PCD) is a highly phylogenetically conserved mechanism by which eukaryotic cells die following a stereotyped series of molecular and cellular events commonly referred to as apoptosis (Glucksmann 1951; Saunders, 1966). Apoptosis has been recognized to be an essential process during development, where it appears to be fundamental for the control of the final numbers of neurons (and glial cells) in the central (CNS) and peripheral nervous system (PNS). Moreover, a growing body of evidence is accumulating to indicate that apoptosis is also responsible for the loss of neurons associated with senescence (Taglialatela *et al,*.1996; Kaufmann *et al,*. 2001). PCD is not exclusive of neurons, since any type of cell in any type of tissue can die independently from a pathological insult. About half of the neurons produced during neurogenesis die before completion of CNS maturation, and nearly all classes of neurons are produced in excess during development. Then these oversized populations of neurons are significantly reduced during the periods of naturally occurring neuronal death (NOND), mainly upon activation of the apoptotic machinery. NOND is involved in the withdrawal of aberrant axon projections or non correct cell position, the target-dependent matching of the sizes of both pre- and post-synaptic neuronal populations, and the refinement of the neural circuits. The neurotrophic theory has clearly established that neurons compete for limited amounts of target-derived trophic factors that have a protective role against apoptosis, and has provided a fundamental conceptual basis for a wide number of studies aiming to elucidate the mechanisms of neuron-to-target interaction eventually leading to cell survival or death. The initial concept of neurotrophic factor based on this theory has now broadened to embrace a large and molecularly heterogeneous group of surviving factors, and parallel studies have disclosed the pro-survival role of electrical activity during the wiring of neural networks.

Still, the general idea beyond this expanded concept is that NOND is closely related to the establishment of proper connections with targets, and under this perspective, little or no attention has been dedicated to apoptosis of neural precursors and/or young neuroblasts at early developmental stages, which seems to be independent from synaptogenesis. In fact, a still broadening array of data is accumulating to indicate that two subsequent periods of cell death occur in several areas of the brain, including the cerebellum and the retina. The first (early NOND) mainly takes place at the onset of neurogenesis and is not apparently related to synapse formation, while the second (late NOND) is linked to the wiring of young post-mitotic neurons. In both periods, cell death is apoptotic (Bähr, 2000; De la Rosa and De Pablo, 2000; Lossi *et al.,* 2002c). Why so many neurons are generated in excess and eliminated shortly thereafter remains to be fully understood: it was in fact surprising that several lines of transgenic animals, in which essential cell death genes have been knocked out and/or apoptosis inhibitors have been over-expressed, display no detectable CNS defects (Lossi and Merighi, 2003).

In neurodegenerative disorders and traumatic neuronal injury, neuronal loss is not only linked to apoptosis, but also to necrosis. In the latter (also referred to as *exogenous* cell death), alterations of the cellular environment eventually result in cell swelling and disruption of the cell membrane, while the main elements of the apoptotic machinery are constitutively expressed or generated by the cell itself as a form of "cell suicide". Although the two types of cell death have initially been considered to be totally independent from each other, they share some cellular and molecular features, and cells can switch from one mode of death to the other upon different conditions.

A great amount of literature in recent years has deepened our knowledge on the cellular and molecular mechanisms of apoptosis in neurons and other cells. Current knowledge about the genetic regulation of apoptosis is mainly based on studies of the nematode worm *C. elegans*. These studies have been extensively reviewed in numerous publications (Hengartner and Horvitz, 1994; Hoffman and Liebermann,1994; Stewart, 1994; Vaux *et al.,* 1994; Yuan, 1995; Fraser *et al.,*1996; White,1996; Meier and Evan,1998; Liu and Hengartner,1999; Rao and White,1997; Selimi *et al,*.1997; Sastry and Rao, 2000; Denecker *et al,*. 2001; Shastry *et al,*.2001). Readers are referred to this material for a more in depth analysis of the literature on the above issues.

Apoptosis is nowadays a central issue not only in neurobiology, but also in many different physiological and pathological processes outside the nervous system. Therefore, we believe that a more general public will find here some useful information to explain the basic mechanisms of cell death which appear to be highly conserved not only in neurons but also in many other types of cells.

Basic Concepts and Definitions

In the mammalian CNS, neurons differentiate from *neuroepithelial stem cells*, which are multipotent cells that give rise to *committed progenitors* of neurons and glial cells. As differentiation proceeds, neuronal progenitors transform into *neuronal precursors* also referred to as *neuroblasts*, which, in vertebrates, are incapable of cell replication. Neuroblasts are thought to establish their future phenotype and thereby transform into differentiated neurons, upon generation by terminal-mode symmetric divisions of committed progenitors (McKonnel, 1995; Yoshikawa, 2000). Thus, differentiated neurons are post-mitotic cells completely devoid of any replicative capability.

Most mammalian CNS neurons reach the post-mitotic state during embryonic life. In doing so, some dividing cells exit from the cell cycle and enter a phase of mitotic quiescence commonly referred to as the G_0 phase. Indeed, post-mitotic neurons are believed to enter an *extended G_0 phase* that, unlike other cell types, is irreversible (Herrup and Busser, 1995).

The intrinsically necessary elimination of specific cell populations during development of multicellular organisms is often, but not always characterized by an apoptotic morphology (Leist and Jaattela, 2001; Schwartz *et al.*,1993). As already mentioned, this process is currently referred to as PCD, a term that is often used as a synonym for apoptosis. However, apoptosis may be only one of several forms of cell death occurring in PCD (Alles *et al.,*1991;

Cohen *et al.,* 1992; Eastman, 1993), and the two terms should not be used as synonyms. We will use here, instead, the term naturally occurring neuronal death (NOND) as a synonym of PCD.

The term apoptosis will be employed to specifically indicate a type of cell death, which is characterized by a very well defined series of morphological and biochemical hallmarks. Specific cell pathways are activated in apoptosis, and, in this type of death, cells are responsible for their own demise, in a form of "cell suicide". For this reason apoptosis is considered a form of *physiological* cell death. In pathology, but now also in the broader field of cellular and developmental neurobiology, the term apoptosis rapidly caught on, in particular to emphasize its *normal* occurrence and independence from pathological insults (Migheli *et al.*,1994; Stewart, 1994; White,1996; Marks and Berg,1999). However, this axiomatic association of apoptosis and physiological cell death turned out to be an oversimplification for several reasons. First, apoptosis, besides being relevant to an array of physiological functions (that, in addition to development, comprise the differentiation and maturation of various types of cells, and several functions of the immune system), is involved with cell injury induced by a spectrum of physical and chemical agents (Boobis *et al.*,1990; Ortiz *et al.*,2001; Dainiak, 2002; Stewart, 1994; Yakovlev and Faden, 2001). Second, apoptosis is also concerned with oncogenesis, tumor homeostasis, and the action of cytotoxic drugs employed in chemotherapy (Hoffman and Liebermann,1994; Stewart, 1994; Singh *et al.*, 2002; Mimeault, 2002). Third, more and more evidence is accumulating to show a role of apoptosis in several neurodegenerative diseases, including Alzheimer's disease (AD) and aging (Marks and Berg, 1999; Sastry and Rao, 2000).

Types of Cell Death

Perhaps the fundamental reason why apoptosis has been so extensively investigated is that, as mentioned above, this type of cell death originally was linked with physiological conditions, unlike necrosis that was instead related to different types of pathologic insults. Interestingly, the existence of these two (and other) types of cell death was initially described on a merely morphological basis. In an ultrastructural study on several embryonic tissues, it was originally proposed that there are three main types of cell death during normal development, based on the role of lysosomes in cell disruption (Schweichel and Merker, 1973). In the first type, cell death occurs without any detectable activation of endogenous lysosomes, and cells are eventually destroyed by secondary lysosome activation in tissue macrophages in a process called *phagocytosis.* This phenomenon has also been referred to as heterophagocytosis. In the second type of cell death (*autophagocytosis*) cells are eliminated after activation of their own lysosomal enzymes. In the third type, there is no obvious lysosome intervention. The first two types are by far more common, and have been described by various authors starting from the sixties. The ultrastructural features of type 1 cell death in the Schweichel and Merker classification correspond to the current definition of apoptosis, type 2 to the definition of autophagy, while at least a variant of type 3 shares several features with necrosis. The issue of the ultrastructural diversity in developmental cell death has been reviewed in detail previously (Clarke, 1990).

Apoptosis

Apoptosis was originally defined as a distinct mode of cell death because of a series of characteristic ultrastructural features such as (in sequence): nuclear and cytoplasmic condensation, cell fragmentation and phagocytosis (Kerr *et al.,* 1972). Initially Kerr and co-authors used the term "shrinkage necrosis" to describe the phenomenon. Subsequently they coined the term "apoptosis" (from the Greek = falling of the leaves), which indicates the dropping of leaves from trees or petals from flowers, to emphasize the role of this mode of cell death in normal tissue turnover.

Apoptosis phases are most evident at the electron microscope level, where changes in the nucleus represent their first unequivocal evidence. Chromatin condensation and segregation into sharply delineated masses that abut on the nuclear envelope are in fact typically observed during the initial phase of apoptotic cell death. Chromatin masses are very electrondense and often show a characteristic crescent-like appearance. High magnification electron micrographs reveal that these masses are made up of closely packed, fine granular material.

This initial condensation of the nucleus eventually leads to true pyknosis. In parallel with nuclear changes, cytoplasm condensation also occurs, and the cell membrane becomes convoluted with the onset of protuberances of various sizes that may give the cell a star-like appearance. As the cytoplasm density increases, some vacuoles may become evident, but the cell organelles remain unaffected, although they become abnormally closely packed, likely because of the loss of cell cytosol. Ribosomes can be detached from the rough endoplasmic reticulum and from polysomes that eventually disappear. As the process continues, the cell and its nucleus assume a more irregular shape and nuclear budding occurs to produce discrete fragments, still surrounded by an intact nuclear envelope. Eventually, the cell is fragmented into membrane-bounded apoptotic bodies, which still display a sharp segregation of condensed chromatin in nuclear fragments and well-preserved organelles. Apoptotic bodies are rapidly cleared out in tissues by macrophages or neighboring cells, and are then degraded within heterophagosomes.

Necrosis

Necrosis (from the Greek = degeneration of the corpse) indicates the death of a cell or group of cells usually because of injury, disease or a pathological state. Therefore, necrosis is traditionally associated with one or more exogenous factors leading to non-physiological cell death. During the necrotic process, two main stages can be distinguished: the *death* of the cell, i.e. an irreversible alteration of the cellular mechanisms that enable the cell to maintain its homeostasis, and the *subsequent degeneration* of the dead cell (Clarke, 1990). In general, necrosis affects large numbers of cells that are often grouped together and is associated with an inflammatory reaction.

In the irreversibly injured cells, the morphological changes characteristic of necrosis consist in clumping of the chromatin without marked changes in its distribution: usually chromatin condenses in multiple small clumps with irregular outlines, and is poorly demarcated from the surrounding nucleoplasm. Sometimes densities in the matrix of abnormally swollen mitochondria, and local membrane disruption become evident. At later stages, there is a more or less pronounced disintegration of cell organelles and membranes, although the cell somehow maintains an overall identity. The chromatin disappears at the end of the process, leaving only "ghost-like" cell residues. Usually, groups of several cells undergo necrosis in tissues, with a marked inflammatory response that eventually leads to the removal of the necrotic debris by cells of the mononuclear phagocytic system.

The Dichotomy of Apoptosis-Versus-Necrosis

The conceptual dichotomy of apoptosis-versus-necrosis has been originally based upon the aforementioned morphological differences between the two modalities of cell death, although some authors started to question such a sharp morphological distinction already at the beginning of the nineties (Clarke,1990).

In more recent times, it has become clear that many intermediate morphological types fill the gap between "classical" apoptosis and necrosis. In these intermediate forms of cell death blebbing may be more or less evident and varying degree of chromatin condensation and margination may be apparent (Leist and Jaattela, 2001). With the more and more in depth dissection of the cellular and molecular mechanisms of apoptosis, it became clear that specific cellular pathways are activated in apoptosis. This led to an enforcement of the concept that, in apoptosis, cells are responsible of their own demise, a reason for which this type of PCD is commonly considered as a sort of "cell suicide".

In addition, a more accurate analysis of the cellular mechanisms of apoptosis suggested that at least some executioners of apoptotic and non-apoptotic (necrotic) cell deaths may be identical (Benchoua *et al.,* 2001; Moroni *et al.,* 2001; Smith, 2001; Cole and Perez-Polo,2002; Fujikawa *et al.,* 2002; Hou and MacManus, 2002; Schwab *et al.,* 2002).

Therefore, as the investigation of cell death in different systems proceeds, we find more and more variations to the classical concept of apoptosis versus necrosis (Lockshin and Zakeri, 2002).

Autophagy

Autophagy is exemplificative of one of these variations. There is no conclusive evidence as regarding the involvement of autophagy in apoptosis, and a causative relationship between autophagy and cell death has not yet been established in full.

Nevertheless, observations of cells with autophagic features in brain areas undergoing PCD have led to define an additional type of cell death commonly referred to as autophagic (also known as cytoplasmic cell death or type II cell death), with morphological features somehow different from those of apoptosis.

It is not known whether autophagic activity in dying cells actually causes death or whether it simply occurs as a process alongside it. However, in many neurological diseases, in certain neuronal cell death pathways and after neuronal injury, increased numbers of autophagosomes have been detected (Amritraj et *al*., 2009).

Apoptotic Pathways in Mammalian Neural Cells

The basis mechanisms of apoptotic cell death are highly conserved not only in different cell types in mammals, but also throughout phylogenesis.

A growing family of related proteases, which are collectively referred to as caspases (caspase = cysteine aspartate protease - Thornberry and Lazebnik, 1998) plays a crucial role in dismantling cell structures and organelles during the course of apoptosis in neural cells. Among the 14 members of the caspase family identified to date, caspases 3, 6, and 7 have a short pro-domain and degrade vital cellular proteins, whereas the others have long pro-domains that mediate protein-protein interactions, and may only in certain circumstances trigger apoptosis (Thornberry and Lazebnik,1998). Each caspase is initially synthesized as a zymogen, and requires processing at specific cleavage sites to generate the active enzyme (Stennicke and Salvesen, 1999). The caspases that are the first to be activated in turn trigger other downstream caspases, giving rise to a proteolytic cascade that culminates with the execution of apoptosis.

A crucial role in apoptosis is also played by a large protein complex called the apoptosome (Adams and Cory, 2002), consisting of apoptosis protease activated factor 1 (APAF 1) (Zou *et al.,* 1997), cytochrome *c* (also referred to as APAF 2) and procaspase 9 (also referred to as APAF 3). Activation of caspases, in fact, requires a number of scaffolding proteins that act in the apoptosome. In particular, biochemical studies have revealed that caspase 3 processing requires not only caspase 9, but also APAF 1 and cytochrome *c* (Liu *et al.*,1996; Li *et al.*,1997). Upon binding to cytochrome *c*, which is released from mitochondria at the onset of apoptosis, a series of conformational changes leads to APAF 1 multimerization and association with procaspase 9 with the generation of the about 1 MDa molecular weight apoptosome (Adams and Cory, 2002).

Another important group of apoptosis-related proteins forms the so-called mammalian B cell lymphoma-2 (BCL-2) protein family (Adams and Cory, 1998; Chao and Korsmeyer, 1998; Newton and Strasser, 1998; Sadoul, 1998; Sanchez and Yuan, 2001). Besides BCL-2, some other protein members of the family (such as BCL-X_L and BCL-W) act as survival factors, whereas others (such as BAX, BAK, BAD, BID) are pro-apoptotic. The anti-apoptotic members of the BCL-2 family protect cells from death by at least two different mechanisms. First, BCL-2 and BCL-X_L (located at the cytoplasmic side of the outer mitochondrial membrane) can prevent the release of cytochrome *c* from mitochondria, avoiding assembling of the apoptosome, and thereby protecting cells from being killed (Li *et al.*,1998; Luo *et al.*,1998). Second, BCL-X_L interacts with APAF 1, and inhibits association of APAF 1 with pro-caspase 9, again with blockage of apoptosome formation and inhibition of caspase 9 activation (Adams and Cory, 2002).

Several different stimuli can initiate the apoptotic death of neurons. However, given that, independently upon the event(s) that trigger(s) apoptosis, common morphological and biochemical alterations are observed, it seems reasonable to hold that most apoptotic pathways converge on a restricted number of common effector pathways (Sastry and Rao, 2000). When considering the relative timing of caspase activation and mitochondrial release of cytochrome *c*, two major pathways can be distinguished. In the first, which is exemplified by activation of death receptors, an effector caspase is activated prior to mitochondrial alterations. In the second, cytochrome *c* is released from the mitochondrial intermembrane space prior to caspase activation.

Apoptosis and Cell Cycle Regulation

Investigation of the mechanisms of NOND has led to the unexpected discovery that in many instances the quiescent and dormant cell cycle machinery is "resuscitated" in neurons undergoing apoptosis. Some data suggest that the uncoordinated expression of molecules that are physiologically implicated in the control of cell cycle progression is a fundamental mechanism of apoptosis, at least in certain neurons (White, 1996; Yoshikawa, 2000; Liu and Greene, 2001).

Three families of proteins are primarily responsible for cell cycle regulation: the cyclins, the cyclin-dependent kinases (CDKs), and the cyclin-dependent kinase inhibitors (CKIs).

Cyclins are a group of proteins whose abundance varies substantially during the cell cycle. They associate with their cognate CDKs acting as activating subunits, eventually leading to formation of active cyclin-CDK complexes. These latter allow progression through the different phases of the cell cycle by phosphorylating their target substrates.

For protection from a variety of different types of insults or stress resulting in DNA damage, eukaryotic cells have developed a system of checkpoints that delay progression to the next phase of the cell cycle and activates DNA repair. When DNA damage is irreparable, checkpoints eliminate potentially hazardous cells by permanent cell cycle arrest or cell death (Bartek and Lukas, 2001). Checkpoints act at two different phases of cell cycle, the $G_2 \rightarrow M$ transition and the mid-to-late G_1.

During normal cell cycle progression, a complex process of activation of the cyclin-dependent protein kinase cell division cycle 2 (Cdc2) kinase triggers initiation of mitosis. For this reason, Cdc2 kinase is the major "engine" that drives the $G_2 \rightarrow M$ transition. Prior to mitosis, the Cdc2-cyclin B1 complex is held in the cytoplasm in an inactive state by Cdc2 phosphorylation. Dephosphorylation is regulated by two cell division cycle 25 (Cdc25) phosphatases, Cdc25B and Cdc25C. In mammals, the checkpoint kinase 1 (Chk1) is the most important molecule that acts upstream of Cdc25C in DNA replication and damage checkpoints (Walworth, 2001). Chk1 phosphorylates Cdc25C and thus blocks mitosis, since dephosphorylation of Cdc25C is required for mitotic entry.

Cell Death in Cerebellar Neurons

Granule Cells

The post-natal cerebellar cortex undergoes significant structural remodeling mainly because of the generation, differentiation, migration and apoptosis of the cerebellar granule cells (CGCs). During early post-natal life, granule cells precursors form the external granular layer (EGL) and subsequently migrate to form the definitive internal granular layer (IGL) (Figure 1). During these phases they proliferate and differentiate to make finally synapses onto the Purkinje cells (PCs); the PC/CGC ratio is highly regulated, yet varies among species

(Lange, 1975). The role of cell death in such a regulation is still under debate, although, starting from the seventies, numerous studies based on counts of pyknotic nuclei in normal and experimental animals, have shown that cells in the external granular layer of the forming cerebellar cortex of several species die during post-natal development. Subsequent work *in vivo* led to the demonstration, in different species of mammals including humans, of extremely high numbers of apoptotic cells in the post-natal cerebellum (Wood et al.,1993; Krueger et al.,1995; Lossi et al.,1998).

The occurrence of cell death in the developing cerebellum has been originally inferred starting from the numerous studies based on counts of pyknotic nuclei in normal and experimental animals that have shown that EGL cells of several species die during post-natal development. Subsequent work *in vivo* led to the demonstration, in different species of mammals including humans, that death is of the apoptotic type (Wood *et al.,1993;* Krueger *et a.l*,1995; Lossi *et al.*, 1998). Cells undergoing apoptosis in the EGL have been mostly identified as immature CGCs and/or their precursors (GCPs), and displayed a consistent correlation with the previously demonstrated pyknotic elements in the same location. *In vivo* analysis of CGC apoptosis is made easier by the fact that this neuronal population comprises billions of cells. Therefore, although NOND is not synchronized throughout the entire cerebellum, and apoptotic CGCs are rapidly removed from tissue, it is still possible to localize these neurons in tissue sections with better chances of success.

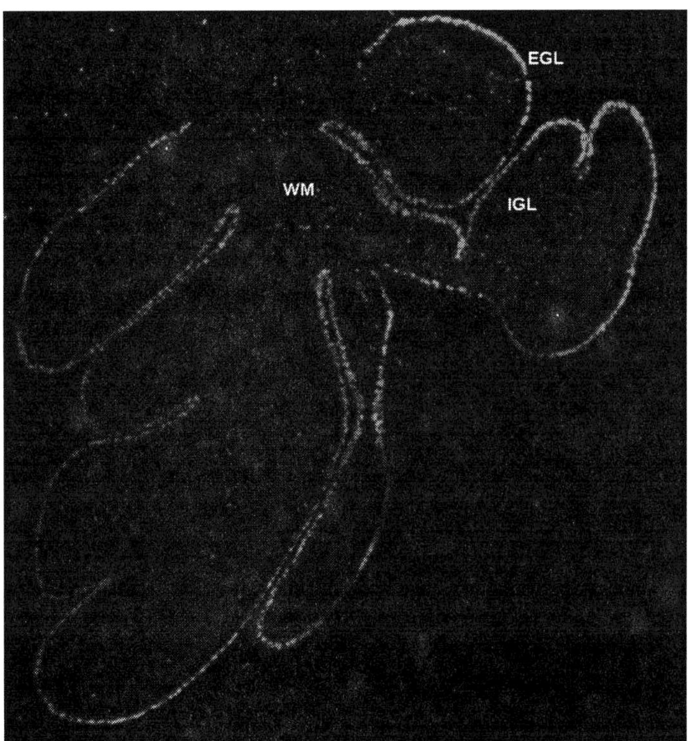

Figure 1. Immunocytochemical localization of BrdU in the post-natal day 10 cerebellum after two hour survival upon tracer administration. Positive nuclei are particularly abundant in the EGL, but are present also in the IGL and white matter. Abbreviations: EGL = external granular layer of the cerebellar cortex; IGL = internal granular layer of the cerebellar cortex; WM = white matter. Original magnification 4x.

Temporal Phases and Modes of Cell Death in CGCs

The existence of two consequent temporal phases of cell death that affect the pre-and post-migratory CGCs, respectively, has been demonstrated in rodents (Lossi *et al.*, 2002b).

The first phase (early NOND) has been neglected until recently, although the existence of NOND in the neuroepithelium at the beginning of neurulation was first described more than 50 years ago (Glucksmann,1951). Early neuronal death was also observed during the formation of the neural crest and neurogenesis, with a widespread distribution throughout the CNS and PNS (De la Rosa and De Pablo, 2000). A number of studies led to the conclusion that these early apoptotic neurons entered the cell cycle shortly before death (Blaschke *et al.*,1996; Galli-Resta and Ensini, 1996; Thomaidou *et al.*,1997).

We have demonstrated that dying cells in early cerebellar NOND are GCPs or newly generated neuroblasts at a stage in which they are totally disconnected from target. What apoptotic pathways are activated *in vivo* during this process is still under debate. We have analyzed the expression of a number of proteins involved in cell cycle control in the post-natal cerebellum, and we have demonstrated that pre-migratory CGCs with typical apoptotic morphologies are stained *in vivo* with antibodies against phospho-Chk1 and two different forms of phospho-Rb (Lossi *et al.,* 2004a). On the other hand, in organotypic cultures transfected with a vector that enables following of the process of caspase 3 activation in living cells, we have proved that caspase 3 is not activated during apoptosis of GCPs and/or pre-migratory CGCs (Lossi *et al.,* 2004b).

Four major conclusions can be drawn from these results. First, the extremely selective localization of phospho Chk1 in early NOND suggests that activation of this effector kinase is associated with the initiation of apoptosis in GCPs and pre-migratory CGCs. Second, hyperphosphorylation of Rb at specific sites is required for CGC apoptosis. Notably, Rb phosphorylation is regulated by G1 phase cyclins and CDKs (Bartek *et al.,*1996), and functional inactivation of Rb requires sequential modifications by at least two distinct cyclin-CDK complexes (Lundberg and Weinberg,1998). Third, apoptosis during early NOND in cerebellum is caspase 3-independent. Four, early NOND occurs well in advance that the time in which the CGCs establish their connections with the PCs.

As mentioned, caspases and members of the BCL-2 family (see below) are very important players in apoptosis. Therefore, distributional studies have focused onto the localization of these proteins and/or their mRNAs mainly in primates (Sohma *et al.*, 1996; Lichnovsky *et al.*,1998; Lossi *et al.*,1998; Bernier *et al.*, 2000; Vinet *et al.*, 2002) and rodents (Castren *et al.*,1994; Frankowski *et al.*,1995; Ishii *et al.*, 1996; Mizuguchi *et al.*,1996; Srinivasan *et al.*, 1998; De Bilbao *et al.*,1999; Mooney and Miller, 2000).

Caspase 3 and 9 are expressed in the telencephalic ventricular zone (VZ) (Sommer and Rao, 2002), and experimental activation of caspase 3 leads to progenitor cell death which is blocked by a pan-caspase inhibitor (D'Sa-Eipper and Roth, 2000). This indicates that neural progenitors can activate a caspase-dependent apoptotic pathway. However, in cerebellum, although caspases represent major actors in apoptosis, some experiments *in vitro* showed that caspase inhibitors were unable to protect CGCs from death (Miller *et al.,*1997; Padmanabhan *et al.,*1999). Since transcriptional inhibitors could block apoptosis, the question of whether it is related to activation of certain components of the cell cycle machinery was raised (Ferrari *et al.,* 1995). These results are consistent with our observations *in vivo* about the independence of early NOND from caspase 3 activation.

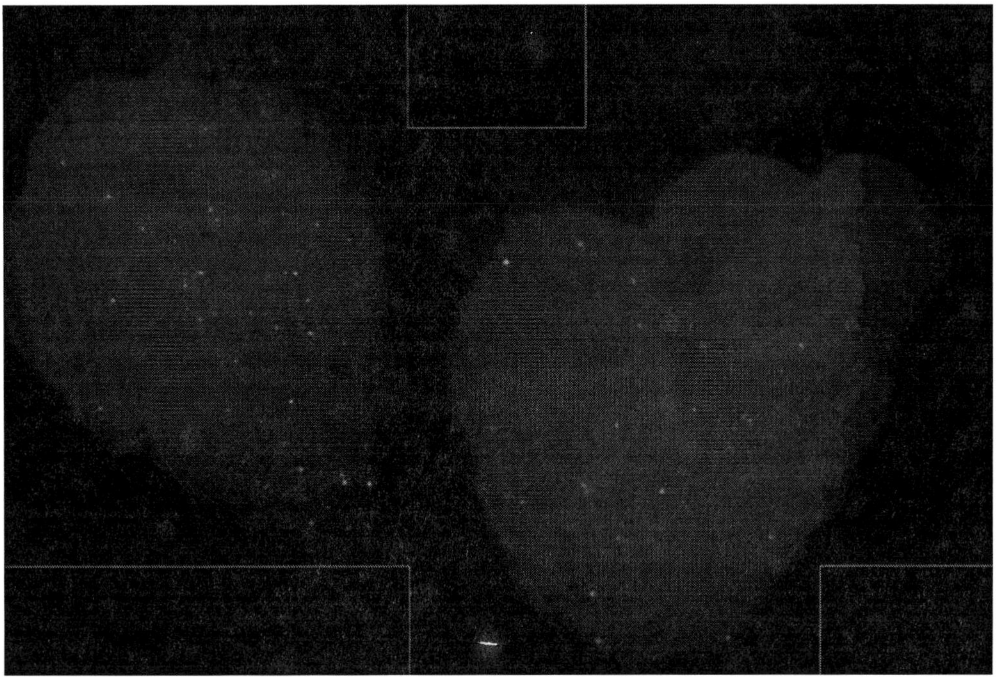

Figure 2. Cerebellar organotypic cultures transfected with human BCL-2 (hBCL-2) protein fused to enhanced yellow fluorescent protein (EYFP). Transfected granule cells appear as intensely fluorescent dots. Original magnification 4x.

The possibility that early NOND of GCPs and/or pre-migratory CGCs is related to the establishment of the correct ratio with the PNs cannot be ruled out by our experiments, but it appears to be very unlikely (Lossi *et al.*, 2002c). An alternative functional explanation for the significance of early NOND is that the massive cell death of the EGL neurons is related to the process of fissuration of the cerebellar cortex that eventually leads to formation of the folia in the mature cortex (Lange 1975)

A second apoptotic wave (late NOND) affects post-mitotic CGCs, and is a widely recognized phenomenon that has been demonstrated to play a crucial role in sculpting and maintaining the architecture of the mature CNS since establishment of the neurotrophic theory (Oppenheim, 1985; Johnson and Oppenheim, 1994; Raff *et al.*, 1994; Miller and Kaplan, 2001). In keeping with this functional explanation, there are several data indicating that the survival of CGCs and PCs is based upon a reciprocal trophic support (Oppenheim, 1985; Johnson and Oppenheim, 1994; Raff *et al.*, 1994; Miller and Kaplan, 2001). Consistently with this idea, we have demonstrated that post-mitotic CGCs in the internal granular layer (IGL) of the forming cerebellar cortex undergo apoptosis *in vivo* very likely as a consequence of failing to make proper synaptic contacts with the PC dendrites (Lossi *et al.*, 2002b).

Because activation of the mitochondrial apoptotic pathway is a well-established phenomenon in several experimental conditions leading to death of post-mitotic neurons, one is led to ask whether or not the same pathway is activated in late cerebellar NOND. We investigated the molecular mechanisms underlying late NOND and obtained evidence for specific cleavage of several caspases and PARP-1, the most biologically relevant substrate of caspase 3 (Smith, 2001). Caspase/PARP-1 cleavage selectively occurs within the IGL, which is known to be populated by post-mitotic, post-migratory CGCs (Lossi *et al.*, 2004). Therefore, in this case, PCD is again apoptotic, but, differently from early NOND that affects the EGL neurons, is caspase-dependent.

Recently, Kirkland e co-authors (2009) have shown an important role for Bax and some reactive oxygen species (ROS) in CGC apoptosis. Their data indicate that Bax lies upstream of most O_2 produced in neurons, and that caspase 3 is required for increased O_2 production during neuronal apoptosis. Also, they showed that caspase 3 is partially involved in O_2 production in non-apoptotic neurons, and that other caspases may also be involved in Bax-dependent O_2 production in non-apoptotic cells other than neurons.

Although the role of BCL-2 is well established in cell death, until recently it remained unclear how it correlates with the two forms of NOND that have been demonstrated to occur in cerebellar cortex. To shed some light on this issue we have employed an organotypic approach that allows for an easier and better-controlled experimental manipulation. It was very recently shown that, similarly to isolated CGCs *in vitro*, differentiation in organotypically cultured CGCs maintained under elevated $[K^+]_e$ is blocked at stage of immaturity, whereas these neurons reach their full maturation in physiological $[K^+]_e$. The process is reversible and, by simply switching from physiological to elevated $[K^+]_e$ or *vice versa*, one can easily interfere with the differentiation state of CGCs. Therefore, we have studied the post-translational regulation of the human BCL-2 (hBCL-2) protein fused to enhanced yellow fluorescent protein (EYFP) in murine organotypic cultures (Lossi *et al.*, 2009) (Figure 2). We have reasoned that the density of EYFPhBCL-2 expressing CGCs, directly monitored under the fluorescence microscope, could be used to analyze the consequences of manipulating the degree of CGC maturation on the regulation of the fusion protein, and that any modifications reflected those occurring to endogenous BCL2. Upon switching from physiological to elevated $[K^+]_e$, we have shown that there are no changes in *bcl-2* mRNA after real time PCR, whereas the levels of the fusion protein and of murine BCL-2 are increased when cultures are maintained in elevated $[K^+]_e$. After using a series of agonists/antagonists for ion channels at the cell membrane or the endoplasmic reticulum (ER), and drugs affecting protein synthesis/degradation, we have related the accumulation of BCL-2 to a reduction in post-translational cleavage by autophagy. Neuronal autophagy is regulated in a quite complex fashion by several signals, among which $[Ca^{2+}]_i$. Our results indicate that the rise in $[Ca^{2+}]_i$ that follows release from ER stores inhibits the autophagic clearance of BCL-2. In keeping with this observation, if autophagy is stimulated by rapamycin inhibition of mTOR, the effect of elevated $[K^+]_e$ on the density of EYFP-tagged CGCs is abolished.

Autophagy has been specifically linked to ERK activation in rat CGCs when switching from elevated to physiological $[K^+]_e$ although the mechanisms of neuronal death involving ERK may be different in mouse. Surprisingly, blocking of ERK did not interfere with the post-translational accumulation of BCL-2 in mouse CGCs. However, when ERK is blocked, the density of EYFP-hBCL-2 tagged CGCs under elevated $[K^+]_e$ normalized to the mean cell density under physiological $[K^+]_e$ is significantly reduced. This suggests that sustained inhibition of ERK (and hence of autophagy) has a detrimental effect on survival both in mature and immature status. Upstream inducers of ERK have also been hypothesized to play a role in neuronal death. Although additional work will be necessary to fully address this issue, our study suggests that $[Ca^{2+}]_i$ regulates the post-translational levels of BCL-2 in CGCs through the activation of the CAMK/ERK pathway, consistently with the observations carried out in other types of neurons.

In brief, the results of our study indicate autophagy as an important mechanism for regulation of BCL-2 levels in neurons. As mentioned earlier, apoptosis is the most common aspect that NOND assumes during post-natal maturation of cerebellar cortex *in vivo*, but autophagy has also been shown to occur in CGCs. In keeping with these observations, previous studies demonstrated that autophagy is constitutively active in neurons and required for neuronal survival, although its role in the post-translational regulation of cellular BCL-2 levels was not previously investigated.

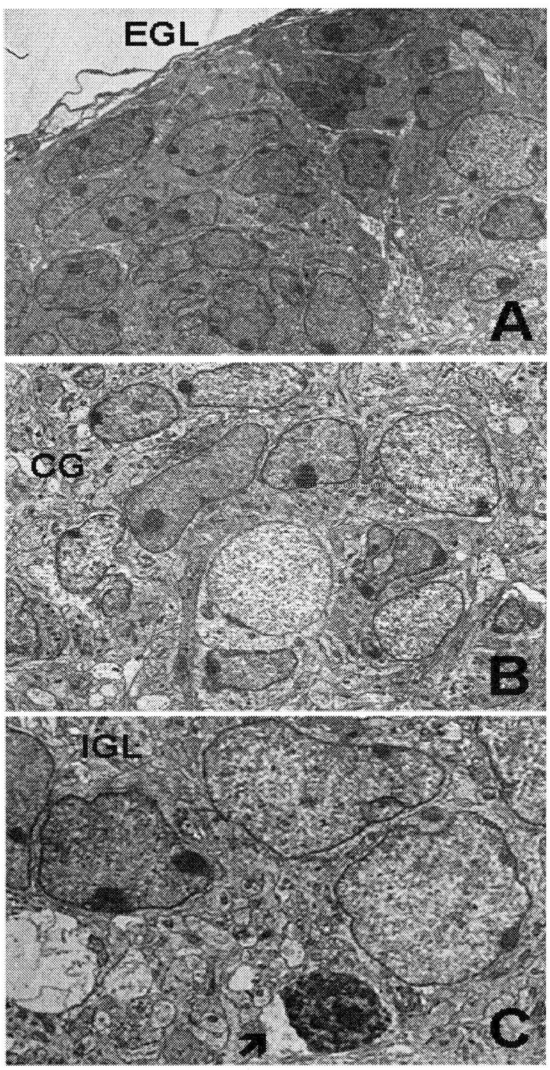

Figure 3. A-B: Electron micrographs of the mouse EGL at post-natal day 10 (P10). Pre-migratory granule cells are densely packed at the subpial surface (A) and at the limit with the Purkinje cell layer (B). C: Post-mitotic granule cells in the mouse IGL at P10 (B). One cell with morphological alterations typical of apoptosis is apparent at centre bottom (C, arrowhead). The cell displays a very condensed nucleus and cytoplasm shrinkage. Abbreviations: EGL = external granular layer of the cerebellar cortex; IGL = internal granular layer of the cerebellar cortex. Original magnification 3000x.

Clearance of Apoptotic CGCs from Cerebellar Cortex

We have described the clearance of apoptotic cells during NOND in the postnatal rabbit cerebellum *in vivo* (Lossi *et al.,* 2002c) after tagging proliferating cells with a modified "window-labeling" technique (Belecky-Adams *et al.,*1996) to follow their fate. We have thus observed that apoptotic CGCs are phagocytosed *in toto* by the microglia, before being fragmented into apoptotic bodies. At early stages of apoptosis, CGCs in the EGL were seen to be contacted by glia that bent around them and became engulfed with the entire apoptotic cell in a fashion very close to that observed in the nematode worm *C. elegans.* Cells with the typical features of late apoptosis were more easily observed in the IGL (Figure 3) and commonly fragmented into several apoptotic bodies inside the heterophagosomes. It was rather easy to spot some of these phagocytic cells engulfed with apoptotic material in close apposition to the blood capillaries, and we have provided ultrastructural evidence for the presence of intraluminal blood monocytes engulfed with heterophagosomes. These latter contained highly condensed nuclear DNA pre-labeled *in vivo* with an S phase marker 24 hours earlier. Proliferation of the CGCs only occurs in the EGL. Therefore our data demonstrate that in the limited span of time between tracer administration and sacrifice some CGCs completed their division, entered the apoptotic program and were cleared by glia. This observation gives an unequivocal demonstration of the remarkable speed by which apoptotic cells are removed from the intact nervous tissue.

Purkinje Cells

PCs are the key neurons of the cerebellar cortex and the only output from it giving rise to all efferent fibers system. Young post-migratory PCs pass through four stages of morphological differentiation before the building of their definitive dendritic trees: (i) simple-fusiform cell, (ii) complex-fusiform cell, (iii) PC with regressive-atrophic dendrites, and (iv) stellate-cell stage.

In vitro PCs undergo apoptosis at P1-P5 (Cheng *et al.,* 2011) However, it is still under debate whether PCs undergo apoptosis during normal development, because usually they do not exhibit the morphological features typical of PCD (Norman *et al.* 2005). However, these cells are by far less numerous than the CGCs, and are not easily recognizable until P4 when they reach their final location and start forming a monocellular layer in the cerebellar cortex (the piriform layer or PC layer of the mature cerebellum). In mouse embryos, pyknotic figures are present in the cortical plate zone where PCs originate and an immunocytochemical study demonstrated active caspase 3 and terminal deoxynucleotidyl transferase dUTP nick-end labeling (TUNEL) positivity in these neurons (Marin-Teva *et al.*, 2004). Obviously, detection of TUNEL-positive and active caspase-3-immunoreactive Purkinje cells is consistent with cell death being due to apoptosis also in this population of cerebellar neurons. In keeping with such a possibility, transgenic mouse over-expressing BCL-2 have 40% more PCs than wild type animals (Zanijani *et al,.* 1996), thus offering an indirect demonstration for the occurrence of apoptosis in these neurons. Recently, (Jankowski *et al.,* 2009) have demonstrated that apoptosis in PCs displays a major variability in the various areas of the cerebellum. This variability may explain the difficulty in assessing apoptosis in PCs in normal animals. Even more recently, (Cheng *et al.*, 2011) detected apoptotic PCs as early as P3 in normal mice.

With increasing age, their density decreased following an exponential curve, probably because of the massive volume expansion of the cerebellum in the P3–5 interval.

A role for neurotrophins in the regulation of PCs' NOND has been hypothesized. However, by using organotypic cultures of P3 cerebellar slices, Dusart and co-workers (2006) demonstrated that at least certain neurotrophins (i.e., BDNF and NT3) and insulin-like growth factor-1 (IGF-1) have minor effects on PCs apoptosis, which, in contrast, is strongly prevented by the inhibition of protein kinase C during the first 5 days *in vitro*.

Not only apoptosis but also other modes of cell loss, such as autophagy, could act simultaneously to regulate PC death during normal development, as demonstrated in mice suffering from spontaneous mutations. In the *Lurcher* mouse PCs degenerate in the post-natal period within four weeks (Selimi *et al.*, 2003). As a properly functioning apoptosis pathway is not required for PC death in *Lurcher* mice and autophagy has been implicated in the execution of type II cell death. This observation suggests that over-activation of autophagy in *Lurcher* may account for the cerebellar neurodegeneration in these mutants.

In the *Toppler* and *Woozy* mutants, PCs undergo apoptosis, sometimes in the presence of an autophagic mechanism but also with the activation of a different death pathway (Dusart *et al.*, 2006). On the other hand, during ageing, these cells die with degenerative changes similar to apoptosis.

In the homozygous *pcd5J* mouse, an autophagosome association with mitochondria is a significant feature of autophagy activation in degenerating PCs (Chakrabarti L. *et al.*, 2009). Electron microscopy of the cerebellum of *pcd5J* homozygous mice shows many PCs undergoing dark cell degeneration, evidence for autophagic degradation of mitochondria, in the form of double-membrane bounded cytoplasmic structures containing mitochondria and/or their fragments.

PCs at the later stages of degeneration display altered mitochondria that eventually become swollen with a reduced electron density. Another common finding is the presence of an abnormally dilated endoplasmic reticulum (ER) with apparent dissociation of ribosomes from the ER segments. Autophagosome-like structures were evident in dendrities and axons in dying PCs likely as a consequence of an over-activation of the autophagic pathway or dysfunctional autophagy. These observations suggest that, in the *pcd* mice, the autophagic degradation of mitochondria, a process known as "mitophagy" could be responsible for the PC demise.

In keeping with these results, a recent study has observed cleaved caspase-3 and Fluoro-Jade C-positive PCs in the cerebellum of Npc1 mice, a model for Niemann-Pick disease (Kim S.J. *et al.*, 2008). In addition, when the ultrastructural features in wild type and Npc1 mice were compared by using conventional electron microscopy, transgenic mice displayed an evident shrinkage of PCs that possessed a very intense electron density. The cell body contained severe cytoplasmic vacuolization and a dark nucleus.

Very interestingly, the number and area occupied by secondary lysosomes with heterogeneous dark content and numerous concentric lamellar bodies were increased. Altogether, the ultrastructural alterations observed in PCs of Npc1 mice are typical of autophagy and indicate an increase of lysosomal activity. This was confirmed by the increase of the autophagy markers LC3-II and beclin-1 in transgenic mice compared to controls. It would be of interest to analyze the cellular levels of BCL-2 in the PCs of these mice, since one expects that they be drastically reduced because of an intense autophagic demolition in keeping with the results described above for CGC autophagy.

Nonetheless one should keep in mind that anti-apoptotic strategies, such as over-expression of BCL-2 or treatment with minocycline (that acts by increasing BCL-2 levels and inhibiting caspase 3), failed to protect neurons in Npc1 mice. It is thus possible that the post-translational regulation of BCL-2 levels by autophagy shows a certain degree of cellular specificity or that redundant apoptotic mechanisms exist in Niemann-Pick pathology.

These observations suggest that an altered autophagy pathway may function as a common theme for PC demise in a variety of neurodegenerative diseases.

Stellate and Basket Cells

Stellate and basket cell progenitors proliferate in the white matter during development and finally reach the molecular layer of the mature cerebellar cortex (Weisheit G *.et al.,* 2006). They are generated post-natally from the VZ at the roof of the fourth ventricle between P2/3 and P16/17, with a peak at P6/7 in rodents (Altman and Bayer 1997). In the first postnatal week, these neurons are present in white matter, whereas in the following days they proliferate and migrate to the molecular layer. Migration is completed at the third week. During the proliferation/migration, the stellate and basket cells undergo apoptosis as demonstrated in the GAD67/GFP mouse (Yamanaka et al. 2004). In these animals, all GABAergic neurons express the fluorescent marker GFP (green fluorescent protein) under the control of the GAD67 promoter that is specifically turned on in neurons synthesizing the inhibitory neurotransmitter GABA.

In these mice, apoptotic cells are present in all layers of the cerebellar cortex between P5 and P21. They are 5–10% of total apoptotic cells in the molecular/PC layer and in the white matter, but less than 2% in the IGL at P10–15.

Stellate/basket cells suffer apoptosis both during migration in the white matter and after their settlement in the molecular layer. Percentage of the GFP-positive cells to total apoptotic cells was rather small, being less than 10%.

Apoptotic GFP-negative cells in GAD67/GFP mice are CGCs and glial cells, as described previously, but might include also some GABAergic neurons that have lost GFP during the apoptotic process (Yamanaka *et al.*, 2004).

Neurons of Deep Cerebellar Nuclei

There are no data showing apoptosis during normal development of deep cerebellar nuclear (CN) neurons. Nonetheless, data from literature demonstrate that CN neurons dye by apoptosis after neonatal rats were treated with ethanol (Dikranian et al., 2005). In other experiments, a unilateral penduculotomy (that causes deafferentation of the hemicerebellum) was perfomed. This type of surgery results in axotomy of the deep CN neurons ipsilateral to the lesion. The effect of taxotomy on CN neurons begun to be evident histologically within 3 hours after lesion, and neurodegeneration occurred within 48 hours with clear apoptotic signs (Sherrard and Bower, 1997).

Cell Death and Circuitry Remodeling in Cerebellum

Cerebellar NOND could be in relation with the final organization of the highly precise pattern of adult cerebellar circuitry and with the intrinsic cerebellar cortex molecular compartmentalization.

Not only apoptosis occurs during the development of PCs, but also of their main primary afferents: the CGCs (as already discussed above) and the inferior olive (IO) neurons. It seems possible that an altered interaction with both types of afferents is responsible for PC apoptosis, although this remains to be established in full.

The relationship with the main output target, the cerebellar nuclear neurons could also determine PC death. Adult cortico-nuclear and nucleo-cortical cerebellar projections are strictly organized. They follow a topographic map, which exactly matches with the sagittal organization of the cerebellar cortex and extracerebellar afferents. Thus, it is conceivable that CN synaptic availability for developing PC axons plays a role on PC death (Madalosso et al., 2005).

CN neurons are born 24 hours before PCs, with an overlapping period of 24 hours on birth dates for the whole populations of both CN neurons and PCs. CN neurons migrate down throughout the migrating clusters of PCs to reach their deep adult location in the cerebellar plate. Early during this migratory period, the axons of the CN neurons leave the cerebellum, while the intermingling between CN and PC migratory neurons could result in the establishment of the proper cortico-nuclear projections.

There is also a temporal coincidence between the occurrence of PC, IO and bulbo-pontine neuron (BPN) NOND in which PC NOND ends earliest. This time-window can be divided into two consecutive and distinct stages. In an early phase, NOND might act independently to delineate the self-organization of the respective cerebellar cortex IO and BPN domains. In a late phase NOND is exclusive of the pre-cerebellar neurons. In this case, IO and BPN NOND is a reflection of the competition for target-derived trophic factors taking place during the final remodeling of the olivo-cerebellar and the ponto-cerebellar projections (Madalosso *et al.,* 2005) Furthermore, PC apoptosis is probably regulated also through the connection with climbing fibers, i.e. the axons of the IO neurons. During the first postnatal week in rats, a single climbing fiber is initially connected to several PCs (Morara *et al.*, 2001; Kapfhammer, 2004). Then, the climbing fiber starts to trim its terminal arbors and begins to focus more on a single PC. The rest of the climbing fiber innervations to other PCs will eventually retract as long as cerebellar development proceeds (Kapfhammer, 2004; Sugihara, 2006). Therefore, the PCs without connections with climbing fibers are presumably eliminated through apoptosis (Chu *et al.*, 2000).

References

Adams J.M. and Cory S. (1998). The Bcl-2 protein family: arbiters of cell survival. *Science* 281, 1322-1326.

Adams J.M. and Cory S. (2002). Apoptosomes: engines for caspase activation. *Curr. Opin. Cell Biol.* 14, 715-720.

Alles A., Alley K., Barrett J.C., Buttyan R., Columbano A., Cope F.O., Copelan E.A., Duke R.C., Farel P.B. and Gershenson L.E. (1991). Apoptosis: a general comment. *FASEB J.* 5, 2127-2128.

Altman J and Bayer SA (1997). Development of the cerebellar system in relation to its evolution, structure and functions. Boca Raton:CRC Press.

Amritraj A, Peake K, Kodam A, Salio C, Merighi A, Vance JE, Kar S. (2009). Increased activity and altered subcellular distribution of lysosomal enzymes determine neuronal vulnerability in Niemann-Pick type C1-deficient mice. *Am. J. Pathol.*

Bähr, M. (2000). Live or let die - retinal ganglion cell death and survival during development and in lesioned CNS. *Trends Neurosci.* 23, 483-490.

Bartek, J. and Lukas, J. (2001). Mammalian G1- and S-phase checkpoints in response to DNA damage. *Curr. Opin. Cell Biol.* 13, 738-747.

Bartek, J., Bartkova, J., and Lukas, J. (1996). The retinoblastoma protein pathway and the restriction point. *Curr. Opin. Cell Biol.* 8, 805-814.

Belecky-Adams, T., Cook, B., and Adler, R. (1996). Correlation between terminal mitosis and differentiated fate of retinal precursor cells *in vivo* and *in vitro*: analysis with the "window-labeling" technique. *Dev. Biol.* 178, 304-315.

Benchoua, A., Guegan, C., Couriaud, C., Hosseini, H., Sampaio, N., Morin, D., and Onteniente, B. (2001). Specific caspase pathways are activated in the two stages of cerebral infarction. *J. Neurosci.* 21, 7127-7134.

Bernier, P. J., Vinet, J., Cossette, M., and Parent, A. (2000). Characterization of the subventricular zone of the adult human brain: evidence for the involvement of Bcl-2. *Neurosci. Res.* 37, 67-78.

Blaschke A.J., Staley K. and Chun J. (1996). Widespread programmed cell death in proliferative and postmitotic regions of the fetal cerebral cortex. *Development* 122, 1165-1174.

Boobis A.R., Fawthrop D.J. and Davies D.S. (1990). Mechanisms of cell toxicity. *Curr. Opin. Cell Biol.* 2, 231-237.

Castren E., Ohga Y., Berzaghi M.P., Tzimaagiorgis G., Thoenen H. and Lindholm D. (1994). bcl-2 messenger RNA is localized in neurons of the developing and adult rat brain. *Neuroscience* 61, 165-177.

Chao D.T. and Korsmeyer S.J. (1998). BCL-2 family: regulators of cell death. *Annu. Rev. Immunol.* 16, 395-419.

Cheng XS, Li MS, Du J, Jiang QY, Wang L, Yan SY, Yu DM, Deng JB. (2011).Neuronal apoptosis in the developing cerebellum *Anat. Histol. Embryol.* 40(1):21-7.

Chu T, Hullinger H, Schilling K, Oberdick J. (2000). Spatial and temporal changes in natural and target deprivation-induced cell death in the mouse inferior olive *J. Neurobiol.* 43(1):18-30.

Clarke P.G.H. (1990). Developmental cell death: morphological diversity and multiple mechanisms. *Anat. Embryol.* 181, 195-213.

Cohen J.J., Duke R.C., Fadok V.A. and Sellins K.S. (1992). Apoptosis and programmed cell death in immunity. *Annu. Rev. Immunol.* 10, 267-293.

Cole K.K. and Perez-Polo J.R. (2002). Poly(ADP-ribose) polymerase inhibition prevents both apoptotic-like delayed neuronal death and necrosis after H(2)O(2) injury. *J. Neurochem.* 82, 19-29.

Dainiak N. (2002). Hematologic consequences of exposure to ionizing radiation. *Exp. Hematol.* 30, 513-528.

De Bilbao F., Guarin E., Nef P., Vallet P., Giannakopoulos P. and Dubois-Dauphin M. (1999). Postnatal distribution of cpp32/caspase 3 mRNA in the mouse central nervous system: an in situ hybridization study. *J. Comp. Neurol.* 409, 339-357.

De la Rosa E.J. and De Pablo F. (2000). Cell death in early neuronal development: beyon the neurotrophic theory. *Trends Neurosci.* 23, 454-458.

Denecker, G., Vercammen, D., Declercq, W., and Vandenabeele, P. (2001). Apoptotic and necrotic cell death induced by death domain receptors. *Cell Mol. Life Sci.* 58, 356-370.

Dikranian K, Qin YQ, Labruyere J, Nemmers B, Olney JW. (2005). Ethanol-induced neuroapoptosis in the developing rodent cerebellum and related brain stem structures. *Brain Res. Dev. Brain Res.* 22, 1-13.

D'Sa-Eipper C. and Roth K.A. (2000). Caspase regulation of neuronal progenitor cell apoptosis. *Dev. Neurosci.* 22, 116-124.

Dusart I, Guenet JL, Sotelo C. (2006). Purkinje cell death: differences between developmental cell death and neurodegenerative death in mutant mice. *Cerebellum* 5, 163-173.

Eastman A. (1993). Apoptosis: a product of programmed and unprogrammed cell death. *Toxicol. Appl. Pharmacol.* 121, 160-164.

Ferrari G., Anderson B.L., Stephens R.M., Kaplan D.R. and Greene L.A. (1995). Prevention of apoptotic neuronal death by G_{M1} ganglioside. Involvement of Trk neurotrophin receptors. *J. Biol. Chem.* 270, 3074-3080.

Frankowski H., Missotten M., Fernandez P.A., Martinou I., Michel P., Sadoul R. and Martinou J.C. (1995). Function and expression of the *Bcl-x* gene in the developing and adult nervous system. *Neuroreport* 6, 1917-1921.

Fraser A., McCarthy N. and Evan G.I. (1996). Biochemistry of cell death. *Curr. Opin. Neurobiol.* 6, 71-80.

Fujikawa D.G., Ke X., Trinidad R.B., Shinmei S.S. and Wu A. (2002). Caspase-3 is not activated in seizure-induced neuronal necrosis with internucleosomal DNA cleavage. *J. Neurochem.* 83, 229-240.

Galli-Resta L. and Ensini M. (1996). An intrinsic time limit between genesis and death of individual neurons in the developing retinal ganglion cell layer. *J. Neurosci.* 16, 2318-2324.

Glucksmann P.D. (1951). Cell deaths in normal vertebrate ontogeny. Biol. Rev. 26, 59-86.

Hengartner M.O. and Horvitz H.R. (1994). Programmed cell death in *Caenorhabditis elegans*. *Curr. Opin. Cell Biol.* 4, 581-586.

Herrup K. and. Busser C.J. (1995) The induction of multiple cell cycle events precedes target-related neuronal death *Development* 121, 2385-2395.

Higashi Y, Murayama S, Pentchev PG, Suzuki K.(1993). Cerebellar degeneration in the Niemann-Pick type C mouse. *Acta Neuropathol.* 85(2):175-84.

Hoffman B. and Liebermann D.A. (1994). Molecular controls of apoptosis: Differentiation/ growth arrest primary response genes, proto-oncogenes, and tumor suppressor genes as positive and negative modulators. *Oncogene* 9, 1807-1812.

Hoffman B. and Liebermann D.A. (1994). Molecular controls of apoptosis: Differentiation/growth arrest primary response genes, proto-oncogenes, and tumor suppressor genes as positive and negative modulators. *Oncogene* 9, 1807-1812.

Hou S.T. and MacManus J.P. (2002). Molecular mechanisms of cerebral ischemia-induced neuronal death. *Int. Rev. Cytol.* 2002. ;221. :93. -148. 221, 93-148.

Ishii N., Wanaka A., Ohno K., Matsumoto K., Eguchi Y., Mori T., Tsujimoto Y. and Tohyama M. (1996). Localization of *bcl-2*, *bax*, and *bcl-x* mRNAs in the developing inner ear of the mouse. *Brain Res.* 726, 123-128.

Jankowski J, Miething A, Schilling K, Baader SL. (2009). Physiological purkinje cell death is spatiotemporally organized in the developing mouse cerebellum *Cerebellum.* Sep;8(3):277-90.

Johnson J. and Oppenheim R. (1994). Neurotrophins: Keeping track of changing neurotrophic theory. The requirement for neurotrophins in the development of specific neuronal populations is endorsed by transgenic mice lacking neurotrophin receptors, but central nervous regions of the mice are paradoxically unaffected. *Curr. Biol.* 4, 662-665.

Kapfhammer J.P. (2004) Cellular and molecular control of dendritic growth and development of cerebellar Purkinje cells *Prog. Histochem. Cytochem.* 39(3):131-82.

Kaufmann, J. A., Bickford, P. C., and Taglialatela, G. (2001) Oxidative-stress-dependent up-regulation of Bcl-2 expression in the central nervous system of aged Fisher-344 rats. *J. Neurochem.* 76, 1099-1108.

Kerr J.F., Wyllie A.H. and Currie A.R. (1972). Apoptosis: a basic biological phenomenon with wide-ranging implications in tissue kinetics. *Br. J. Cancer* 26, 239-257.

Kim SJ, Lim MS, Kang SK, Lee YS, Kang KS. (2008). Impaired functions of neural stem cells by abnormal nitric oxide-mediated signaling in an in vitro model of Niemann-Pick type C disease. *Cell Res.* Jun;18(6):686-94.

Kirkland R.A., Saavedra G.M., Cumming B.S. Franklin J.L. (2010). Bax Regulates Production of Superoxide in Both Apoptotic and Nonapoptotic Neurons: Role of Caspases *J.Neurosci.* 30(48):16114 –16127.

Krueger B.K., Burne J.F. and Raff M.C. (1995). Evidence for large-scale astrocyte death in the developing cerebellum. *J. Neurosci.* 15, 3366-3374.

Lange W. (1975). Cell number and cell density in the cerebellar cortex of man and some other mammals. *Cell Tiss. Res.* 157, 115-124.

Leist M. and Jaattela M. (2001). Four death and a funeral: from caspases to alternative mechanisms. *Nat. Rev. Mol. Cell. Biol.* 8, 324-326.

Li P., Nijhawan D., Budihardjo I., Srinivasula S.M., Ahmad M., Alnemri E.S. and Wang X. (1997). Cytochrome c and dATP-dependent formation of Apaf-1/caspase-9 complex initiates an apoptotic protease cascade. *Cell* 91, 479-489.

Lichnovsky V., Kolar Z., Murray P., Hlobilkova A., Cernochova D., Pospisilova E., Vojtesek B. and Nenutil R. (1998). Differences in p53 and Bcl-2 expression in relation to cell proliferation during the development of human embryos. *Mol. Pathol.* 51, 131-137.

Liu Q.A. and Hengartner M.O.(1999). The molecular mechanism of programmed cell death in C. elegans. *Ann. N. Y. Acad. Sci.* 887, 92-104.

Liu X., Kim C.N., Yang J., Jemmerson R. and Wang X. (1996). Induction of apoptotic program in cell-free extracts: requirement for dATP and cytochrome c. *Cell* 86, 147-157.

Liu, D. X. and Greene, L. A. (2001). Neuronal apoptosis at the G1/S cell cycle checkpoint. *Cell Tissue Res.* 305, 217-228.

Lockshin, R. A. and Zakeri, Z. (2002). Caspase-independent cell deaths. *Curr. Opin. Cell Biol.* 14, 727-733.

Lossi L., Zagzag D., Greco M.A. and Merighi A. (1998). Apoptosis of undifferentiated progenitors and granule cell precursors in the post-natal human cerebellar cortex correlates with expression of BCL-2, ICE and CPP-32 proteins. *J. Comp. Neurol.* 399, 359-372.

Lossi L, Mioletti S, Aimar P, Bruno R. and Merighi A (2002a). In vivo analysis of cell proliferation and apoptosis in the CNS. In: Merighi A and Carmignoto G, editors. Cellular and Molecular Methods in Neuroscience Research.New York:Springer Verlag. p. 235-258.

Lossi L., Mioletti S. and Merighi A. (2002b). Synapse-independent and synapse-dependent apoptosis of cerebellar granule cells in postnatal rabbits occur at two subsequent but partly overlapping developmental stages. *Neuroscience* 112, 509-523.

Lossi L. and Merighi A. (2003). In vivo cellular and molecular mechanisms of neuronal apoptosis in the mammalian CNS. *Progr. Neurobiol.* 69, 287-312.

Lossi L., Gambino G., Mioletti S. and Merighi A. (2004a). In vivo analysis reveals different apoptotic pathways in pre- and post-migratory cerebellar granule cells of rabbit. *J. Neurobiol.* 60, 437-452.

Lossi L., Tamagno I. and Merighi A. (2004b). Molecular morphology of neuronal apoptosis: activation of caspase 3 during postnatal development of mouse cerebellar cortex. *J. Mol. Histol.* 35, 621-629.

Lossi L., Alasia S., Salio C. and Merighi A. (2009.) Cell death and proliferation in acute slices and organotypic cultures of mammalian CNS. *Prog. Neurobiol.* 88(4):221-45.

Lundberg, A. S. and Weinberg, R. A. (1998) Functional inactivation of the retinoblastoma protein requires sequential modification by at least two distinct cyclin-cdk complexes. *Mol. Cell Biol.* 18, 753-761.

Luo X., Budihardjo I., Zou H., Slaughter C. and Wang X. (1998). Bid, a Bcl2 interacting protein, mediates cytochrome c release from mitochondria in response to activation of cell death surface receptors. *Cell* 94, 481-490.

Maddalossoa S., Perz-Villegasa E., Amengola J.A. (2005). Naturally occurring neuronal death during the postnatal development of Purkinje cells and their precerebellar afferent projection *Brain Research Review* 49 267-279.

Marin-Teva JL, Dusart I, Colin C, Gervais A, van Rooijen N, Mallat M. (2004). Microglia promote the death of developing Purkinje cells. *Neuron* 19, 535-547.

Marks N. and Berg M.J. (1999). Recent advances on neuronal caspases in development and neurodegeneration. *Neurochem. Int.* 35, 195-220.

McKonnel, S. K. (1995). Constructing the cerebral cortex: neurogenesis and fate determination. *Neuron* 15, 761-768.

Migheli A., Cavalla P., Piva R., Giordana M.T. and Schiffer D. (1994). bcl-2 protein expression in aged brain and neurodegenerative diseases. *Neuroreport* 5, 1906-1908.

Miller T.M., Moulder K.L., Knudson C.M., Creedon D.J., Deshmukh M., Korsmeyer S.J. and Johnson E.M., Jr. (1997). Bax deletion further orders the cell death pathway in cerebellar granule cells and suggests a caspase-independent pathway to cell death. *J. Cell Biol.* 139, 205-217.

Miller, F. D. and Kaplan, D. R. (2001). Neurotrophin signalling pathways regulating neuronal apoptosis. *Cell Mol. Life Sci.* 58, 1045-1053.

Mimeault M. (2002). New advances on structural and biological functions of ceramide in apoptotic/necrotic cell death and cancer. *FEBS Lett.* 530, 9-16.

Mizuguchi M., Sohma O., Takashima S., Ikeda K., Yamada M., Shiraiwa N. and Ohta S. (1996). Immunochemical and immunohistochemical localization of Bcl-x protein in the rat central nervous system. *Brain Res.* 712, 281-286.

Mooney S.M. and Miller M.W. (2000). Expression of bcl-2, bax, and caspase-3 in the brain of the developing rat. *Brain Res. Dev. Brain Res.* 123, 103-117.

Morara S., Van Der Want J.L., De Weerd,H., Provini L., Rosina A.(2001). Ultrastructural analysis of climbing fiber–Purkinje cell synaptogenesis in the rat cerebellum *Neuroscience.* 2001;108(4):655-71.

Moroni F., Meli E., Peruginelli F., Chiarugi A., Cozzi A., Picca R., Romagnoli P., Pellicciari R. and Pellegrini-Giampietro D.E. (2001). Poly(ADP-ribose) polymerase inhibitors attenuate necrotic but not apoptotic neuronal death in experimental models of cerebral ischemia. *Cell Death. Differ.* 8, 921-932.

Morrison, J. H. and Hof, P. R. (1997). Life and death of neurons in the aging brain. Science 278, 412-419.

Newton K. and Strasser A. (1998). The Bcl-2 family and cell death regulation. *Curr. Opin. Genet. Dev.* 8, 68-75.

Oppenheim R.W. (1985). Naturally occurring cell death during neural development. *Trends Neurosci.* 8, 487-493.

Ortiz A., Lorz C., Justo P., Catalan M.P. and Egido J. (2001). Contribution of apoptotic cell death to renal injury. *J. Cell Mol. Med.* 5, 18-32.

Padmanabhan J., Park D.S., Greene L.A. and Shelanski M.L. (1999). Role of cell cycle regulatory proteins in cerebellar granule neuron apoptosis. *J. Neurosci.* 19, 8747-8756.

Raff, M. C., Barres, B. A., Burne, J. F., Coles, H. S. R., Ishizaki, Y., and Jacobson, M. D. (1994). Programmed cell death and the control of cell survival. *Philos. Trans. R. Soc.* Lond. [Biol] 345, 265-268.

Rao, L. and White, E. (1997). Bcl-2 and the ICE family of apoptotic regulators: making a connection. *Curr. Opin. Genet. Dev.* 7, 52-58.

Sadoul, R. (1998). Bcl-2 family members in the development and degenerative pathologies of the nervous system. *Cell Death Differ.* 5, 805-815.

Sanchez, I. and Yuan, J. (2001). A convoluted way to die. *Neuron* 29, 563-566.

Sastry P.S. and Rao K. (2000). Apoptosis in the nervous system. *J. Neurochem.* 74, 1-20.

Saunders J.V. (1966). Death in embryonic systems. *Science* 154, 604-612.

Schwab B.L., Guerini D., Didszun C., Bano D., Ferrando-May E., Fava E., Tam J., Xu D., Xanthoudakis S., Nicholson D.W., Carafoli E. and Nicotera P. (2002). Cleavage of plasma membrane calcium pumps by caspases: a link between apoptosis and necrosis. *Cell Death Differ.* 9, 818-831.

Schwartz L.M., Smith S.W., Jones M.E. and Osborne B.A. (1993). Do all programmed cell death occur via apoptosis? *Proc. Natl. Acad. Sci. USA* 90, 980-984.

Schweichel, J. U. and Merker, H. J. (1973) The morphology of various types of cell death in prenatal tissues. *Teratology* 7, 253-266.

Selimi, F., Mariani, J., and Martinou, J. C. (1997) Caenorhabditis elegans et la mort neuronale chez les mammiferes. *Rev. Neurol.* (Paris) 153, 478-483.

Selimi F, Lohof AM, Heitz S, Lalouette A, Jarvis CI, Bailly Y, Mariani J Lurcher (2003). GRID2-induced death and depolarization can be dissociated in cerebellar Purkinje cells. *Neuron.* Mar 6;37(5):813-9.

Shastry, P., Basu, A., and Rajadhyaksha, M. S. (2001) Neuroblastoma cell lines - a versatile in vitro model in neurobiology. *Int. J. Neurosci.* 108, 109-126.

Sherrard RM, Bower AJ. (1997) Acute neuronal and vascular changes following unilateral cerebellar pedunculotomy in the neonatal rat. *J. Anat.* 191, 177-189.

Singh R.P., Dhanalakshmi S. and Agarwal R. (2002). Phytochemicals as cell cycle modulators--a less toxic approach in halting human cancers. *Cell Cycle* 1, 156-161.

Smith S. (2001). The world according to PARP. *Trends Biochem. Sci.* 26, 174-179.

Sohma O., Mizuguchi M., Takashima S., Yamada M., Ikeda K. and Ohta S. (1996). High espression of Bcl-x protein in the developing human cerebellar cortex. *J. Neurosci. Res.* 43, 175-182.

Sommer L. and Rao M. (2002). Neural stem cells and regulation of cell number. *Prog. Neurobiol.* 66, 1-18.

Srinivasan A., Roth K.A., Sayers R.O., Schindler K.S., Wong A.M., Fritz L.C. and Tomaselli K.J. (1998). In situ immunodetection of activated caspase-3 in apoptotic neurons in the developing nervous system. *Cell Death Differ.* 5, 1004-1016.

Stennicke H.R. and Salvesen G.S. (1999). Catalytic properties of the caspases. *Cell Death. Differ.* 6, 1054-1059.

Stewart B.W. (1994). Mechanisms of apoptosis: Integration of genetic, biochemical, and cellular indicators. *JNCI* 86, 1286-1296.

Sugihara I. (2006) Organization and remodeling of the olivocerebellar climbing fiber projection. *Cerebellum.* 2006;5(1):15-22.

Taglialatela, G., Gegg, M., Perez-Polo, J. R., Williams, L. R., and Rose, G. M. (1996). Evidence for DNA fragmentation in the CNS of aged Fischer-344 rats. *Neuroreport* 7, 977-980.

Thomaidou D., Mione M.C., Cavanagh J.F. and Parnavelas J.G. (1997). Apoptosis and its relation to the cell cycle in the developing cerebral cortex. J. Neurosci. 17, 1075-1085.

Thornberry N.A. and Lazebnik Y. (1998). Caspases: enemies within. *Science* 281, 1312-1316.

Vaux D.L., Haecker G. and Strasser A. (1994). An evolutionary perspective on apoptosis. *Cell* 76, 777-779.

Vinet J., Bernier P.J. and Parent A. (2002). Bcl-2 expression in thalamus, brainstem, cerebellum and visual cortex of adult primate. *Neurosci. Res.* 42, 269-277.

Walworth, N. C. (2001). DNA damage: Chk1 and Cdc25, more than meets the eye. *Curr. Opin. Genet. Dev.* 11, 78-82.

Weisheit G, Gliem M, Endl E, et al. (2006) Postnatal development of the murine cerebellar cortex: formation and early dispersal of basket, stellate and Golgi neurons. *Eur. J. Neurosci;* 24:466–478.

White, E. (1996) Life, death,and the pursuit of apoptosis. *Genes. Dev.* 10, 1-15.

Wood K.A., Dipasquale B. and Youle R.J. (1993). In situ labeling of granule cells for apoptosis-associated DNA fragmentation reveals different mechanisms of cell loss in developing cerebellum. *Neuron* 11, 621-632.

Yakovlev A.G. and Faden A.I. (2001). Caspase-dependent apoptotic pathways in CNS injury. *Mol. Neurobiol.* 24, 131-144.

Yamanaka H., YanagawaY., Obata K., (2004) Development of stellate and basket cells and their apoptosis in mouse cerebellar cortex. *Neuroscience Research* 50, 1: 13-22.

Yoshikawa K. (2000). Cell cycle regulators in neural stem cells and postmitotic neurons. *Neurosci. Res.* 37, 1-14.

Yuan J. (1995). Molecular control of life and death. *Curr. Opin. Cell Biol.* 7, 211-214.

Zanjani H.S., Vogel M.W., Delhaye-Bouchaud N., Martinou J.C. and Mariani J. (1996). Increased cerebellar Purkinje cell numbers in mice overexpressing a human bcl-2 transgene. *J. Comp Neurol.* 374, 332-341.

Zou H., Henzel W.J., Liu X., Lutschg A. and Wang X. (1997). Apaf-1, a human protein homologous to C. elegans CED-4, participates in cytochrome c-dependent activation of caspase-3. *Cell* 90, 405-413.

In: Neuronal Cell Apoptosis
Editors: C. J. Schmid and J. L. Wolfe

ISBN: 978-1-61324-633-7

Chapter V

Fundamental Mechanisms and Implications of Neuronal Apoptosis

Binu Tharakan[1], Indrajit Chowdhury[2] and Ganapathy Bhat[2*]

[1]Center for Cell Death and Differentiation,
Texas A&M Health Science Center College of Medicine and Scott and White Memorial Hospital, Temple, Texas, USA
[2]Department of Obstetrics and Gynecology,
Morehouse School of Medicine, Atlanta, Georgia, USA

Abstract

Brain cells fundamentally differ, structurally, biochemically and functionally from other cell types of the body. In the brain, neuronal apoptosis is a normal physiological process that is an essential and integral part of neurogenesis. The massive overproduction of neurons that occur during the development of the brain is followed by a programmed cell death of roughly one-half of the originally formed neurons. This precisely controlled neuronal demise is highly conserved in diverse organisms. However, apoptosis when becomes aberrant has been implicated in various neurological disorders as diverse as Parkinson's disease, Alzheimer's disease, Huntington's disease, and amyotrophic lateral sclerosis, spinal muscular atrophy, diabetic encephalopathy, and ischemic and traumatic brain injuries. At the fundamental level, the main focus of the recent research has been in the area of understanding the functional mechanisms of the death receptor mediated 'extrinsic' and mitochondria mediated 'intrinsic' apoptotic pathways, and the roles of caspases and protein members of the Bcl-2 family. Neuronal apoptosis is classically viewed as an active and highly regulated form of neuronal death, whereas neuronal necrosis is considered as an accidental form of cell death. Also, the recent concept of a programmed necrosis has considerable scope in neuronal cell death while autophagy, a caspase independent apoptosis related mechanism in which the cytoplasm is destroyed by

* Corresponding Author, Ganapathy Bhat, email: gbhat@msm.edu

lysosomes has been implicated in cell death as well as in cell protection. The main purpose of this chapter is to provide the basic details and update the current understanding of neuronal apoptosis and its implications in neurodegenerative, ischemic and traumatic conditions. Thus, this chapter focuses on two major areas (a) the fundamental mechanisms of neuronal apoptosis and (b) the pathophysiological implications of neuronal apoptosis.

Introduction

This chapter is intended to provide an account on the fundamental concepts and the pathophysiological implications of programmed cell death of neurons, with special emphasis to the most recent developments in the field. The first part of this chapter is devoted to the coverage of the various intracellular mechanisms involved in neuronal cell death in considerable detail and the second part briefly addresses the implications of neuronal apoptosis in various but selected neurological disorders. Some of the most recent findings in the field of apoptosis that are yet to be validated in neurons are also discussed in the first section. Once fully established in neurons, such findings will provide enormous opportunities for future research in the field of neuronal apoptosis.

1. Basic Mechanisms of Neuronal Apoptosis

At the fundamental level, the main focus of the recent research on neuronal apoptosis has been in the area of understanding the functional mechanisms of the two pathways, the mitochondria mediated 'intrinsic' apoptotic pathway and the death receptor mediated 'extrinsic' pathway (Figure 1).

The induction of extrinsic pathway is initiated by the binding of the death receptors Fas (Apo-1 or CD95), tumor necrosis factor receptor-1 (TNFR-1/p55/CD120a), and interferon (IFN) and TRAIL (TNF-related apoptosis-inducing ligand or Apo2-L) receptors to their ligands in the plasma membrane of the cell. The signal transduction of cell death from TNF-R1 is via its cytoplasmic death domain (DD) by the activation of caspase 8 alone to activate caspase 3 or by the activation of the mitochondria-dependent amplification loop to cause apoptosis. TNF-R2 directly recruits TNF receptor-associated factors (TRAF), induces gene expression and intensively cross-talks with TNF-R1. The mitochondria-dependent path is triggered by activated caspase 8 through the cleavage of Bid to truncated Bid (tBid), which activates intrinsic apoptotic pathway. The intrinsic or mitochondrial pathway, results in mitochondrial outer membrane permiabilization (MOMP), permiability transition pore (PTP) and cristae remodeling, subsequent release of cytochrome *c* (Cyt *c*) from the mitochondrial intermembrane space (IMS) into the cytosol, followed by the formation of the apoptosome (a complex formed by cytochrome c + Apaf1 + procaspase-9) and activation of initiator caspase-9. This pathway is regulated by the members of the B-cell-lymphoma (Bcl)-2 family proteins, mainly by a balance in the cell of the anti-apoptotic (Bcl2, BclxL) and pro-apoptotic (Bax, Bak) proteins. The release of other IMS proteins, including Smac, Diablo, endonuclease G (Endo G), apoptosis inducing factor (AIF) and HtrA2/Omi, in to the cytosol following MOMP results in the execution of apoptosis through the caspases 3 and 7.

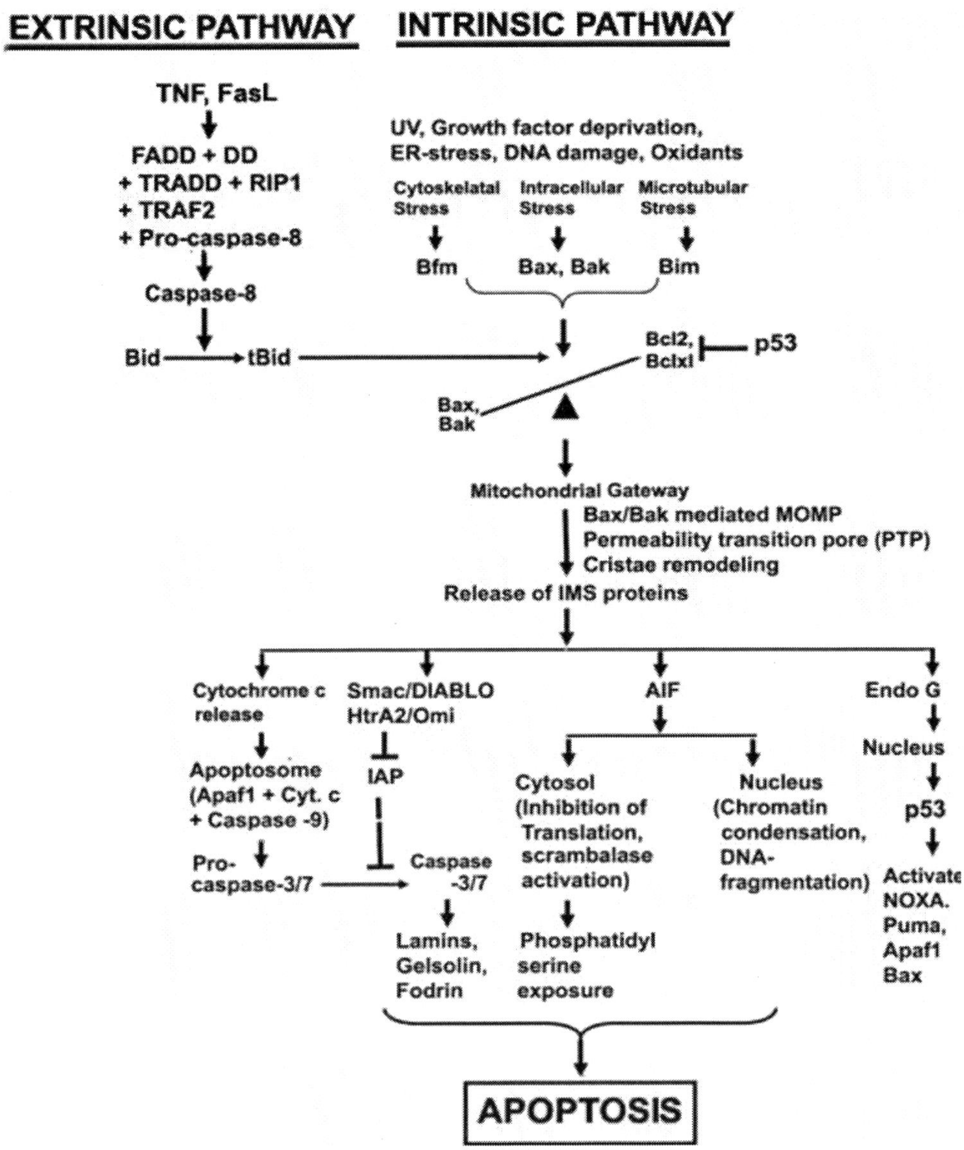

Figure 1. Schematic representation of extrinsic and intrinsic apoptotic signaling pathways, the critical regulators, caspase activation, key events involved and the cross-talk in these pathways.

The blunt-headed bold lines in the figure indicate the roles of the inhibitors regulating the intrinsic pathways.

1.1. Intrinsic Pathway

Mitochondria are key players in the induction and regulation of apoptotic neuronal cell death. Various apoptotic stimuli such as reactive oxygen species (ROS) and Fas signaling lead to a mitochondrial mediated neuronal death. The process of neuronal cell death occurs due to the release of several pro-apoptotic proteins into the cytosol that are normally located in the mitochondrial intermembrane space. Permeabilization of the mitochondrial membrane is key to the release of the pro-apoptotic molecules into the cytosol. Several mechanisms have been proposed to explain the permeabilization of the mitochondrial membrane. Induction of the mitochondrial permeability transition (MPT) is the first step in this process and this involves the opening of a mitochondrial permeability transition pore (PTP) [1]. The PTP opening causes swelling of the mitochondrial matrix, which results in mitochondrial uncoupling, rupturing of the mitochondrial outer membrane, and release of pro-apoptotic proteins into the cytosol [2]. The other important mechanism that regulates the outer mitochondrial membrane permeabilization involves members of the Bcl-2 family of proteins that are either pro- apoptotic or anti-apoptotic proteins. The proteins that are released from the mitochondria, that play major role in apoptosis are cytochrome *c*, second mitochondria-derived activator of caspases (Smac/Diablo), apoptosis inducing factor (AIF), and endonuclease G (endoG); most recently serine protease high temperature requirement protein A2 (HtrA2/OMI) have also been involved in caspase activation [3]. These pro-apoptotic proteins, once released from the mitochondria into the cytosol, trigger caspase-dependent or -independent apoptotic death pathways [4]. Among the mitochondrial apoptogenic factors released in to the cytosol, cytochorme *c* is considered the most important molecule that triggers further apoptotic signaling by activating a series of caspases, in various cell types including neuronal cells. Cytochrome *c* is a small heme protein loosely attached to the inner mitochondrial membrane and is an essential component of the mitochondrial respiratory chain, where it carries one electron. Caspase activation is the main function of the cytosolically located cytochrome *c*. The caspases are normally inactive and require proteolytic processing for their activity. This is achieved by the formation of large protein complexes known as the DISC (death inducing signaling complex) and the apoptosome. In the DISC complex, the central scaffold protein is a mammalian CED-4 homologue known as apoptotic protease-activating factor-1 (Apaf-1). In the caspase dependent pathway of apoptosis, cytochrome *c* binds to Apaf-1 and forms the apoptosome complex [4-7]. In the apoptososme complex, pro-caspase 9 is recruited and activated in the presence of ATP or dATP. Caspase-9 is an initiator caspase that is present in all mammals. The caspase-9 activation is followed by caspase 9 mediated clevage and activation of the effector caspases, procaspase 3 and/or 7, which process substrates like caspase-activated DNase (ICAD) or PARP, and leads to DNA fragmentation and neuronal death [8]. The effector caspase, the caspase-3 is a member of the cysteine-aspartic acid protease family and plays a central role in the execution of apoptosis. Proteolytic activation of initiator procaspases is a crucial step in the activation of caspase-3. Other models have been proposed for the caspase activation mechanism, namely the oligomerization or induced proximity model and the allosteric regulation model. The oligomerization holds that procaspases become activated upon proper oligomerization by an adaptor protein. The allosteric regulation model states that the adaptor is an allosteric regulator for procaspases which has been applied for the activation of procaspase-9 by Apaf-1 in an oligomeric complex, the apoptosome. Inhibitors such as X-chromosomal inhibitor of apoptosis protein (XIAP), c-IAP1, c-IAP2, and survivin, regulate the intracellular activity of caspases. These inhibitors bind through their Baculovirus IAP repeat (BIR) domain to suppress the catalytic activity of caspases [9]. Proteins like Smac and Omi/HtrA2 (that are discussed above) are released from the mitochondria and act specifically on IAPs to block their effect [4, 10]. A more detailed account on the role of caspases in neuronal apoptosis is described later in one of the following sections. In a caspase-independent pathway, AIF translocates from the mitochondria to the nucleus and induces DNA fragmentation and chromatin condensation [7]. AIF is a mitochondrial flavoprotein and is the main mediator of caspase independent apoptosis. It is released into the cytosol upon permeabilization of the outer mitochondrial membrane. The translocation of endoG from mitochondria into the nucleus leads to internucleosomal DNA fragmentation [11]. The release of the apoptogenic proteins from mitochondria into the cytosol is regulated by members of the Bcl-2 family of proteins that are either pro-apoptotic or anti-apoptotic.

According to their structure and activity, the Bcl-2 family of proteins is divided into three groups that comprises of over 30 members. These members are the anti-apoptotic proteins (Bcl-2, Bcl-xL, Bcl-w, Mcl-1, A1/Bfl-1, NR-13, Boo/Diva/Bcl-2-L-10, and Bcl-B), the pro-apoptotic proteins (Bax, Bak, Bok/Mtd, and Bcl-Xs), and the BH3-only proteins (Bid, Bad, Noxa, Puma, Bmf, BimL/Bod, Bik/Nbk, Blk, Hrk/DP5, Bnip3, and Bnip3 L) [7]. The anti-apoptotic proteins have four Bcl-2 homology domains (BH 1-4). The pro-apoptotic proteins contain three BH domains. The BH3-only proteins induce apoptosis by activating proapoptotic or by inhibiting anti-apoptotic proteins. The BH3-only proteins are activated by cytotoxic signals that promote activation of pro-apoptotic proteins of the Bax-like family, which induce their oligomerization. The oligomeric form of pro-apoptotic proteins has the ability to form pores and to penetrate different cellular membranes. Bcl-2 family of proteins regulates apoptosis by controlling mitochondrial membrane permeability. The permeabilization of the outer mitochondrial membrane by proapoptotic Bax/Bak occurs through direct pore formation or through interaction with MPT pore components, in particular, with the voltage-dependent anion channel (VDAC) [12]. The VDAC plays a crucial role in apoptotic changes of the mitochondria and its activity is directly regulated by some Bcl-2 family members, including Bcl-2/Bcl-x(L) and Bax/Bak but not Bid. It has been shown that Bim has an ability to activate directly the VDAC, which plays an important role in apoptosis of mammalian cells. Accumulating evidence implicates that the VDAC functions in mitochondria-mediated apoptosis and as a critical player in the release of apoptogenic proteins, such as cytochrome. VDAC oligomerization is involved in mitochondria-mediated apoptosis and may represent a general mechanism common to numerous apoptogens acting via different initiating cascades. Thus, targeting the oligomeric status of VDAC, and hence apoptosis, offers a therapeutic strategy for combating cancers and neurodegenerative diseases [13]. The anti-apoptotic members of the Bcl-2 family play major role in maintaining the mitochondrial membrane integrity. They prevent the release of various mitochondrial proteins such as cytochrome *c* and Smac by interacting with, and inhibiting the pro-apoptotic proteins such as Bax/Bak and the BH3-only proteins. The ratio between pro-and anti-apoptotic proteins thus regulates the susceptibility of a neuronal cell to apoptotic death. The intrinsic apoptotic signaling pathway has also been depicted in Figure 1. A detailed account on caspases and their role on neuronal apoptosis is given in one of the later sections.

1.2. Extrinsic Pathway

Extrinsic apoptotic pathway comprises mainly of a death receptor pathway. The death receptors belong to the super family of tumor necrosis factor (TNF)/nerve growth factor receptor. These proteins are transmembrane in nature and possess an extracellular ligand-binding domain, a transmembrane domain, and a C-terminal domain. They are involved in transmitting the death signal from the neuronal surface to intracellular pathway upon substrate binding [14, 15]. The binding of a ligand to the death receptor initiates the death receptor-mediated apoptosis. This process causes receptor trimerization and recruitment of adaptor proteins at the cytosolic death domain (DD) of the receptor. Fas (CD95 or APO-1), TNF receptor 1(TNFR1), TNF-related apoptosis inducing ligand receptor 1 (TRAIL-R1) (death receptor 4, DR4), and TRAIL receptor 2 (TRAIL-R2) (death receptor 5, DR5) [11] are the best-characterized death receptors that play major role in apoptosis. Fas ligand is a type II transmembrane protein belonging to the TNF family. Fas- and TRAIL-associated death domain proteins further bind pro-caspases 8 and/or 10 and form the DISC complex (described earlier) where the initiator caspases are activated. A soluble Fas decoy receptor, DcR3, and three decoys for TRAIL receptors, TRAIL-R3 (decoy receptor 1, DcR1), TRAIL-R4 (decoy receptor 2, DcR2), and OPG (osteoprotegerin) have been characterized, but they are unable to trigger apoptosis because of the absence or truncation of the cytoplasmic death domain [7, 16, 17]. OPG acts as a soluble antagonist receptor capable of inhibiting TRAIL-mediated cell death. Neuronal damage mediated by the TRAIL-system is considered to be involved in the pathogenesis of neuroinflammatory diseases of the central nervous system. Depending on the nature of the downstream-mediated signaling, cells are generally divided into two categories. They are Type I cells and Type II cells. Type I cells exhibit efficient DISC formation and strong activation of caspase 8 that directly activates the downstream executioner caspases 3 and 7 [7, 18]. In type II cells, DISC formation is weak and therefore insufficient to induce apoptosis by the catalyzing effect of caspase-8. Here it initiatives a different signaling pathway by initiating the cleavage of a pro-apoptotic protein, BCL-2 Interacting Domain (Bid). Bid serves as a regulator of mitochondrial mediated apoptotic signaling [7, 19]. The role of Bid as a link between the extrinsic and the intrinsic pathways is described in more detail, in section 1.3 of this chapter. The extrinsic apoptotic signaling pathway has also been depicted in Figure 1.

The role of death receptors in mediating neuronal apoptosis is not a very well studied field of research. In the CNS, the expression of Fas-Fas ligand system is relatively weak, including glial cells and neurons, and is associated with the maintenance of immune suppression and prevention of an inflammatory response [7, 20]. In the developing brain, Fas plays a role in neurite remodeling and also mediates neuronal branching through its death domain, without caspase activation [21]. Neurite remodeling is a significant early step in the formation of axons and dendrites. Increased production of TNF-α and expression of TNF-α receptors were shown to contribute to neuronal cell death [22, 23]. TNFR1 and TNFR2 are present in human neuronal progenitors and microglial cells [24, 25]. A constitutive expression of apoptosis-mediating (TRAIL-R1 and R2) and nonapoptotic-mediating (TRAIL-R3 and -R4) receptors are present in human brain. These receptors are specifically localized in neurons, astrocytes, and oligodendrocytes but without TRAIL expression [26]. Increased TRAIL production occurs in brain macrophages or T cells during neurodegeneration and contributes to death receptor-mediated neuronal apoptosis [27].

1.3. Relationship between Intrinsic and Extrinsic Pathways

There is a relationship between the extrinsic and intrinsic death pathways that is dependent on Bid cleavage [28-30]. As mentioned in the above section, Bid is a BH3-only

Bcl-2 family member that is inactive in its full length form and bears a caspase-8 recognition site. When processed by caspase-8, Bid becomes active in its ability to antagonize anti-apoptotic Bcl-2 family members at the level of the mitochondria and initiate interaction with the intrinsic pathway. Full-length Bid is localized in cytosol whereas truncated Bid (tBid) translocates to the mitochondria. Caspase-8 in the extrinsic pathway can therefore, by cleaving Bid, lead to cytochrome *c* release and activation of caspase-9 of the intrinsic pathway. The caspase-8-mediated proteolytic cleavage of Bid and subsequent mitochondrial translocation has been detected in various neuronal cell death conditions. It has been shown that, granzyme B, a serine protease that is released by cytotoxic T cells and natural killer cells, can also cleave and activate Bid [31]. This is one mechanism used by these lymphocytes to induce apoptosis in their target cells. Pharmacological inhibition of Bid provides a promising therapeutic strategy in neurological diseases where programmed cell death is prominent [32].

1.4. Role of Caspases

Evidence obtained over the last decade shows that the cellular and biochemical features of apoptosis are all a consequence of the cleavage of a subset of proteins by proteases of the caspase family [5, 33]. Caspases are cysteine proteases that mediate apoptotic death in a variety of cellular systems, including neurons. As mentioned in one of the above sections, caspases are activated through intrinsic or extrinsic pathways. The former pathway is used by most neurons in most situations. In this pathway, release of mitochondrial cytochrome *c* into the cytoplasm induces formation of the apoptosome, which leads to the activation of caspase 9 and subsequently other caspases. To date, 14 mammalian caspases have been described. The caspases exist within the cells as inactive pro-forms. Following induction of apoptosis, these inactive enzymes (zymogens) will be cleaved to form active caspases. Caspases are highly conserved through the evolutionary process as their homologues can be found in diverse organisms such as insects, worms, sponges and hydra [5, 34]. The specificity of caspase is based on the recognition of a stretch of four amino acid residues, the last one of them corresponding to the aspartic residue after which the cleavage occurs [35]. Although eleven members of the caspase family have been described in humans, all of them do not play a role in apoptosis [36-38]. Members of the caspase family have been divided into pro-apoptotic and pro-inflammatory sub-families. The pro-apoptotic sub-family is further classified as activator caspases (caspase-2, -8, -9, -10, and -12) and executioner caspases (caspase- 3, -6, and -7). A feed-back loop has been demonstrated between caspase-3 and caspase-9 that lead to maximal activity of the proteases [33]. Upon APAF-1 and dATP-mediated oligomerization of caspase-9 following cytochrome *c* release from the mitochondria, caspase-9 is activated to a certain extent. This leads to partial processing and activation of caspase-3. Caspase-3 can then further process caspase-9 augmenting its activation allowing caspase-9 in turn to fully activate caspase-3 [5, 39]. The absence of caspase-3 and -7 greatly delayed UV-induced cytochrome *c* release that is upstream of the caspases [33]. This finding has considerable significance because cytochrome *c* release was not known to be under the regulation of downstream executioner caspases. Thus, executioner caspases control caspase-9 activation not only by being able to cleave caspase-9 but also by regulating the extent of cytochrome *c* release from mitochondria. A list of various caspases and their role in apoptosis has been given in Table 1 (for details, see Chowdhury et al [5]).

Table 1. Mammalian caspase (Cysteinyl aspartate-specific proteinase or Cysteinyl aspartate-specific proteinase) family and their selected characteristics

Subfamily	*Group*	*Member name*	*Structure*	*Substrate specificity*
ICE	*I (inflammatory mediator)*			*Hydrophobic proteolytic specificity*
		Caspase-1 (ICE)	1CARD+ p20+p10	WEHD
		Caspase-4 (ICErelII, TX, ICH-2)	1CARD+ p20+p10	(W/L) EHD
		Caspase-5 (ICErelIII, TY)	1CARD+ p20+p10	WEHD
		Caspase-11 (murine)	p20+p10	
		Caspase-12	1CARD+ p20+p10	WEHD
		Caspase-13 (ERICE)	1CARD+ p20+p10	
		Caspase-14 (MICE)	p20+p10	WEHD
CED-3	*Apoptotic*			
	II (initiator/ apoptotic activator)			*Aliphatic proteolytic specificity*
		Caspase-2 (ICH-1)	1CARD+ p20+p10	DEHD
		Caspase-8 (MACH, FLICE, Mch5)	2DED+ p20+p10	
		Caspase-9 (ICE-LAP6, Mch6)	1CARD+ p20+p10	
		Caspase-10 (Mch-4)	2DED+ p20+p10	
	III (effector/ apoptotic executioner)			*Asp proteolytic specificity*
		Caspase-3 (CPP32, Apopain, Yama)	p20+p10	DEHD
		Caspase-6 (Mch-2)	p20+p10	VEHD
		Caspase-7 (Mch3, ICE-LAP-3, CMH-1)	p20+p10	DEVD

The nascent caspases are originally synthesized as catalytically-dormant tripartite proenzyme (a single polypeptide chain of 32–55 kDa representing 3 domains in common) with a 17–21 kDa central large internal domain (p20) containing a large catalytic subunit (active site), a 10–13 kDa small C-terminal domain (p10) also called small catalytic subunit, and a 3–24 kDa NH_2-terminus prodomain called death domain (DD). In a number of procaspases, the p20 and p10 subunits are separated by a small linker sequence. The large N-terminal prodomains of procaspases contain 80–100 residue long structural motifs called death domain superfamily, involved in the transduction of apoptotic signals. Death domain (DD) is the member of TNF-receptor family. The DD has two sub-domains, the death effector domain (DED) and the caspase-recruitment domain (CARD).

1.5. Caspase Activation That Does Not Lead To Neuronal Apoptosis

There is increasing evidence that caspase-3 activation does not always lead to neuronal apoptosis. Pro-apoptotic caspases such as caspase-3 can mediate nonapoptotic functions, including activation of cell survival pathways and cell differentiation [40-44]. One of the areas in which caspase-3 plays a significantly different role in the brain is learning and memory. The brain appears to have adapted a core component of cell death machinery to serve a unique role, in learning and memory [45]. In the brain N-methyl-D-aspartate (NMDA) receptor-dependent synaptic modifications, such as long-term potentiation (LTP) and long-term depression (LTD), are essential for brain development and function. The activation of caspase-3 via mitochondria is required for LTD and AMPA receptor internalization in hippocampal neurons [46]. In the zebra finch auditory forebrain, caspase-3 activity is regulated at postsynaptic sites in brain following stimuli associated with memory (neural activation and subsequent response habituation), evidenced by the increase in the concentration of caspase-3 active sites briefly within minutes after exposure to tape-recorded birdsong [45].

1.6. Role of Mitochondrial ROS

Reactive oxygen species (ROS) are highly reactive molecules containing oxygen that plays a significant role in normal cell signaling and apoptosis. ROS has distructive actions on both DNA and proteins. Mitochondria are a major source of ROS within the cell. Mitochondrial ROS play an important role in apoptosis induction under both physiologic and pathologic conditions and they directly or indirectly mediate cytochrome *c* release from mitochondria that triggers caspase activation. In the mitochondria, the energy supply is achieved by oxidative phosphorylation of ADP to ATP using the electrochemical proton gradient generated by the mitochondrial respiratory chain complex. Cytochrome *c* is one of the components of the respiratory chain and transfers electrons from Complex III to Complex IV. Impaired mitochondrial electron transport system as occurs with oxidative challenge, can enhance ROS production. Within the mitochondria, cytochrome *c* is bound to the outer surface of the mitochondrial inner membrane by its association with cardiolipin (cardiolipin is an anionic phospholipid present predominantly in the mitochondria). The increased production of mitochondrial ROS and subsequent oxidative stress can lead to the oxidation of cardiolipin and the subsequent release of cytochrome *c* from the mitochondria to the cytosol [47]. However, the precise role of cardiolipin and its relationship to the mitochondrial ROS in regulating the release of cytochrome *c* from mitochondria during apoptosis is still unclear [47]. The mitochondrial redox compartment, as distinct from the cytosolic compartment, maintains mitochondrial integrity. Thus, the depletion of matrix glutathione (GSH), regardless of the cytosolic GSH status, is a key event in sensitization of cells to oxidant-induced apoptosis. GSH is a tripeptide antioxidant that is known to protect cellular components against the adverse effects of ROS. The mitochondrial transition pore that is critical for cytochrome *c* release is modulated by thiol oxidation. In such situations functional divalent thiol reactive agents induce formation of disulfide cross links which block Bcl-2 function in preventing permeability transition [48]. Oxidative stress induces mitochondrial DNA (mtDNA) damage leading to apoptosis [49]. Mitochondrial genome encodes several components of the mitochondrial respiratory chain complex. The mtDNA damage by ROS will decrease gene expression of key respiratory proteins and further alter mitochondrial ROS levels [50]. The disruption in electron flow eventually enhances mitochondrial ROS formation, which leads to mitochondrial dysfunction that includes decreased ATP production and loss of MTP that in turn will lead to apoptosis [51]. The oxidative mtDNA damage in the mitochondrial complex I of mitochondrial electron transport has been implicated in neurodegerative conditions. The involvement of mitochondrial ROS formation, mtGSH redox and mtDNA in regulation of neuronal apoptosis needs to be further established. A detailed account on the role of apoptosis in various neurodegerative diseases has been discussed in one of the following sections.

1.7. Programmed Necrosis

As seen in the above sections, neuronal apoptosis is classically viewed as an active and highly regulated form of cell death. Neuronal necrosis is considered as an accidental form of cell death. The most recent concept of a programmed necrosis has considerable scope in neuronal cell death. Recent studies from diverse organisms have begun to suggest that necrosis may in fact be "programmed" and not a default "accidental" pathway as previously believed.

These studies show that necrosis follows a stereotypical series of cellular and molecular events that includes swelling of organelles, increases in ROS formation and cytoplasmic calcium, a decrease in ATP, activation of calpain and cathepsin proteases, and finally rupture of organelles and plasma membrane [52, 53].

Genetic and chemical manipulations demonstrate that necrosis can be inhibited. This indicates that necrosis can indeed be controlled and that a possible 'program' exists in its functioning. Recent studies conducted in organisms as diverse as *C. elegans*, yeast, dictyostelium, drosophila, and mammals provide evidence for conserved mechanisms of necrosis [54]. As necrosis is prominent in ischemic and traumatic injury and various conditions of neurodegeneration, further detailed studies of this process may have potential therapeutic significance [7, 52].

1.8. Mitochondrial Fission and Fusion

Mitochondria are highly dynamic organelles that play a very important role in apoptosis. In normal cells, the mitochondria continuously undergo two opposite processes, fission and fusion. These mitochondrial dynamics influence mitochondrial morphology, biogenesis, mitochondrial distribution within the cell, cell bioenergetics, and cell injury or cell death. Mitochondrial fission and fusion are linked to synaptic activity and are implicated in the regulation of apoptotic cell death in many other cell types [55]. In neurons, the imbalance of mitochondrial fission/fusion can influence neuronal physiology, such as synaptic transmission and neuronal plasticity, and affect neuronal survival. This has considerable implications in the process of learning and memory. Recently, important regulatory components of the mitochondrial fission/fusion machinery have been identified through genetic studies in various model organisms [56]. The important players in this process are, the GTPase family of proteins, including fission proteins, dynamin related protein 1 (Drp1), mitochondrial fission 1 (Fis1), and fusion proteins (Mfn1, Mfn2 and Opa1). All these molecules are essential to maintain mitochondrial fission and fusion balance in neurons. The Drp1 plays a role in determining shape, size, distribution, remodeling, and maintenance of mitochondria. Drp1 is associated with several cellular functions, including mitochondrial and peroxisomal fragmentation, phosphorylation, SUMOylation, ubiquitination, and cell death. It has been hypothesized that mutant proteins in several neurodegenerative diseases interact with Drp1, activate mitochondrial fission machinery, fragment mitochondria excessively, and impair mitochondrial transport and mitochondrial dynamics, ultimately causing mitochondrial dysfunction and neuronal apoptosis [57]. The rate of fission exceeds the rate of fusion in healthy neuronal processes, and, therefore, the fission/fusion ratio alone is insufficient to explain mitochondrial morphology at steady state [55]. The antiapoptotic protein Bcl-xL, increases the rates of both fusion and fission, but more importance is given for explaining the longer organelle morphology induced by Bcl-xL is its ability to increase mitochondrial biomass. Deficits in these Bcl-xL-dependent mechanisms may be critical in neuronal dysfunction during the earliest phases of neurodegeneration, long before commitment to cell death [55]. Mutations of Mfn2 and Opa1 lead to neuropathies such as Charcot-Marie-Tooth disease type 2A and autosomal dominant optic atrophy. Moreover, both Aβ peptide and mutant huntingtin protein induce mitochondrial fragmentation and neuronal cell death. In addition, mutants of Parkinson's disease-related genes also show abnormal mitochondrial morphology [56]. These studies establish mitochondrial dynamics as a new paradigm for neurodegenerative disease research. Compounds that modulate mitochondrial fission/fusion could have potential therapeutic value in disease intervention [58].

2. Implications of Neuronal Apoptosis

In the central nervous system (CNS), neuronal apoptosis is a normal physiological process that is an essential and integral part of neurogenesis. The massive overproduction of neurons that occurs during the development of the brain is followed by a programmed cell death of roughly one-half of the originally formed neurons. However, neuronal apoptosis when becomes aberrant leads to various neurological disorders. Sustained progression of neuronal cell death results in the loss of the brain tissue. Neuronal apoptosis has been implicated following ischemic and hemorrhagic stroke, traumatic brain injuries and in various neurological disorders as diverse as Parkinson's disease, Alzheimer's disease, Huntington's disease, and amyotrophic lateral sclerosis and diabetic encephalopathy. A brief account on the role of neuronal apoptosis in normal development and in some of the selected brain abnormalities has been described below. The original list of neurological disorders and associated conditions, where, neuronal apoptosis plays direct or indirect role, is quite lengthy and is a subject of detailed discussion and thus not discussed in this chapter.

2.1. Neuronal Apoptosis and Normal Brain Development

Neuronal cell death in the developing brain was first identified almost a century ago. Since then its biological significance in normal brain development has been a major focus of research. During brain development, apoptosis has been widely demonstrated to play a crucial role in the formation and maintenance of the architecture of the mature brain [59]. It is assumed that about half of the neurons produced during development die before brain maturation, and nearly all neuronal classes are produced in excess during development [60]. Neuronal apoptosis is the preferred mechanism by which these oversized neuronal populations are eliminated. In such situations, the neurons are eliminated either by early death of proliferating precursors and young postmitotic neuroblast cells or by the late death of postmitotic neurons. This precisely controlled process is a highly conserved cellular mechanism in diverse organisms, ranging from invertebrate species such as the nematode, *Caenorhabditis elegans,* and insects, to nearly all of the studied vertebrate species [7]. A high number of dying neurons with characteristic apoptotic morphology has been detected in the developing postnatal rat cerebellum [61]. In addition, DNA fragmentation was detected in the postnatal mouse [62], rat [63], rabbit [64], and human cerebellum [65], and during prenatal development of human cerebral cortex [66]. In recent years, neuronal apoptosis during prenatal and early postnatal development has been extensively studied in the cerebral cortex of the mouse [67] and rat [68]. Apoptosis of cerebellar Purkinje cells has been also studied in this context. During normal cerebellar development, dying Purkinje cells were highly localized within the vermis specifically; however, the apoptosis of Purkinje cells in the cerebellar hemispheres was low with relatively minor variability. Although there are substantial data regarding cerebellar neuronal apoptosis, the relationship between apoptosis and cerebellum development is not well elucidated, in particular during neuronal proliferation and migration. A relationship with the intrinsic pathway has been known in brain development. Naturally occurring neuronal death in the rat brain is accompanied by a decreased Bcl-2 to Bax ratio. This is associated with an increase in the expression of active caspase 3 [7]. A proper balance of pro- and anti-apoptotic factors was found to be crucial for the normal development of the nervous system [69]. For instance, deficiency in pro-apoptotic proteins or overexpression of anti-apoptotic proteins leads to a variety of brain abnormalities, such as nervous system hypertrophy, aberrant structural organization in different parts of the brain, and reduced susceptibility to certain apoptotic stimuli. On the other hand, deficiency in antiapoptotic proteins leads to decreased neuronal survival or exacerbation of neuronal apoptosis [70].

2.2. Ischemic Stroke

Ischemic stroke occurs due to the abrupt occlusion of cerebral blood vessels and generally results in neuronal cell death within the affected area. Effective stroke therapies require recanalization of occluded blood vessels. However, reperfusion can cause blood brain barrier break down, cerebral edema, brain hemorrhage, and neuronal death. Activation of the 'intrinsic' mitochondrial apoptotic pathway and an excess production of mitochondrial reactive oxygen species in the reperfusion phase is a major trigger for neuronal apoptosis and also is a limiting factor in the benefits of stroke therapies [71]. Cerebral ischemia triggers both the intrinsic pathway and extrinsic pathway of apoptosis. Although many of the key apoptotic proteins have been identified, our understanding of the complex underlying mechanisms of apoptosis remains poor in stroke patients. In addition to the simplistic concept that stroke-induced-caspase-mediated apoptosis occurs predominantly in neurons, new evidence suggests that apoptosis is prevalent in non neuronal cells and that caspase-independent mechanisms also play a major role. Although the ischemic penumbra is under threat of infarction, it is potentially salvageable and thus represents an opportunity for therapeutic intervention [72]. Mitochondria are centrally involved in the development of ischemic stroke induced tissue injury. This is due to the modifications of their major role in supplying ATP and to changes in their properties that can contribute to apoptotic and necrotic cell death. In ischemic stroke, early restoration of blood flow is important and can result in substantial recovery of energy-related metabolites throughout the affected area. However, glucose oxidation is markedly decreased due both to lower energy requirements in the post-ischemic tissue and limitations on the mitochondrial oxidation of pyruvate. A secondary deterioration of mitochondrial function subsequently develops that may contribute to significant progression of the condition to neuronal death. Mitochondrial release of multiple apoptogenic proteins has been identified in ischemic and post-ischemic brain, mostly in neurons. Caspase-dependent and caspase-independent apoptosis and the mitochondrial permeability transition are important contributors to tissue damage, particularly when induced by short periods of temporary focal ischemia [73]. Formation of mitochondrial ROS during ischemic stroke-reperfusion and subsequent activation of caspase mediated neuronal apoptosis are known [73]. At the mitochondrial levels, recent studies have shown a role for cyclophilin D (CypD), a regulator of the PTP, to play major role in ischemic stroke-induced neuronal apoptosis [74]. Cyclophilin D enhances Ca(2+)-induced mitochondrial permeabilization and cell death in the brain. At interventional levels, the inhibition of pro-apoptotic Bax markedly reduces caspase activation and brain injury in immature neurons but is ineffective in adult [75]. The extrinsic pathway also has a major impact on the outcome in ischemic stroke. Fas apoptotic inhibitory molecule 2 (Faim2) is an evolutionary conserved, neuron-specific inhibitor of Fas/CD95-mediated apoptosis. Recent findings show that Faim2 is a novel neuroprotective molecule in the context of cerebral ischemia [76].

2.3. Traumatic Brain Injury

Traumatic brain injury (TBI) is a cause of high mortality and morbidity. Apoptosis of neurons and glia contribute to the overall pathology of traumatic brain injury (TBI) in both humans and animals [77]. Recent studies have shown that the inhibition of apoptosis reverses many of the deleterious effects of TBI [78]. A major role for apoptosis in delayed neuronal loss after both acute and chronic brain injury is emerging in experimental animals and in humans. This includes the identification of neuronal apoptosis after TBI, molecules that initiate the process of apoptosis, key modulators of apoptosis such as the Bcl-2 family members, key executioners of apoptosis such as the members of caspase family and potential therapeutic interventions for blocking neuronal apoptosis after TBI [78]. One of the most important and recent observations in TBI relates to the potential role of apoptosis in secondary brain injury. Apoptosis was shown in brain tissue of all traumatic contusions from the pericontusional zone (PCZ) of patients and in most of PCZ in post-mortem specimens [79]. *In vivo* samples showed higher expression of antiapoptotic proteins Bcl-2and Bcl-XL than post-mortem samples. In autopsies, the expression of Fas and Bim were higher in PCZ than in the zone distal from the contusion. *In vitro* studies showed that apoptotic rate was an independent factor associated with mortality. Cerebral apoptosis is a prominent form of cell death in the PCZ of human traumatic cerebral contusions [79].

2.4. Neurodegerative Diseases

2.4.1. Parkinson's Disease

Parkinson's disease (PD) is one of most common age-related neurodegenerative disorder. It is the second most diagnosed neurodegerative disorder, after Alzherimer's disease. The progressive degeneration of dopaminergic neurons in the substantia nigra pars compacta, resulting in typical signs of slowness of voluntary movements, resting tremors, rigidity, and gait imbalance is a characteristic of PD. The formation of Lewy bodies, intraneuronal proteinaceous cytoplasmic inclusions composed of misfolded and aggregated α-synuclein protein etc are hall mark of PD [80]. PD cases are generally of sporadic origin, and only less than 5% are familial in nature. A well-maintained balance of the processes of mitochondrial dynamics is essential for neuronal signaling, plasticity and neurotransmitter release. In PD, oxidative stress and mitochondrial dysfunction underlie the development of neuropathology [33]. The recessive Parkinsonism-linked genes PTEN-induced kinase 1 (PINK1) and Parkin maintain mitochondrial integrity by regulating diverse aspects of mitochondrial function, including membrane potential, calcium homeostasis, cristae structure, respiratory activity, and mtDNA integrity [81]. Experimental results from genetic, cellular, and toxicological studies indicate that many of the etiological factors associated with PD may be related to the autophagic process in various model systems [7, 82]. In addition, Parkin is crucial for autophagy-dependent clearance of dysfunctional mitochondria. In the absence of PINK1 or Parkin, cells often develop fragmented mitochondria. Whereas excessive fission may cause apoptosis, coordinated induction of fission and autophagy is believed to facilitate the removal of damaged mitochondria through mitophagy. A better understanding of the relationship between the specific changes in mitochondrial dynamics and cell death will be important to identify potentially neuroprotective pathways steering PINK1-deficient cells towards survival [83].

2.4.2. Alzheimer's Disease

Neuronal apoptosis is a well studied and established mechanism of neuronal loss in Alzheimer's disease (AD). The involvement of apoptosis in AD brains was suggested by findings of DNA fragmentation as detected by TUNEL staining with concomitant up regulation of pro-apoptotic Bax protein [84], c-Jun protein [85], and effector caspase 3 [86]. The mitochondrial and extrinsic apoptotic pathways were activated in AD hippocampal sections, wherein active caspase 9 was found co-localized with active caspase 8 [86]. Further more, the intracellular amyloid β-peptide1-42 has been shown to induce human neuronal cell apoptosis through Bax activation that resulted in cytochrome *c* release [87]. In addition to this, the distribution of caspase-cleaved fragments of tau suggests that the activation of caspases preceded the formation of neurofib-rillary tangles in brains of AD patients. The interaction between oxidative stress and apoptosis in AD is still debated. Considerable evidence of oxidative stress has been reported alongside neuronal apoptosis in AD. The expressions of other antioxidant enzymes such as catalase, glutathione peroxidase, and glutathione reductase were elevated in hippocampus and inferior parietal lobule, indicating a possible protective gene response to increased peroxidation in the brain regions affected by AD [88]. Adversely affecting lipid peroxidation products, like 4-hydroxynonenal (HNE), were found in higher amounts in the brain and cerebrospinal fluid of AD patients [89, 90]. Oxidized DNA bases, indicative of ROS mediated DNA damage, have been shown in certain regions of AD brains [88].

2.4.3. Huntington's Disease

Huntington's disease (HD) is a genetic and progressive neurodegenerative disorder characterized by selective neuronal loss in the brain striatum and the cerebral cortex leading to uncontrolled movements, general motor impairment, dementia, and personality changes. Neuronal loss due to apoptosis is an important mechanism in the pathophysiology of HD. Mitochondrion is a target in HD pathology. The deleterious effects of the mutant Huntington protein on lymphoblast mitochondria from HD patients have been known [91]. The mutant Huntington decreases the levels of two subunits of mitochondrial complex II, resulting in decreased complex II activity in the HD striatum [92]. In Huntington gene, the expansion of CAG repeats in exon 1 is involved in inheritable autosomal dominant disease. These repeats are translated into an expanded polyglutamine tract near the N-terminus of the Huntington protein [91]. In rat hippocampal neurons, overexpression of Huntington with CAG expansion leads to induction of JNK activation and apoptotic neuronal death. It has been suggested that the expansion of the polyglutamine tract in the Huntington protein affects its interaction with pro-apoptotic factors. Transolcoation of the pro-apoptotic factors Smac/Diablo and Htra from mitochondria into the cytosol followed by a decrease in cytosolic IAP1 and XIAP have been known in HD [93]. Decreased levels of IAP1 and XIAP have also been shown in HD postmortem tissues [93]. Mutant Huntington gene induces the elevation of Bax and Puma proteins [94] and promotes apoptosis in cultures of primary rat neurons by caspase activation [95]. Caspase inhibition effectively attenuated the progression of the disease in a transgenic mouse model of HD [96]. The role of neuronal apoptosis in the pathogenesis is further supported by the findings of activation of caspases 1, 3, 8, and 9, and cytochrome *c* release in patients who had HD [97, 98].

2.4.4. Amyotrophic Lateral Sclerosis

Amyotrophic lateral sclerosis (ALS; Lou Gehrig's disease) is a motor neuron disease characterized by selective and progressive degeneration of upper and lower motor neurons. Nearly 10% of ALS cases are familial, usually with autosomal dominant inheritance. The pathophysiology and etiology of motor neuron degeneration in ALS is not well understood. Several cellular mechanisms have been postulated to be involved in the pathology of ALS. Prominent among them are oxidative stress, glutamate excitotoxicity, mitochondrial damage, defective axonal transport, glia cell pathology, aberrant RNA metabolism and apoptosis [7, 99]. Mitochondria are an immediate target in ALS. Superoxide dismutase 1 (SOD1) encoded by SOD1 gene plays an important role in the pathophysiology of ALS. Mutant SOD1 was found to be preferentially associated with mitochondria leading to mitochondrial dysfunction associated with ALS. Morphological and functional defects in mitochondria were found in mouse models of ALS (mutant SOD1 over expressing) and in ALS patients. Recent studies suggest that axonal transport of mitochondria along microtubules and mitochondrial dynamics may also be disrupted pointing to the significance of neuronal apoptosis in ALS [99].

2.4.5. Diabetic Encephalopathy

Diabetes mellitus is associated with the occurrence of well characterized microvascular complications, including retinopathy, nephropathy and peripheral neuropathy. Diabetic encephalopathy is a recently recognized complication associated with type 1 diabetes. In diabetic encephalopathy, a slow and progressive alteration in brain function and structure occurs along with diabetes [100]. Though less well studied, cognitive performance in diabetic encephalopathy has recently gained attention and remains an important area to be investigated [101, 102]. In diabetic patients and diabetic animal models, alterations in cognition, neuropsychology and neurobehavior have been known with elevated neuronal apoptosis primarily at the level of the brain hippocampus [103]. Impairment of cognitive function in type 1 diabetic individuals is accompanied by changes in neuronal structure and function [104, 105], and is characterized by mild to moderate slowing of mental speed and a diminished mental flexibility, without evident impairment of learning and memory [106, 107]. Experimental studies have suggested that neuronal apoptosis may play a crucial role in apoptosis, neuronal loss and impaired cognitive function associated with the disease particularly in diabetic animal models [103]. In animal models, apoptosis was associated with decreased neuronal densities, and learning and cognitive deficits [98, 103]. The patients suffering from neurodegenerative disorders such as AD, PD, and HD generally exhibit a higher prevalence of diabetes [108], indicating common pathways for neuronal loss in these diseases.

Conclusion

At the fundamental level, the main focus of the recent research on neuronal apoptosis has been in the area of understanding the functional mechanisms of the mitochondrial mediated 'intrinsic' and the death receptor mediated 'extrinsic' apoptotic signaling pathways. The activation of the intrinsic pathway is associated with the release of several pro-apoptotic proteins into the cytosol that are normally located in the mitochondrial intermembrane space.

The mitochondrial membrane permeabilization and release of such factors involves members of the Bcl-2 family of proteins that are either pro- apoptotic or anti-apoptotic proteins. The extrinsic pathway consists of the death receptors belonging to the super family of tumor necrosis factor (TNF)/nerve growth factor receptor and the binding of a ligand to the death receptor initiates this pathway of apoptosis. The cellular and biochemical features of apoptosis are all a consequence of the cleavage of a subset of proteins by proteases of the caspase family. Mitochondrial ROS plays an important role in neuronal apoptosis and they directly or indirectly mediate cytochrome *c* release from mitochondria that triggers caspase activation. In neurons, the imbalance of mitochondrial fission/fusion can influence neuronal physiology, such as synaptic transmission and neuronal plasticity, and affect neuronal survival. In the brain, neuronal apoptosis is a normal physiological process that is an essential and integral part of neurogenesis and brain development. Neuronal apoptosis when becomes aberrant leads to various neurological disorders. Sustained progression of neuronal cell death results in functional deficits following stroke, traumatic brain injury and a variety of neurodegenerative diseases.

References

[1] Chowdhury I, Tharakan B, Bhat GK. Current concepts in apoptosis: the physiological suicide program revisited. *Cell Mol. Biol. Lett,* 11:506-25, 2006.

[2] Yang JC, Cortopassi GA. Induction of the mitochondrial permeability transition causes release of the apoptogenic factor Cytochrome *c*. *Free Radic. Biol. Med,* 24:624–631, 1998.

[3] Saelens X, Festjens N, Vande Walle L, van Gurp M, van Loo G, Vandenabeele P. Toxic proteins released from mitochondria in cell death. *Oncogene,* 23:2861-74, 2004.

[4] Jiang X, Wang X. Cytochrome *c* promotes caspase-9 activation by inducing nucleotide binding to Apaf-1. *J. Biol. Chem,* 275:31199–31203, 2000.

[5] Chowdhury I, Tharakan B, Bhat GK. Caspases - an update. *Comp. Biochem. Physiol. B Biochem. Mol. Biol,* 151:10-27, 2008.

[6] Krantic S, MechawarN, Reix S, Quirion R. Apoptosis-inducing factor: A matter of neuron life and death. *Progress in Neurobiology,* 81: 179–196, 2007.

[7] Okouchi M, Ekshyyan O, Maracine M, Aw TY. Neuronal apoptosis in neurodegeneration. *Antioxidant and Redox Signaling,* 9: 1059-1096, 2007

[8] Robertson JD, Orrenius S, Zhivotovsky B. Review: nuclear events in apoptosis. *J. Struct. Biol,* 129: 346–358, 2000.

[9] Deveraux QL, Takahashi R, Salvesen GS, Reed GC. X-linked IAP is a direct inhibitor of cell-death proteases. *Nature,* 388: 300–304, 1997.

[10] Suzuki Y, Imai Y, Nakayama H, Takahashi K, Takio K, Takahashi R. A serine protease, Htra2, is release from the mitochondria and interacts with XIAP, inducing cell death. *Mol. Cell,* 8: 613–621, 2001.

[11] Van Gurp M, Festjens N, van Loo G, Saelens X, Vandenabeele P. Mitochondrial intermembrane proteins in cell death. *Biochem. Biophys. Res. Commun,* 304: 487–497, 2003.

[12] 12. Antonsson B, Conti F, Ciavatta A, Montessuit S, Lewis S, Martinou I, Bernasconi N, Bernard A, Mermod JJ, Mazzei G, Maundrell K, Gambale F, Sadoul R, Martinou JC. Inhibition of Bax channel-forming activity by Bcl-2. *Science,* 277: 370–372, 1997.

[13] Keinan N, Tyomkin D, Shoshan-Barmatz V. Oligomerization of the mitochondrial protein voltage-dependent anion channel is coupled to the induction of apoptosis. *Mol. Cell Biol,* 30:5698-709, 2010.

[14] Degterev A, Boyce M, Yuan J. A decade of caspases. *Oncogene,* 22: 8543–8567, 2003.

[15] Nagata S. Apoptosis by death factor. *Cell,* 88: 355–365, 1997.

[16] Ashkenazi A. Targeting death and decoy receptors of the tumornecrosis factor superfamily. *Natl. Rev. Cancer,* 2: 420–430, 2002.

[17] Sheridan JP, Marsters SA, Pitti RM, Gurney A, Skubatch M, Baldwin D. Control of TRAIL-induced apoptosis by a family of signaling and decoy receptors. *Science,* 277: 818–821, 1997.

[18] Peter ME, Krammer PH. The CD95 (APO-1/Fas) DISC and beyond. *Cell Death Differ,* 10: 26–35, 2003.

[19] Li S, Zhao H, He X, Kim TH, Kuharsky DK, Rabinowich H, Chen J, Du C, Yin XM. Relief of extrinsic pathway inhibition by the Bid-dependent mitochondrial release of Smac in Fas-mediated hepatocyt apoptosis. *J. Biol. Chem,* 277: 26912–26920, 2002.

[20] Raoul C, Henderson CE, Pettmann B. Programmed Cell death of embryonic motoneurons triggered through the Fas death receptor. *J. Cell Biol,* 29: 1049–1061, 1999.

[21] Zuliani C, Kleber S, Klussmann S, Wenger T, Kenzelmann M, Schreglmall N, Martinez A, Del Rio JA, Soriano E, Vodrazka P, Kuner R, Groene Hj, Herr I, Krammer PH, Martin-Villarba A. Control of neuronal branching by the death receptor CD95 (Fas/APO-1). *Cell Death Differ,* 13: 31–40, 2006.

[22] Li Y, Ji A, Weihe E, Schafer M. Cell-specific expression and lipoplysaccharide-induced regulation of tumor necrosis factor á (TNF-á) and TNF receptors in rat dorsal root ganglion. *J. Neurosci,* 24: 9623–9631, 2004.

[23] Veglianese P, Lo Coco D, Bao Cutrona M, Magnoni R, Pennachini D, Pozzi B, Gowing G, Julien JP, Tortarolo M, Bendotti C. Activation of the p38MAPK cascade is associated with upregulation of TNF alpha receptors in the spinal motor neurons of mouse models of familial ALS. *Mol. Cell Neurosci,* 31: 218–231, 2006.

[24] Dopp JM, Mackenzie–Graham A, Otero GC, Merrill JE. Differential expression, cytokine modulation, and specific functions of type-1 and type-2 of tumor necrosis factor receptors in rat glia. *J. Immunol,* 75: 104–112, 1997.

[25] Figiel I, Dzwonek K. TNF_ and TNF receptor 1 expression in the mixed neuronal-glial culturesof hippocampal dentate gyrus exposed to glutamate or trimethyltin. *Brain Res,* 1131: 17–28, 2007.

[26] Dorr J, Bechmann I, Waiczies S, Aktas O, Walczak H, Krammer PH, Nitsch R, Zipp F. Lack of tumor necrosis factor-related apoptosis inducing ligand but presence of its receptors in the the human brain. *J. Neurosci,* 22: 1–5, 2002.

[27] Spierings DC, De Vries EG, Vellenga E, Van den Heuvel F, Koornstra JJ, Wesseling J, Hollema H, de Jong S. Tissue distribution of the death ligand TRAIL and its receptors. *J. Histoche Cytochem,* 52: 821–831, 2004.

[28] Burton TR, Dibrov A, Kashour T, Amara FM. Anti-apoptotic wild-type Alzheimer amyloid precursor protein signaling involves the p38 mitogen-activated protein kinase/MEF2 pathway. *Brain Res. Mol. Brain Res,* 108: 102–120, 2002.

[29] Kim SK, Park HJ, Hong HS, Baik EJ, Jung MW, Mook-Jung I. ERK1/2 is an endogenous negative regulator of the gamma-secretase activity. *FASEB J,* 20: 157–159, 2006.

[30] Klivenyi P, St Clair D, Wermer M, Yen HC, Oberley T, Yang L, Flint Beal M. Manganese superoxide dismutase. *Neurobiol. Dis*, 5: 253-258, 1998.

[31] Danial NN, Korsmeyer SJ. Cell death: critical control points. *Cell,* 116: 205–219, 2004.

[32] Culmsee, C., Zhu, C., Landshamer, S., Becattini, B.,Wagner, E., Pellecchia, M., Blomgren, K., Plesnila, N. Apoptosis-inducing factor triggered by poly(ADP-ribose) polymerase and Bid mediates neuronal cell death after oxygen-glucose deprivation and focal cerebral ischemia. *J. Neurosci*, 25, 10262–10272, 2005.

[33] Bulat N, Widmann C. Caspase substrates and neurodegenerative diseases. *Brain Research Bulletin,* 80: 251–267, 2009.

[34] Oberst A, Bender C, Green DR. Living with death: the evolution of the mitochondrial pathway of apoptosis in animals. *Cell Death Differ*, 15: 1139–1146, 2008.

[35] Nicholson DW, Thornberry, NA. Caspases: killer proteases. *Trends Biochem. Sci,* 22:299–306, 1997.

[36] Shi Y. Mechanisms of caspase activation and inhibition during apoptosis, *Mol Cell,* 9:459-470, 2002.

[37] Siegel RM. Caspases at the crossroads of immune-cell life and death. *Nat. Rev. Immunol*, 6: 308-317, 308–317, 2006.

[38] Yi CH, Yuan J. The Jekyll and Hyde functions of caspases. *Dev. Cell,* 16: 21-34, 2009.

[39] Li ZG, Zhang W, Sima AA. C-peptide enhances insulin-mediated cell growth and protection against high glucose-induced apoptosis in SH-SY5Y cells. *Diabetes Metab. Res. Rev,* 19: 375–385, 2003.

[40] Abraham MC, Shaham S. Death without caspases, caspases without death. *Trends Cell Biol,* 14, 184-193, 2004.

[41] Algeciras-Schimnich A, Barnhart BC, Peter ME. Apoptosis-independent functions of killer caspases. *Curr. Opin. Cell Biol,* 14: 721-726, 2002.

[42] Launay S, Hermine O, Fontenay M, Kroemer G, Solary E, Garrido C. Vital functions for lethal caspases. *Oncogene,* 24: 5137–5148, 2005.

[43] Yang J.-Y, Michod D, Walicki J, Widmann C. Surviving the kiss of death. *Biochem. Pharmacol*, 68: 1027-1031, 2004.

[44] Yi CH, Yuan J. The Jekyll and Hyde functions of caspases. *Dev. Cell,* 16: 21-34, 2009.

[45] Huesmann GR, Clayton DF. Dynamic role of postsynaptic caspase-3 and finch BIRC4 in zebra song-response habituation. *Neuron,* 52:1067-1072, 2006.

[46] Li Z, Jo J, Jia JM, Lo SC, Whitcomb DJ, Jiao S, Cho K, Sheng M. Caspase-3 activation via mitochondria is required for long-term depression and AMPA receptor internalization. *Cell,* 28: 141: 859-871, 2010.

[47] Orrenius S. Reactive oxygen species in mitochondria-mediated cell death. *Drug Metab. Rev*, 39:443-55, 2007.

[48] Constantini P, Belzacq AS, Vieira H, larochette N, de Pablo M, Zamzami N, Susin SA, Brenner C, Kroemer G. Oxidation of a critical thiol residue of the adenine nucleotide

translocator enforces Bcl-2-independent permeability transition pore opening and apoptosis. *Oncogene,* 19: 307–314, 2000.
[49] Ozawa T. Oxidative damage and fragmentation of mitochondrial DNA in cellular apoptosis. *Biosci. Rep,* 17: 237–250, 1997.
[50] Ballinger SW, Patterson C, Yan CN, Doan R, Burow DL, Young CG, Yakes FM, Van Houten B, Ballinger CA, Freeman BA, Runge MS. Hydrogen peroxide- and peroxynitrite-induced mitochondrial. DNA damage and dysfunction in vascular endothelial and smooth muscle cells. *Circ. Res,* 86: 960–966, 2000.
[51] Goldsmit Y, Erlich S, Pinkas–Kramarski R. Neuregulin induces sustained reactive oxygen species generation to mediate neuronal differentiation. *Cell Mol. Neurobiol,* 21: 753–769, 2001.
[52] Golstein P, Kroemer G. Cell death by necrosis: towards a molecular definition. *Trends Biochem. Sci.* 32:37-43, 2007.
[53] McCall K. Genetic control of necrosis - another type of programmed cell death. *Curr. Opin. Cell Biol. 22*:882-8, 2010.
[54] Golstein, P., Aubry, L., Levraud, J.P. Cell-death alternative model organisms: why and which? *Nat. Rev. Mol. Cell Biol,* 4: 798–807, 2003.
[55] Berman SB, Chen YB, Qi B, McCaffery JM, Rucker EB, Goebbels S, Nave KA, Arnold BA, Jonas EA, Pineda FJ, Hardwick JM. Bcl-x L increases mitochondrial fission, fusion, and biomass in neurons. *J. Cell Biol,* 184:707-719, 2009.
[56] Cho DH, Nakamura T, Lipton SA. Mitochondrial dynamics in cell death and neurodegeneration. *Cell Mol. Life Sci,* 67:3435-47, 2010.
[57] Chang CR, Blackstone C. Dynamic regulation of mitochondrial fission through modification of the dynamin-related protein Drp1. *Ann. N. Y. Acad. Sci,* 21201:34-9, 2010.
[58] Lu B. Mitochondrial dynamics and neurodegeneration. *Current Neurol. Rep*, 9:212-219, 2009.
[59] Miller FD, Kaplan DR. Neurotrophin signalling pathways regulating neuronal apoptosis. *Cell Mol. Life Sci*, 58:1045-53, 2001.
[60] Lossi L, Merighi A. In vivo cellular and molecular mechanisms of neuronal apoptosis in the mammalian CNS. *Prog. Neurobiol*, 69:287-312, 2003.
[61] Jankowski J, Miething A, Schilling K, Baader SL. Physiological Purkinje cell death is spatiotemporally organized in the developing mouse cerebellum. *Cerebellum,* 8:277-290, 2009.
[62] Wood KA, Dipasquale B, Youle RJ. In situ labeling of granule cells for apoptosis-associated DNA fragmentation reveals different mechanisms of cell loss in developing cerebellum. *Neuron,* 11: 621–632, 1993.
[63] Krueger BK, Burne JF, Raff MC. Evidence for large-scale astrocyte death in the developing cerebellum. *J. Neurosci.* 15: 3366–3374, 1995.
[64] Lossi L, Gambino G, Mioletti S, Merighi A. *In vivo* analysis reveals different apoptotic pathways in pre- and postmigratory cerebellar granule cells of rabbit. *J. Neurobiol,* 60: 437–452, 2004.
[65] Lossi L, Zagzag D, Greco MA, Merighi A. Apoptosis of undifferentiated progenitors and granule cell precursors in the postnatal human cerebellar cortex correlates with expression of BCL- 2, ICE, and CPP32 proteins. *J. Comp. Neurol,* 399: 359–372, 1998.

[66] Voiculescu B, Nat R, Lin E, Iosef C. Apoptosis in human embryo development: 1. Cerebral cortex. *J. Cell Mol. Med,* 4: 284–288, 2000.

[67] Blaschke AJ, Staley K, Chun J. Widespread programmed cell death in proliferative and postmitotic regions of the fetal cerebral cortex. *Development*, 122:1165-74, 1996.

[68] Thomaidou D, Mione MC, Cavanagh JF, Parnavelas JG. Apoptosis and its relation to the cell cycle in the developing cerebral cortex. *J. Neurosci*, 17:1075-85, 1997.

[69] Yoshida H, Kong YY, Yoshida R, Elia AJ, Hakem A, Hakem R, Penninger JM, Mak TW. Apaf1 is required for mitochondrial pathways of apoptosis and brain development. *Cell,* 94: 739–750, 1998

[70] Lossi L, Merighi, A. *In vivo* cellular and molecular mechanisms of neuronal apoptosis in the mammalian CNS. *Prog. Neurobiol,* 69: 287–312, 2003.

[71] Jung JE, Kim GS, Chen H, Maier CM, Narasimhan P, Song YS, Niizuma K, Katsu M, Okami N, Yoshioka H, Sakata H, Goeders CE, Chan PH. Reperfusion and neurovascular dysfunction in stroke: from basic mechanisms to potential strategies for neuroprotection. *Mol. Neurobiol,* 41:172-179, 2010.

[72] Broughton BR, Reutens DC, Sobey CG. Apoptotic mechanisms after cerebral ischemia. *Stroke*, 40:e331-9, 2009.

[73] Sims NR, Muyderman H. Mitochondria, oxidative metabolism and cell death in stroke. *Biochim. Biophys. Acta*, 1802: 80-91, 2010.

[74] Wang X, Carlsson Y, Basso E, Zhu C, Rousset CI, Rasola A, Johansson BR, Blomgren K, Mallard C, Bernardi P, Forte MA, Hagberg H. Developmental shift of cyclophilin D contribution to hypoxic-ischemic brain injury. *J. Neurosci*, 29:2588-96, 2009.

[75] Wang X, Han W, Du X, Zhu C, Carlsson Y, Mallard C, Jacotot E, Hagberg H. Neuroprotective effect of Bax-inhibiting peptide on neonatal brain injury. *Stroke,* 41:2050-2055, 2010.

[76] Reich A, Spering C, Gertz K, Harms C, Gerhardt E, Kronenberg G, Nave KA, Schwab M, Tauber SC, Drinkut A, Harms K, Beier CP, Voigt A, Göbbels S, Endres M, Schulz JB. Fas/CD95 regulatory protein Faim2 is neuroprotective after transient brain ischemia. *J. Neurosci*, 31:225-33, 2011.

[77] Wong J, Hoe NW, Zhiwei F, Ng I. Apoptosis and traumatic brain injury. *Neurocrit. Care,* 3:177-82, 2005.

[78] Zhang X, Chen Y, Jenkins LW, Kochanek PM, Clark RS. Bench-to-bedside review: Apoptosis/programmed cell death triggered by traumatic brain injury. *Crit. Care*, 9:66-75, 2005.

[79] Miñambres E, Ballesteros MA, Mayorga M, Marin MJ, Muñoz P, Figols J, López-Hoyos M. Cerebral apoptosis n severe traumatic brain injury patients: an in vitro, in vivo, and postmortem study. *J. Neurotrauma*, 25:581-591, 2008.

[80] Spillantini MG and Goedert M. The alpha-synucleinopathies: Parkinson's disease, dementia with Lewy bodies, and multiple system atrophy. *Ann. N. Y. Acad. Sci.* 920: 16–27, 2000.

[81] Gasser T. Genetics of Parkinson's disease. *Curr. Opin. Neurol,* 18: 363–369, 2005.

[82] Banerjee R, Beal MF, Thomas B. Autophagy in neurodegenerative disorders: pathogenic roles and therapeutic implications. *Trends Neurosci,* 33:541-9, 2010.

[83] Büeler H. Mitochondrial dynamics, cell death and the pathogenesis of Parkinson's disease. *Apoptosis,* 15:1336-53, 2010

[84] Su JH, Deng G, Cotman CW. Bax protein expression is increased in Alzheimer's brain: correlations with DNA damage, Bcl-2 expression, and brain pathology. *J. Neuropathol. Exp. Neurol,* 56: 86–93, 1997.

[85] Anderson AJ, Pike CJ, Cotman CW. Differential induction of immediate early gene proteins in cultured neurons by beta-amyloid (A beta): association of c-Jun with A beta-induced apoptosis. *J. Neurochem,* 65: 1487–1498, 1995.

[86] Rohn TT, Rissman RA, Davis MC, Kim YE, Cotman CW, Head E. Caspase-9 activation and caspase cleavage of tau in the Alzheimer's disease brain. *Neurobiol. Dis,* 11: 341–354, 2002.

[87] Zhang Y, McLaughlin R, Goodyer C, LeBlanc A. Selective cytotoxicity of intracellular amyloid peptide1-42 through p53 and Bax in cultured primary human neurons. *J. Cell Biol,* 156: 519–529, 2002.

[88] Lyras L, Cairns NJ, Jenner A, Jenner P, Halliwell B. An assessment of oxidative damage to proteins, lipids, and DNA in brain from patients with Alzheimer's disease. *J. Neurochem,* 68: 2061–2069, 1997.

[89] Lovell MA, Xie C, Markesbery WR. Decreased glutathione transferase activity in brain and ventricular fluid in Alzheimer's disease. *Neurology,* 51: 1562–1566, 1998.

[90] Marcus DL, Thomas C, Rodriguez C, Simberkoff K, Tsai JS, Strafaci JA, Freedman ML. Increased peroxidation and reduced antioxidant enzyme activity in Alzheimer's disease. *Exp. Neurol,* 150: 40–44, 2003.

[91] The Huntington's Disease Collaborative Research Group. A novel gene containing a trinucleotide repeat that is expanded and unstable on Huntington's disease chromosomes. *Cell,* 72: 971–983, 1993

[92] Benchoua A, Trioulier Y, Zala D, Gaillard MC, Lefort N, Dufour N, Saudou F, Elalouf JM, Hirsch E, Hantraye P, Deglon N, Brouillet E. Involvement of mitochondrial complex II defects inneuronal death produced by N-terminus fragment of mutated huntingtin. *Mol. Biol. Cell,* 17: 1652–1663, 2006.

[93] Goffredo D, Rigamonti D, Zuccato C, Tartari M, Valenza M, Cattaneo E. Prevention of cytosolic IAPs degradation: a potential pharmacological target in Huntington's Disease. *Pharmacol. Res,* 52: 140–150, 2005.

[94] Bae BI, Xu H, Igarashi S, Fujimuro M, Agrawal N, Taya Y, Hayward SD, Moran TH, Montell C, Ross CA, Snyder SH, Sawa A. p53 mediates cellular dysfunction and behavioral abnormalities in Huntington's disease. *Neuron,* 47: 29–41, 2005.

[95] Rigamonti D, Sipione S, Goffredo D, Zuccato C, Fossale E, Cattaneo E. Huntingtin's neuroprotective activity occurs via inhibition of procaspase-9 processing. *J. Biol. Chem,* 276: 14545–14548, 2001.

[96] Ona VO, Li M, Vonsattel JP, Andrews LJ, Khan SQ, Chung WM, Frey AS, Menon AS, Li XJ, Stieg PE, Yuan J, Penney JB, Young AB, Cha JH, Friedlander RM. Inhibition of caspase-1 slows disease progression in a mouse model of Huntington's disease. *Nature,* 399: 263–267, 1999.

[97] Kiechle T, Dedeoglu A, Kubilus J, Kowall NW, Beal MF, Friedlander RM, Hersch SM, Ferrante RJ. Cytochrome C and caspase- 9 expression in Huntington's disease. *Neuromolec. Med,* 1: 183–195, 2002.

[98] Sanchez I, Xu CJ, Juo P, Kakizaka A, Blenis J, Yuan J. Caspase-8 is required for cell death induced by expanded polyglutamine repeats. *Neuron,* 22: 623–633, 1999.

[99] Shi P, Gal J, l Kwinter DM, Liu X, Zhu H. Mitochondrial dysfunction in amyotrophic laterasclerosis. *Biochim. Biophys. Acta,* 1802:45-51, 2010.

[100] Biessels GJ, Kamal A, Ramakers GM, Urban IJ, Spruijt BM, Erkelens DW, Gispen WH. Place learning and hippocampal synaptic plasticity in streptozotocin-induced diabetic rats. *Diabetes,* 45: 1259–1266, 1996.

[101] Biessels GJ, Gispen WH. The impact of diabetes on cognition: what can be learned from rodent models? *Neurobiol. Aging,* 26 Suppl 1:36–41, 2005.

[102] Li ZG, Sima AA. C-peptide and central nervous system complications in diabetes. *Exp. Diabesity Res,* 5: 79–90, 2004.

[103] Li ZG, Zhang W, Sima AA. The role of impaired insulin/IGF action in primary diabetic encephalopathy. *Brain Res,* 1037: 12–24, 2005.

[104] Makimattila S, Malmberg–Ceder K, Hakkinen AM, Vuori K, Salonen O, Summanen P, Yki–Jarvinen H, Kaste M, Heikkinen S, Lundbom N, Roine RO. Brain metabolic alterations in patients with type 1 diabetes-hyperglycemia-induced injury. *J. Cereb. Blood Flow Metab,* 4:1393–1399, 2004.

[105] Wessels AM, Simsek S, Remijnse PL, Veltman DJ, Biessels GJ, Barkhof F, Scheltens P, Snoek FJ, Heine RJ, Rombouts SA. Voxel-based morphometry demonstrates reduced grey matter density on brain MRI in patients with diabetic retinopathy. *Diabetologia,* 49: 2474–2480, 2006.

[106] Brands AM, Biessels GJ, de Haan EH, Kappelle LJ, Kessels RP. The effects of type 1 diabetes on cognitive performance: a metaanalysis. *Diabetes Care,* 28:726–735, 2005

[107] Ryan CM. Neurobehavioral complications of type I diabetes. Examination of possible risk factors. *Diabetes Care,* 11: 86–93, 1988.

[108] Ristow M. Neurodegenerative disorders associated with diabetes mellitus. *J. Mol. Med,* 82: 510–529, 2004.

In: Neuronal Cell Apoptosis
Editors: C. J. Schmid and J. L. Wolfe

ISBN: 978-1-61324-633-7

Chapter VI

How Can We Prevent Neuronal Apoptosis?

*G. H. Doherty**
School of Biology, University of St. Andrews,
United Kingdom

Abstract

The cells that make up the nervous system die at two distinct times during the lifetime of an organism. These periods of neuron loss occur during embryonic development and with advancing age, occurring at an accelerated rate in neurodegenerative disorders. The major risk factor for many neurodegenerative diseases is ageing. During development, neuronal apoptosis is essential for the correct shaping and hard-wiring of the nervous system. However, in a number of neurological conditions, apoptosis causes a devastating, and currently irreversible, loss of neurons. This review considers the extracellular and intracellular signals that determine which neurons live, and which neurons die, during development. It then asks the question of whether our understanding of apoptosis during neurodevelopment can lead to novel treatments to protect the nervous system from neuron loss due to pathology.

Abbreviations

Aβ –	amyloid beta
BDNF –	brain derived neurotrophic factor
BFCN –	basal forebrain cholinergic neurons
MPP+ -	1-Methyl-4-phenylpyridinium
MPTP -	1-methyl-4-phenyl-1,2,3,6-tetrahydropyridine

* Tel. +44 1334 463611, e-mail: ghm@st-andrews. ac.uk.

NBM - nucleus basalis of Meynert
NGF – nerve growth factor
NT-3 – neurotrophin -3
NT-4/5 – neurotrophin -4/5

Introduction

The body has two major control systems. The endocrine system controls activities that require duration rather than speed, such as growth or metabolic regulation. In contrast, the nervous system allows us to rapidly respond to situations in our environment in a precise manner, whether escaping a predator or reaching forward to turn the pages of this book. Structurally the nervous system consists of collections of cells gathered together in the brain, spinal cord and in numerous ganglia; with processes extending out from them allowing communications with targets, both within, and outwith, the nervous system.

To attain this structure, and the requisite precise connections between neurons and their targets, entails numerous cellular processes during development. Cells proliferate, differentiate, innervate and consolidate the connections that they make. In the midst of all this, vast numbers of neurons die. It is estimated that at least 50% of the neurons that are generated during the ontogeny of the nervous system die before development is complete (Oppenheim, 1991).

After the mature conformation of the nervous system is established, neuronal death is rarely considered until an organism is aged. Advanced age is by the far the greatest risk factor for a number of devastating neurodegenerative conditions including Alzheimer's Disease and other dementias, and Parkinson's Disease. The large scale degeneration of neurons within specific regions of the brain are a hallmark of these disorders, and with the loss of these neurons comes the loss of function that makes these conditions so distressing for both sufferers and their carers.

Taken together, the observation that neuronal death is a feature of both normal neuronal development, and of neurodegenerative disorders, raises some intriguing questions. Can a better understanding of cell death during neurodevelopment enhance our understanding of what goes wrong in neurodegenerative diseases? And more importantly, can we learn any lessons from neurodevelopment that would allow us to prevent cell death in conditions such as Alzheimer's and Parkinson's, and thus go some way to alleviating the symptoms of these disorders?

Apoptosis and Cell Death

The subject of this review, and indeed of this book, is apoptosis: a particular form of cell death. The term apoptosis is derived from Greek meaning literally falling off, as of leaves. However, it must be noted that neurons can die by a number of different pathways and mechanisms, of which apoptosis is just one.

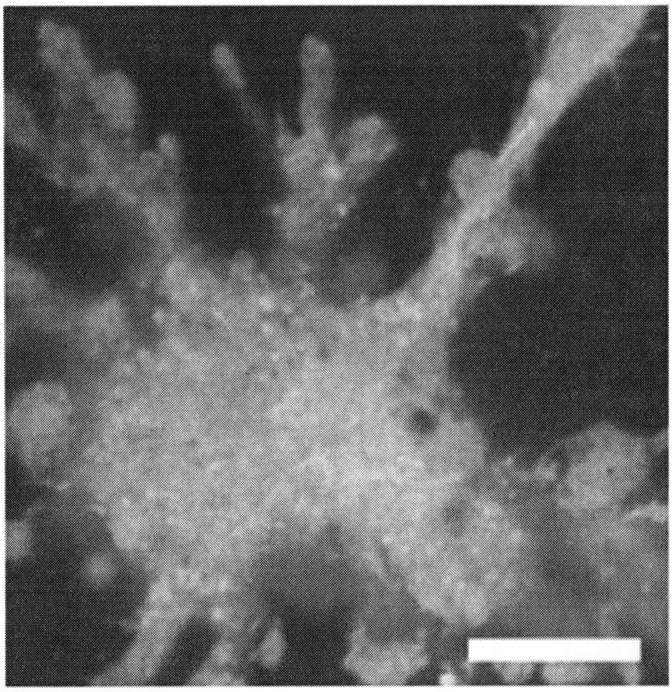

Figure 1. A fluorojade B- labelled apoptotic murine cortical neuron *in vitro*. Clear evidence of blebs forming on the soma can be seen. Scale bar represent 5μm. Fluorojade B dye specifically labels dying neurons.

Apoptotic cell death triggers a characteristic cellular morphology (Figure 1). Early on in the apoptotic cascade nuclear DNA condenses around the perimeter of the nucleus. As the process continues, bulges, known as blebs, appear in the cellular membrane, the nuclear DNA condenses and the nucleus is known as pyknotic. Finally the cellular contents are packaged up into membrane-bound apoptotic bodies that are engulfed by other cells and digested via a lysosomal pathway (Kerr et al., 1972).

The morphological changes that are observed microscopically when cells undergo apoptosis are underpinned by a well-defined molecular pathway (Figure 2). A number of triggers can initiate an apoptotic pathway: both extracellular stimuli and cues from the cell's own internal environment. Activation of death receptors via ligand binding can trigger apoptosis. In this situation, caspases 8 and 10 are activated and these, in turn, cleave, and thus activate, a number of other caspases that target cellular products to dismantle the cell (Alenzi et al., 2010). Thus caspases represent the machinery that breaks down the cell during apoptosis. Such death receptors include TNF- R1, Fas and p75 (Bunone et al., 1997; Ashkenazi and Dixit, 1998; Raoul et al., 1999). In many other situations, the initial control of the entry into apoptosis is at the behest of the members of the Bcl-2 protein family, which can be subdivided into pro-survival and pro-death family members dependent upon molecular and biological function.

Pro-survival members of this family, such as Bcl-2 and Bcl-x_L, share four homologous domains that allow interaction between family members. Three of these domains: BH1, BH2 and BH4 define a hydrophobic groove within the molecule. The fourth domain, termed BH3, is a short region that can bind within that groove (Kuwana et al., 2002).

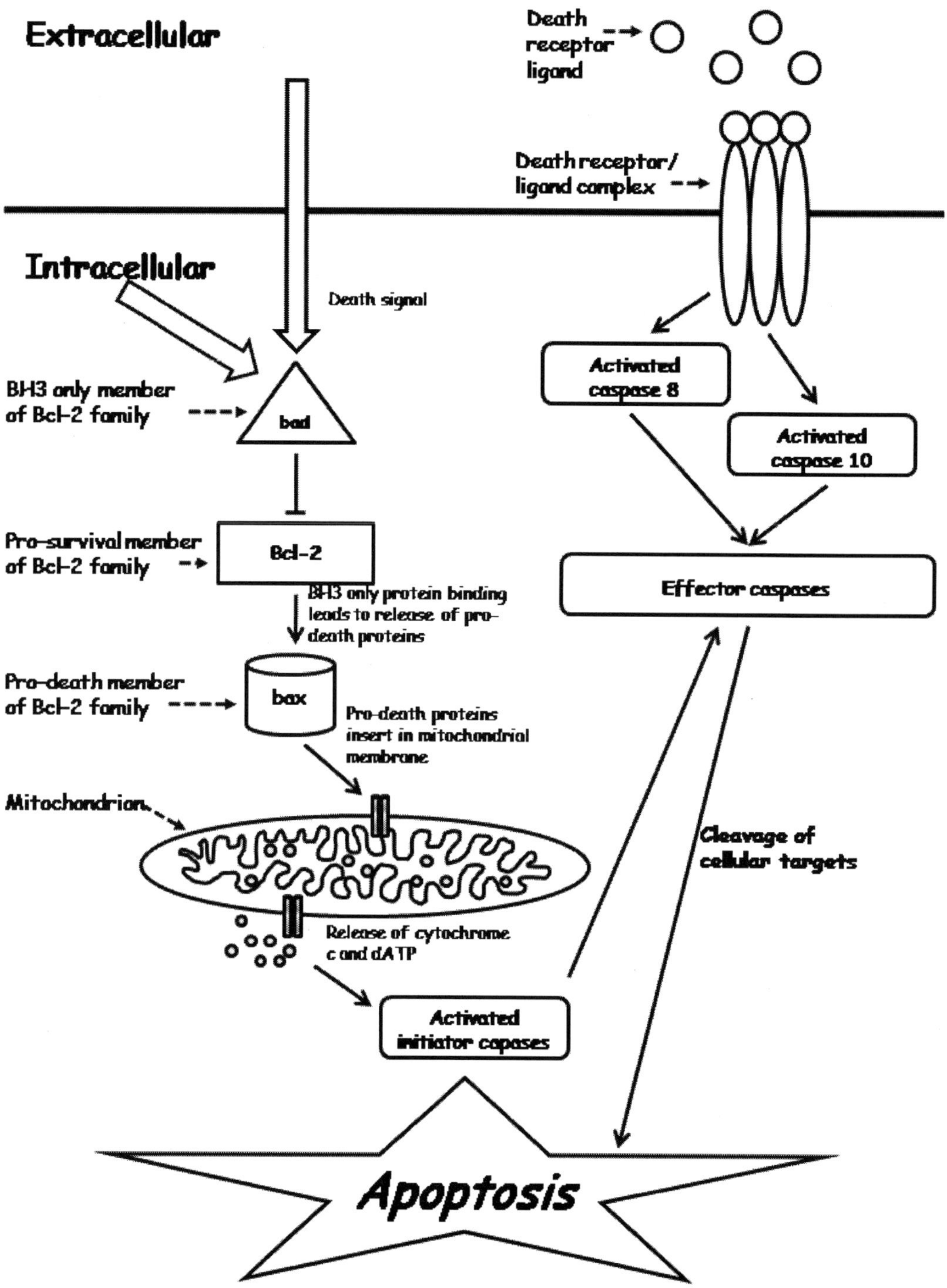

Figure 2. Schematic diagram outlining the events in apoptosis. Representative members of each sub-group of the Bcl-2 family are shown (BH3 only, pro-survival and pro-death). It is well established that other proteins within each category can carry out the same function.

The BH regions allow interaction with the pro-death family members, bax and bak, that possess BH1, BH2 and BH3 but not BH4. In addition to interacting with pro-survival proteins, bax and bak can also form homo-oligomers that insert into the mitochondrial membrane. These form a pore allowing the release of molecules from within the inner mitochondrial environment including cytochrome c and dATP (Liu et al., 1996). The remaining members of the Bcl-2 protein family possess only the BH3 domain, and these proteins (such as bid, bad and bim) all promote apoptosis (Willis et al., 2007).

It is now well established that the relative ratios of the members of this protein family determine whether or not a cell undergoes apoptosis (Middleton et al., 2000). Thus, when pro-survival family members such as Bcl-2 and Bcl-x_L are in excess, they can bind bax and bak preventing mitochondrial translocation. However, when the levels of bax and/ or bak outweigh the levels of the pro-survival proteins, they can homo-oligomerise and insert into the mitochondrial membrane. Since the BH3 domain of the BH3-only family members can bind to the groove created by BH1, BH2 and BH4 found within pro-survival Bcl-2 proteins, these molecules can alter the ratio of pro-survival to pro-death family members. In so doing they promote the formation of pores in the mitochondrial membrane. This is achieved by binding to the pro-survival proteins and thus preventing their binding to bax or bak, leaving these proteins free to form mitochondrial membrane pores.

As mentioned above, mitochondrial pore formation by pro-death members of the Bcl-2 protein family leads to the release of mitochondrial molecules including cytochrome c and dATP (Figure 2). These two molecules activate caspase 9 that exists bound to the scaffold protein apaf-1 in the cytoplasm of the apoptotic cell. caspase 9 can then activate a number of downstream caspases that cleave a variety of proteins that bring about the characteristic morphology of an apoptotic cell. Notably, caspase cleavage of the rho-kinase, ROCK1, is responsible for the occurrence of membrane blebbing (Coleman et al., 2001). Contribution to the nuclear changes observed in apoptosis is made via cleavage of DFF45 by caspase 3, which has been demonstrated to translocate to the nucleus during apoptosis (Janicke et al., 1998a, 1998b; Nakagawara et al., 1997; Zhivotovsky et al., 1999).

As mentioned earlier, it is important to remember that whilst this review focuses on apoptosis, cell death can occur by other mechanisms, and these pathways may be important during neurodevelopment and neurodegeneration. For instance, autophagic neurons have been identified in Alzheimer's Disease, Parkinson's Disease and Huntington's Disease (Cuervo, 2004; Nixon, 2006; Chu, 2006). Autophagy, meaning self-eating, is a tightly regulated form of cell death whereby autophagic vacuoles engulf portions of cytoplasm, which are then degraded. As well as leading to cell death, autophagy may be an important mechanism for recycling cellular constituents during times of stress, thus cells showing morphological features of autophagy may not actually die. Necrosis has been observed in a number of pathological conditions in the nervous system (Behl et al., 1994; Stadelmann et al., 1998; Jellinger et al., 2000). This is the primary mechanism of cell death for instance *in vitro* when cerebellar granule cells are exposed to homocysteine (Foister et al., 2005) and occurs in situations where the apoptotic pathway is blocked (Kruman et al., 2000).

Neurodevelopmental Cell Death

The first event in organogenesis is the formation of the neural tube that is the precursor of the central nervous system, and that also contributes to certain structures of the peripheral nervous system. Inductive signals from the developing notochord and mesoderm cause the thickening of an area of ectoderm giving rise to the neurepithelium. During the development of the nervous system this rolls up to form the neural tube (Figure 3). The neural tube then undergoes a process of expansion, bending and shaping to form the brain and spinal cord; whilst neural crest cells detach from the crests of the elevating neural folds and migrate to contribute to a host of body structures including the sympathetic and dorsal root ganglia, Schwann cells, dermis in the face and neck and 'C' cells of the thyroid gland.

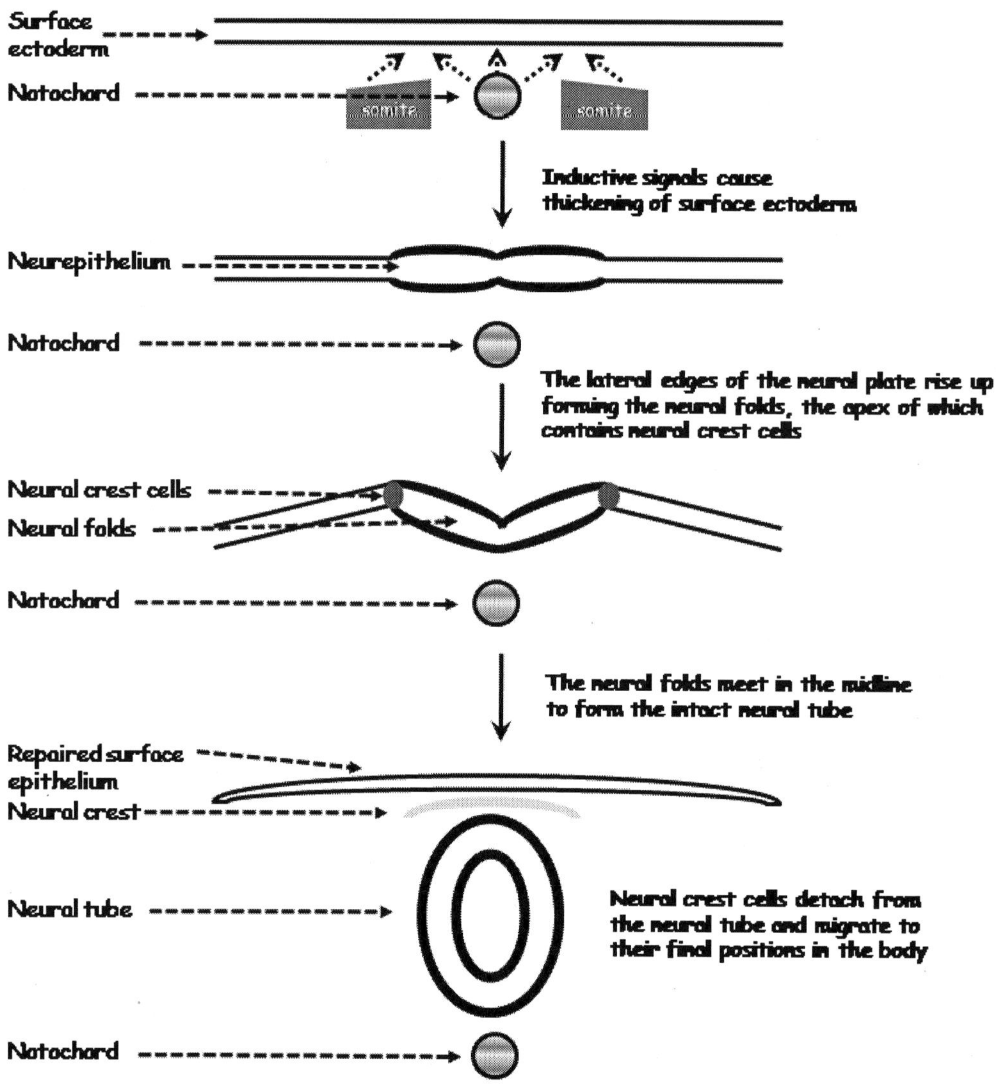

Figure 3. Schematic diagram of neurulation. Inductive signals from the underlying notochord and developing somites cause thickening of the overlying ectoderm to form the neurepithelium. This rolls up to form the neural tube that sinks below the surface of the embryo. Neural crest cells bud off from the apex of the neural fold and migrate along well-defined pathways to establish a variety of body structures.

During the very earliest stages of neurulation, apoptosis has been observed in the neurepithelium (Glucksmann, 1951), and it has been postulated that this loss of developing neurons is necessary to allow correct closure of the neural tube (Weil et al., 1997). In support of this, treatment of developing chick embryos with caspase inhibitors that block apoptosis prevents neural tube closure (Weil et al., 1997). However, evidence from both knock-out mice lacking key components of the apoptotic cascade, and from murine embryos treated with caspase inhibitors, reveals that apoptosis is not required for neural tube closure in mammalian embryos (Massa et al., 2009).

Thus it remains unclear why apoptosis occurs at such an early stage of neurodevelopment. It may serve to help control cell number by establishing equilibrium between apoptosis and mitosis, or this process may eliminate cells that have developed abnormally. One final possibility is that apoptosis helps in cellular selection to generate diversity and thus assist morphogenesis (Voyvodic, 1996; Gilmore et al., 2000; Middleton, 2004).

A vast number of studies investigating apoptosis during the development of the nervous system focus on the neuronal loss that occurs during target field innervation (Figure 4). At this time, neurons are generated in excess, and as they differentiate and mature, they extend processes towards their target field for innervation.

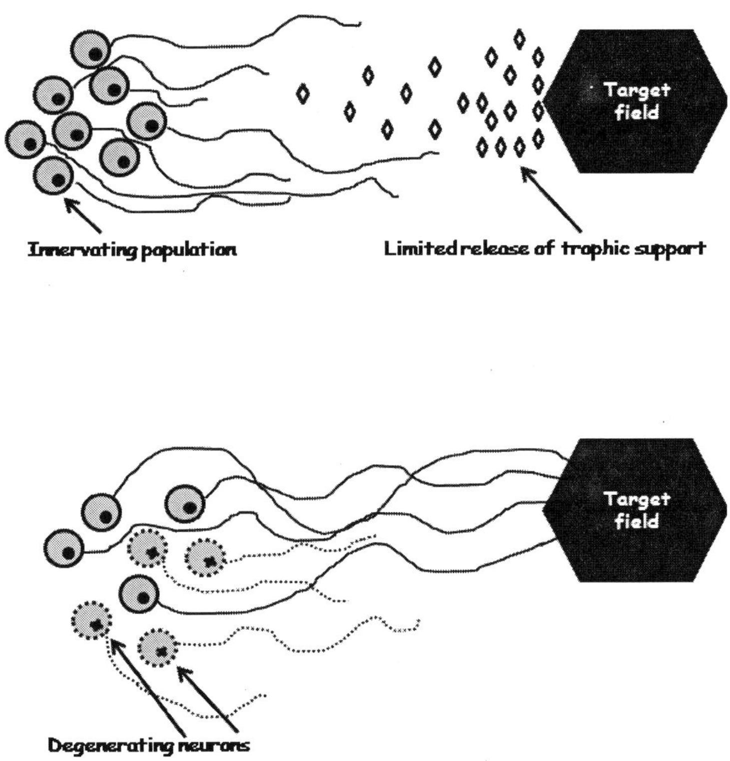

Figure 4. Death of neurons during target field innervation. Neuronal precursors proliferate, migrate and differentiate before sending out processes towards their target field. This, in turn, releases limiting supplies of neurotrophic factors that guide and support the viability of the developing neurons. Whilst neurons that procure adequate neurotrophin supplies live, those that do not undergo apoptosis.

This concept is termed the neurotrophic hypothesis (Middleton, 2004). This system presents the developing organism with great flexibility. It permits size matching between neuronal target fields and the neurons that innervate them. Thus, if an evolutionary change alters the size of the target field, making it smaller, it will release less trophins and therefore a smaller number of innervating neurons will survive. Because of this, the neural population will be of the appropriate size for the target field. Conversely, an increase in target field size, with a corresponding increase in neurotrophin release, will result in the rescue of more of the innervating population (Middleton, 2004).

The first experiments in the neurotrophic hypothesis involved the removal of the peripheral target field, i.e. the limb bud, prior to innervation of the structure by the neurons of the dorsal root ganglia. In this experiment, removal of the limb bud markedly enhanced the magnitude of cell death in the innervating dorsal root ganglia. Conversely, increasing the size of the target field by grafting in an exogenous limb bud decreased the amount of cell death in the innervating dorsal root ganglia (Hamburger and Levi-Montalcini, 1949). These experiments demonstrated that a target-derived factor regulates the amount of cell death in the dorsal root ganglia.

The nature of the signal released by the target field varies between neuronal populations. The trophic factor responsible for the observations made by Hamburger and Levi-Montalcini was identified as nerve growth factor (NGF) (Cohen, 1960). NGF is the founding member of a family of structurally related neurotrophins including brain derived neurotrophic factor (BDNF) (Barde et al., 1982), neurotrophin-3 (NT-3) (Hohn et al., 1990, Maisonpierre et al., 1990; Rosenthal et al., 1990), and neurotrophin 4/5 (NT-4/5) (Berkemeier et al., 1991; Hallbook et al., 1991; Ip et al., 1992). The neurotrophins bind to specific receptor tyrosine kinases of the Trk family and also to p75, a receptor that will be discussed in more detail later. In addition to the NGF family of neurotrophins, a number of other molecules have been identified that support neuronal survival during target field innervation including members of the glia derived neurotrophic factor (GDNF) family (Tomac et al., 1995), hepatocyte growth factor (HGF) (Barres et al., 1993), and the neurotrophic cytokines (Barbin et al., 1984).

The neuronal death that is observed during target field innervation has proven to be largely apoptotic. Characteristic pyknotic neuclei are observed in the ganglia (Middleton et al., 1998), and activated caspase 3, one of the initiator caspases can be detected immunohistochemically. Downstream signalling events following trophic factors binding to their receptors have been well elucidated and important roles for proteins that suppress apoptotic signalling have been identified. Thus pro-survival signalling by members of the Bcl-2 protein family including Bcl-2 (Pinon et al., 1997), Bcl-x_L and Bcl-w (Middleton et al., 2001) plays an important role in preventing apoptosis in neurons that survive the period of naturally occurring cell death. In contrast to these findings, mice that have null mutations in key caspases do not show decreased loss of neurons during target field innervation or beyond implying that other pathways to cell death can be activated if apoptosis is blocked after the point where cytochrome c is released from the mitochondria (Oppenheim et al., 2001).

Beyond the time of target field innervation, apoptosis can elimate cells whose processes that have formed incorrectly (O'Leary et al., 1986; Catsicas et al., 1987). The refinement of connections at this point can occur due to interruptions in electrical activity or axonal transport in presynaptic neurons leading to postsynaptic neuronal loss (Middleton, 2004). Differential rates of neuronal loss also account for sexual dimorphism in certain nuclei such as subcomponents of the medial preoptic nucleus, which play an important role in reproductive behaviour (Arai et al., 1996) and here steroid hormones play a critical role in regulating neuronal viability (Nordeen et al., 1985; Davis et al., 1995).

Much of this brief review of neurodevelopmental cell death has focused on the idea that the lack of an external signal, whether neurotrophin or activity input, induces the neuron to undergo cell death by apoptosis. Thus these external factors prevent apoptosis. However, there are also signals that have been identified that actively trigger neurons to undergo apoptosis during neurodevelopment. As such, TNF-α has been demonstrated to induce cell death during the development of the peripheral nervous system, contributing to the loss of NGF-dependent sensory and sympathetic neurons following NGF deprivation (Barker et al., 2001). In developing dopaminergic ventral mesencephalic neurons (Doherty, 2007), TNF-α is also able to induce neuronal death. Interestingly, TNF-α can also act in a neuroprotective manner during the ontogeny of the nervous system with its effects dependent upon cell type and circumstance (Doherty, 2007; Oldreive and Doherty, 2010). Further examples of active cell killing during neurodevelopment have been reported for the Fas receptor (Raoul et al., 1999) and the NGF receptor, p75 (Bunone et al., 1997).

Apoptosis in Neurodegeneration

With advancing age, comes a decline in neuronal numbers, a process that is accelerated in particular regions of the nervous system during neurodegenerative conditions. Thus in Parkinson's Disease there is a marked loss of dopaminergic neurons in several brain regions including the substantia nigra pars compacta (Figure 5), whilst in Alzheimer's Disease large scale neuronal death is observed in the hippocampus and cortex. As might be expected there have been reports of multiple forms of cell death in such pathological conditions. However, an important role for apoptosis has been identified and it is this that will be discussed here.

Apoptotic neurons have been identified in the substantia nigra of *post mortem* samples taken from Parkinson's patients at a higher frequency than in age-matched controls (Tompkins et al., 1997; Mizuno et al., 1998). Accordingly, ablation of the pro-apoptotic member of the Bcl-2 protein family bax protects mice from dopaminergic neuron loss in an *in vivo* model of Parkinson's Disease (Vila et al., 2001). Furthermore, paraquat is known to induce dopaminergic neuron death and this effect is dependent upon the pro-apoptotic Bcl-2 family member bak (Fei et al., 2008), giving more weight to the hypothesis that Bcl-2 proteins are important in dopaminergic neuronal loss. Further evidence for a role for apoptosis in Parkinson's Disease comes from the observation that *post mortem* samples taken from patients exhibit higher levels of activated caspase 3 than do samples from age-matched controls (Hartmann et al., 2000).

There is some evidence that dopaminergic neurons may be particularly vulnerable to caspase activity. Parkin protein is important for the survival of the neurons that degenerate in Parkinson's disease and studies have shown that this protein can be cleaved by caspases. This suggests that activation of caspases can impair the function of parkin removing a pathway designed to protect against apoptosis (Kahns et al., 2002).

Neuronal death in neurodegenerative disorders is a complex issue. It would be fair to say that neurons will not simply die without a fight.

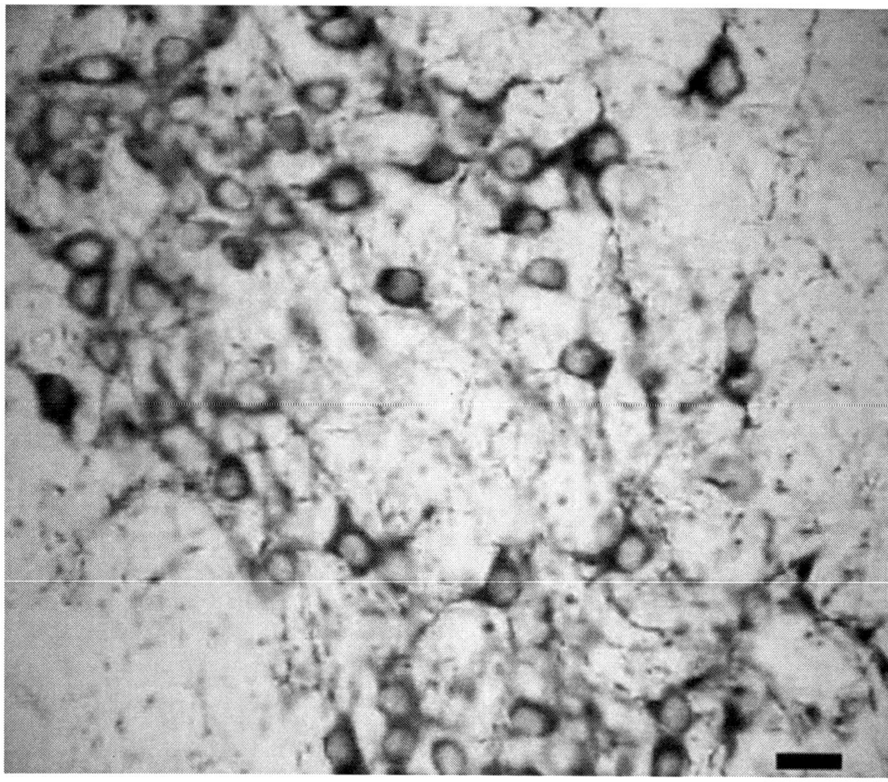

Figure 5. The dopaminergic neurons of the substantia nigra pars compacta. Image of the murine midbrain immunohistochemically stained with an antibody to tyrosine hydroxylase (the rate limiting enzyme in dopamine biosynthesis) to identify dopaminergic neurons. Scale bar represents 50μm.

Thus in Alzheimer's Disease the pro-apoptotic protein bax is upregulated (Su et al., 1997) together with other pro-apoptotic Bcl-2 family members bak and bad (Kitamura et al., 1998); and yet, the expression of the anti-apoptotic protein Bcl-2 increases with disease severity (Satou et al., 1995)

Taken together these data suggest that neurons are attempting to prevent cell death induced by pro-apoptotic Bcl-2 family proteins through upregulation of a pro-survival protein. In support of this, it has been demonstrated that in Alzheimer's Disease patients the upregulation of Bcl-2 expression is associated with neurons that have DNA damage (Su et al., 1996). Similarly, Bcl-x_L, another pro-survival member of the Bcl-2 protein family, is upregulated *in vitro* in neurons exposed to sublethal levels of toxins linked to Alzheimer's Disease, again suggesting that the cells are activating a protective mechanism (Luetjens et al., 2001).

The amyloid β (Aβ) peptide has been widely implicated in the pathogenesis of Alzheimer's Disease (Glenner and Wong, 1984; Bahmanyar et al., 1987; Delabar et al., 1987; Higgins et al., 1998). It has been established that this peptide downregulates Bcl-2 and upregulates bax *in vitro* suggesting that neurons are being pushed towards an apoptotic pathway (Paradis et al., 1996). In addition, Aβ can activate caspases such as caspase 8 (Ivins et al., 1999) and caspase 2 (Troy et al., 2000). Again we see evidence that an apoptotic pathway could, at least in part, contribute to the degenerative changes seen in Alzheimer's Disease.

Discussion

From the evidence presented above it is clear that apoptosis is a hallmark of both neuronal loss during development and in the degenerative disorders associated with ageing. We must now consider whether prevention of apoptosis could be of benefit to patients that are afflicted by these disorders.

It has been estimated that up to 80% of the neurons of the substantia nigra have already been lost at the time of diagnosis of Parkinson's Disease (Bennet, 1988). Therefore much current research, looking for a cure, is focused on replacing these neurons that have already died. Whilst this, of course, is the ultimate aim, such therapies may be some time before they are of use in a clinical setting. An alternative view would be that although the symptoms of Parkinson's are debilitating and depressing at the time of diagnosis, they become far more so as the disease progresses and more neurons, together with their associated networks and pathways, degenerate. Thus, even in symptomatic patients, a viable therapeutic target is the prevention of further cell death, even in the absence of strategies to replace the already-lost neurons. This has the potential to halt disease progression. To this ends, targeting apoptosis could present a real treatment option.

So, how could we prevent further neuronal loss in Parkinson's patients? One option would be to target the machinery that disassembles the cell during apoptosis – the caspases. Cultured rat mesencephalic neurons are protected from the toxin MPP+, that specifically targets dopamienrgic neurons and is thus used to model Parkinson's Disease, by inhibition of caspases (Bilsland et al., 2002). Similarly, in a rat model of Parkinson's Disease *in vivo*, where dopaminergic neuronal death was induced by administration of the dopaminergic neurotoxin 6-hydroxydopamine, the pan-caspase inhibitor, z-VAD-fmk, was able to prevent apotosis in response to this stimulus (Cutillas et al., 1999). This implies that caspase inhibitors can be neuroprotective for midbrain dopaminergic neurons. However, in this experiment, the caspase-inhibiting drug was administered directly into the substantia nigra at the time of insult. This raises obvious problems for human treatment – administration directly into the brain is a risky clinical procedure. The development of systemically active and available pan-caspase inhibitors such as Q-VD-OPH may go some way to assist in drug delivery. This compound, administered intraperitoneally, offers protection from a range of neurotoxins in mice models of neurodegeneration, including MPTP-induced Parkinsonian degeneration (Yang et al., 2004). Furthermore, administration of a caspase inhibitor is potentially dependent upon the temporal point along the apoptotic pathway at which the dopeminergic neurons find themselves. For instance a role for caspase 8 has been identified in Parkinson's Disease but blockade of this protein pushes the neurons towards cell death by a necrotic pathway (Hartmann et al., 2001). Thus in an attempt to block apoptosis by caspase inhibition we may simply push the neurons into an alternate mechanism of cell death.

A major concern with using systemically available and active caspase inhibitors is the lack of tissue specificity with regards to their action. Apoptosis is not simply a hallmark of degenerating neurons, apotosis is required for normal tissue homeostasis. The anti-apoptotic protein, Bcl-2, was identified at the breakpoint of the translocation between chromosomes 14 and 18 found in a high proportion of human follicular B-cell lymphomas (Tsujimoto and Croce, 1986). This translocation moves Bcl-2 from its usual position at 18q21 into juxtaposition with the immunoglobulin heavy chain locus at 14q32 resulting in constitutive expression of Bcl-2. Thus Bcl-2 induces tumorogenesis by suppressing signals that would normally lead to apoptosis (Nunez et al., 1990; Otani et al., 1993). It is therefore clear that systemic blockade of capases could produce severe side effects and as such if caspase inhibitors are to be developed as potential treatments for neurodegeneration, ways of getting these compounds safely and specifically into the brain must be found.

Looking back at ontogeny, adequate supplies of neurotrophins are required to promote the survival of developing neurons. The midbrain dopaminergic neurons of the substantia nigra that are lost in Parkinson's Disease are no exception to the theory, and are supported during their embryonic development by growth factors including BDNF (Hyman et al., 1991) and GDNF (Lin et al., 1993). This raises the question of whether neurotrophins can protect these neurons under pathological circumstances and if so, whether they represent viable treatment options for Parkinson's Disease.

It has been determined that BDNF can protect adult rat dopaminergic neurons of the substantia nigra from 6-hydroxydopamine toxicity (Levivier et al., 1995). Similarly, BDNF administration protects these neurons from MPTP toxicity in Japanese monkeys (Tsukahara et al., 1995). Both of these toxins specifically target dopaminergic neurons and thus are useful models for neuronal loss in Parkinson's Disease. Interestingly, Parkinson's Disease patients have reduced expression of BDNF in the substantia nigra (Parain et al., 1999; Howells et al., 2000) suggesting that depletion of neurotrophin levels is part of the pathology of the disease, although whether or not this contributes to neuronal apoptosis remains to be determined. Interestingly, it has also been demonstrated that a single nucleotide polymorphism in the promoter region of the BDNF gene is linked to familial but not sporadic Parkinson's Disease (Parsian et al., 2004) hinting at a role for this protein in the etiology of certain familial forms of the disorder. What is clear is that the remaining neurons of the substantia nigra of Parkinson's patients retain normal levels of trkB mRNA (Benisty et al., 1998) thus expression of the receptor required to respond to BDNF is maintained in these neurons. These data suggest that BDNF *per se* may have a beneficial effect on Parkinson's patients as it is able to protect midbrain doapminergic neurons and we know that these cells possess the machinery to respond to BDNF in Parkinson's patients. In addition, upregulation of BDNF has been observed in response to a number of molecules that are beneficial to the neurons afflicted in this disorder further suggesting that BDNF may have a central neuroprotective role for these neurons. Thus cystamine, an agent that has proven protective in models of a number of neurodegenerative conditions, can protect against MPTP-induced neuron loss in a murine model of Parkinson's Disease and as one of its effects upregulates the expression of BDNF (Gibrat et al., 2010). Similarly, the anti-Parkinsonian drug selegiline, a monoamine oxidase B inhibitor, upregulates BDNF expression in the mouse brain (Gyarfas et al., 2010). It is of interest to note that acupuncture is frequently used as an alternative or complementary therapy by Parkinson's Disease patients. In a study on mice using 6-hydroxydopamine to model neuronal loss in Parkinson's Disease, acupuncture markedly reduced apoptosis with a concurrent increase in trkB expression (Park et al., 2003). Long-term high-frequency electro-acupuncture upregulates BDNF mRNA in the substantia nigra in a rat model of Parkinson's (Liang et al., 2002). Thus anti-apoptotic signalling via BDNF/ trkB may underpin the beneficial effects reported for acupuncture in these patients.

GDNF has anti-apoptotic effects on midbrain dopaminergic neurons during ontogeny and has also been tested as a potential therapy for Parkinson's. Thus infusion of GDNF can aid in the recovery of motor function after a partial lesioning of the rat striatum that destroys dopeminergic midbrain neurons (Kirik et al., 2001). In these experiments GDNF was applied 2 weeks after lesioning and inhibited further degeneration of nigral neurons that was observed in vehicle controls, implying that its anti-apoptotic functions may be important in this result. Either single or continuous administration of GDNF four weeks post-6-hydroxydopamine lesioning in rats significantly enhances the recovery of both behavioural and biochemical effects associated with this insult (Aoi et al., 2000). The exciting point about this research data is that GDNF administration some time after injury was demonstrating beneficial effects on the remaining neurons and their physiology. This suggests that GDNF has potential for intervention in patients that have already lost a significant number of neurons. This is in contrast to much of the research data on anti-apoptotic molecules that are applied at the time of lesioning and thus may not be useful once the disorder has taken hold. Initial studies in patients, however, did not reveal any motor benefits of intraventricular injection of GDNF and the treatment induced a number of unwanted side effects including nausea and vomiting, anorexia and psychotic episodes (Nutt et al., 2003; Kordower et al., 1999). Further studies delivering via an intraputaminal route rather than intraventricularly showed greater success with lesser side effects (Gill et al., 2003). However, around 10% of patients in a clinical trial of GDNF therapy developed anti-GDNF antibodies, although the bioactivity and toxicity of these antibodies remain to be determined (Lang et al., 2006) and larger-scaled randomised control trials have not reinforced the beneficial effects of GDNF (Lang et al., 2006). Thus at this time, GDNF has not developed into an effective therapy for patients with Parkinson's Disease although it may do in the future.

One further area in Parkinson's therapy where lessons from development may aid in the discovery of therapies for this disorder, is in the field of fetal nigral cell transplantation, where midbrain tissue from aborted fetuses is transplanted into the depleted substantia nigra of Parkinson's patients. However, to date this therapy has proven disappointing in treated subjects at least in part due to difficulties in supporting the grafted tissue in the host environment to prevent apoptosis (Barker et al., 1996). To this end, inhibition of caspases may prove useful and also practical, as strategies to prevent caspase activity can be employed at the time of transplantation. Caspase inhibition with the broad spectrum inhibitor Ac-YVAD-cmk has been shown to protect grafted dopaminergic neurons in experiments on rats (Schierle et al., 1999), yet specific targeting of caspases 1or 3 was not neuroprotective (Hurelbrink et al., 2001) perhaps suggesting that other caspases are active in this situation and that pan-caspase inhibition would be necessary. Other researchers have investigated the possibility of using growth factors either alone or in combination with caspase inhibitors to support grafted tissue. Thus, a combination of GDNF, together with the caspase inhibitor BOC-ASP-CH_2F protected grafted dopaminergic neurons from apoptosis when transplanted into a rat host (Helt et al., 2001). What may be important when designing strategies to support graft tissue is to consider donor tissue age and thus developmental stage. Thus BDNF exhibits optimal ability to protect grafted tissue when BDNF administration was timed to coincide with the point where BDNF would appear in the striatum during the *in vivo* development of the graft (Yurek et al., 1998). In addition, the response of dopaminergic neurons to exogenous factors alters as they mature and thus TNF-α is a trophic factor for early embryonic dopaminergic neurons but is neurotoxic at later developmental stages (Doherty, 2007).

It was noted above that steroid hormones can influence neuronal survival during neurodevelopment. Given that aged men have a greater incidence of Parkinson's Disease than aged women (Mayeux et al., 1995; van den Eeden et al., 2003), this raises the possibility that steroid hormones may have important pro- or anti-apoptotic functions in dopaminergic neurons. Interestingly, androgens induce apoptosis in dopaminergic cell lines (Cunningham et al., 2009), whilst estrogens can protect against apoptosis in dopaminergic neurons (Sawada et al., 2000). These findings imply that males endure the double problem of lacking the protective effects of estrogen whilst being exposed to the pro-apoptotic effects of androgens and this may account for the gender bias observed in Parkinson's Disease patients. Thus estrogen or related steroid compounds may prove beneficial to some Parkinson's patients.

Research into anti-apoptotic agents in Alzheimer's Disease has been somewhat less exhaustive than for Parkinson's Disease. One of the reasons for this is that it is much easier to recreate a Parkinson's phenotype in laboratory animals than it is to model Alzheimer's Disease. And yet whilst none of the Parkinson's models discussed above accurately exhibit all the features of the disease observed in patients, they are easily established in the laboratory and can lend some clues as to the disease process.

The population of neurons lost in Alzheimer's Disease that seems to be most critical for the observed decline cognitive function is that of basal forebrain cholinergic neurons (BFCNs) of the nucleus basalis of Meynert (NBM) (Perry et al., 1978a and 1978b). During embryonic development, a critical neurotrophin for the survival of these neurons is NGF (Levi-Montalcini and Hamburger, 1951). Thus it is of interest that in Alzheimer's Disease patients there is reduced NGF immunoreactivity in BFCNs in comparison with age-matched controls (Mufson et al., 1995). In adult rats, administration of NGF is protective of BFCNs targeted via injury or that exhibit a decline in function due to age (Hefti, 1986; Williams et al., 1986). As a result of empirical data suggesting that NGF had beneficial effects on the BFCNs lost in Alzheimer's Disease, administration via direct cerebral infustion was tested in patients. Unfortunately, no obvious clinical improvements were observed prior to termination of the study due to adverse side effects such as pain and weight loss (Olson et al., 1992; Eriksdotter Jonhagen et al., 1998).

In addition to NGF, a potential role for BDNF in Alzheimer's Disease has been investigated. It is well known that BDNF regulates synaptic plasticity therefore playing a central role in memory formation (Hellweg and Jockers-Scherubl, 1994). Both the mRNA and protein expression of BDNF is decreased in the hippocampus and cortex of the brains of Alzheimer's Disease patients (Murer et al., 2001; Siegel and Chauhan, 2000). Serum BDNF levels are elevated in Alzheimer's patients with poor scores in verbal and visual memory although there was no correlation with disease severity (O'Bryant et al., 2011), further implicating this neurotrophin in Alzheimer's Disease. Other studies have linked elevated circulating levels of BDNF to a slower rate of cognitive decline in Alzheimer's patients (Laske et al., 2010). Thus neurotrophins or memtic drugs based on their pro-survival and cognitive enhancing properties may be on benefit to Alzhiemer's patients.

Quite a large volume of work has accumulating considering the role of steroid hormones in Alzheimer's Disease. It has been demonstrated that hormone replacement therapies (HRT) reduces the occurrence of Alzheimer's Disease in women by up to a third (Miller et al., 2001; Brann et al., 2007). Estrogen is known to boost the expression of synaptic proteins and to increase synaptic plasticity underpinning its reported beneficial effects on learning and memory (Stone et al., 1998; Li et al., 2004). In addition, estrogen has been shown to protect against neuronal death induced by Aβ and oxidative stress, both of which have been implicated in Alzheimer's Disease (Tang et al., 1996; Xu et al., 1998; Li et al., 2000). Thus it is possible that estrogen and related steroids may present a possible stream for therapeutic development to counteract Alzheimer's Disease.

Conclusion

It is clear that there are many parallels to be drawn between neurodevelopment and the neurodegenerative diseases that mainly afflict the elderly. Most notably, neurons die in large numbers. In development it is well established that this cell death is largely apoptotic. In neurodegeneration it is less clear which mechanism of cell death is employed. Whilst there is a large volume of evidence to support the occurrence of apoptosis, it is also obvious that other cell death mechanisms come into play. The relative contribution of each mechanism is difficult to ascertain due to cross-talk between mechanisms and also due to the fact that the *post mortem* tissue that is analysed gives a momentary snap shot of what is occurring in the degenerative process. Unlike developmental cell death which is a rapid process that occurs in defined periods, over days or hours, neurodegenerative cell death occurs over years and indeed decades.

Nonetheless it is clear that an understanding of apoptosis during neurodevelopment has opened up some interesting leads in the search for treatments for Alzheimer's and Parkinson's Diseases. Potential roles for caspase inhibitors, neurotrophins and steroid hormones have all been identified and the realisation of the potential importance of these molecules lies in our understanding of neurodevelopment. And yet neurodevelopment is not fully understood, and over the coming years and decades more mechanisms controlling cell survival and cell death will be discovered. It is likely that these too will proffer potential leads in the search for pharmacological reagents to aid in the fight against neurodegeneration.

Acknowledgment

Dr Doherty holds an Alzheimer's Society personal research fellowship (grant number 93) with support from the Henry Smith Charity. Alzheimer's Society is a charity (registration no. 296645) and a company registered in England and Wales (registration no. 2115499). The University of St Andrews is a charity registered in Scotland (registration no. SC013532).

References

Alenzi FQ, Lotfy M, Wyse R., 2010, Swords of cell death: caspase activation and regulation. *Asian Pac. J. Cancer Prev.*, 11(2):271-80.

Aoi M, Date I, Tomita S, Ohmoto T., 2000, Single or continuous injection of glial cell line-derived neurotrophic factor in the striatum induces recovery of the nigrostriatal dopaminergic system. *Neurol. Res,* 22(8):832-6.

Arai Y, Sekine Y, Murakami S., 1996, Estrogen and apoptosis in the developing sexually dimorphic preoptic area in female rats. *Neurosci. Res.*, 25(4):403-7.

Ashkenazi A, Dixit VM., 1998, Death receptors: signaling and modulation. *Science.*, 281(5381):1305-8.

Bahmanyar S, Higgins GA, Goldgaber D, Lewis DA, Morrison JH, Wilson MC, Shankar SK, Gajdusek DC., 1987, Localization of amyloid beta protein messenger RNA in brains from patients with Alzheimer's disease. *Science.*, 237(4810):77-80.

Barbin G, Manthorpe M, Varon S., 1984, Purification of the chick eye ciliary neuronotrophic factor. *J. Neurochem.*, 43(5):1468-78.

Barde YA, Edgar D, Thoenen H., 1982, Purification of a new neurotrophic factor from mammalian brain. *EMBO J.*, 1(5):549-53.

Barker RA, Dunnett SB, Faissner A, Fawcett JW., 1996, The time course of loss of dopaminergic neurons and the gliotic reaction surrounding grafts of embryonic mesencephalon to the striatum. *Exp. Neurol.*, 141(1):79-93

Barker V, Middleton G, Davey F, Davies AM., 2001, TNFα contributes to the death of NGF-dependent neurons during development. *Nature Neuroscience4*(12), 1194-1198

Barres BA, Schmid R, Sendnter M, Raff MC., 1993, Multiple extracellular signals are required for long-term oligodendrocyte survival. *Development.*, 118(1):283-95.

Behl C, Davis JB, Klier FG, Schubert D., 1994, Amyloid beta peptide induces necrosis rather than apoptosis. *Brain Res.*, 645(1-2):253-64.

Benisty S, Boissiere F, Faucheux B, Agid Y, Hirsch EC., 1998, trkB messenger RNA expression in normal human brain and in the substantia nigra of parkinsonian patients: an in situ hybridization study. *Neuroscience.*, 86(3):813-26.

Bennet JP., 1988, Biochemical pathology and pharmacology of Parkinson's disease. In: Stern MB, Hurtig HI, eds. The comprehensive management of Parkinson's disease. New York: PMA Publishing, 63-76.

Berkemeier LR, Winslow JW, Kaplan DR, Nikolics K, Goeddel DV, Rosenthal A., 1991, Neurotrophin-5: a novel neurotrophic factor that activates trk and trkB. *Neuron.*, 7(5):857-66.

Bilsland J, Roy S, Xanthoudakis S, Nicholson DW, Han Y, Grimm E, Hefti F, Harper SJ., 2002, Caspase inhibitors attenuate 1-methyl-4-phenylpyridinium toxicity in primary cultures of mesencephalic dopaminergic neurons. *J. Neurosci.*, 22(7):2637-49.

Brann DW, Dhandapani K, Wakade C, Mahesh VB, Khan MM., 2007, Neurotrophic and neuroprotective actions of estrogen: basic mechanisms and clinical implications. *Steroids.*, 72(5):381-405.

Bunone G, Mariotti A, Compagni A, Morandi E, Della Valle G., 1997, Induction of apoptosis by p75 neurotrophin receptor in human neuroblastoma cells. *Oncogene.* 1997, 14(12):1463-70.

Catsicas S, Thanos S, Clarke PG., 1987, Major role for neuronal death during brain development: refinement of topographical connections. *Proc. Natl. Acad. Sci. USA*, 84(22):8165-8.

Chu CT., 2006, Autophagic stress in neuronal injury and disease., *J. Neuropathol. Exp. Neurol.*, 65(5):423-32.

Cohen S, 1960, Purification of nerve growth factor-promoting protein from the mouse salivary gland and its neurocytotoxic antiserum. *Proc. Natl. Acad. Sci. USA.*,46: 302-311.

Coleman ML, Sahai EA, Yeo M, Bosch M, Dewar A, Olson MF., 2001, Membrane blebbing during apoptosis results from caspase-mediated activation of ROCK I. *Nat. Cell Biol.*, 3(4):339-45.

Cuervo AM., 2004, Autophagy: many paths to the same end. *Mol. Cell Biochem.* 2004, 263(1-2):55-72.

Cunningham RL, Giuffrida A, Roberts JL., 2009, Androgens induce dopaminergic neurotoxicity via caspase-3-dependent activation of protein kinase Cdelta. *Endocrinology.*, 150(12):5539-48.

Cutillas B, Espejo M, Gil J, Ferrer I, Ambrosio S., 1999, Caspase inhibition protects nigral neurons against 6-OHDA-induced retrograde degeneration. *Neuroreport.*, 10(12):2605-8.

Davis EC, Shryne JE, Gorski RA., 1995, A revised critical period for the sexual differentiation of the sexually dimorphic nucleus of the preoptic area in the rat. *Neuroendocrinology.*, 62(6):579-85.

Delabar JM, Goldgaber D, Lamour Y, Nicole A, Huret JL, de Grouchy J, Brown P, Gajdusek DC, Sinet PM., 1987, Beta amyloid gene duplication in Alzheimer's disease and karyotypically normal Down syndrome. *Science.*, 235(4794):1390-2.

Doherty G., 2007, Developmental switch in the effects of TNFα on dopaminergic midbrain neurons. *Neuroscience Research*, 57, 296-305.

Eriksdotter Jönhagen M, Nordberg A, Amberla K, Bäckman L, Ebendal T, Meyerson B, Olson L, Seiger, Shigeta M, Theodorsson E, Viitanen M, Winblad B, Wahlund LO., 1998, Intracerebroventricular infusion of nerve growth factor in three patients with Alzheimer's disease., *Dement. Geriatr. Cogn. Disord.*, 9(5):246-57.

Fei Q, McCormack AL, Di Monte DA, Ethell DW., 2008, Paraquat neurotoxicity is mediated by a Bak-dependent mechanism. *J. Biol. Chem.*, 283(6):3357-64.

Foister NS, Oldreive CE, Mackie JB, Doherty GH., 2005, Embryonic cerebellar granule cells are resistant to necrosis induced by homocysteine. *Brain Res. Dev. Brain Res.*, 160(1):85-9.

Gibrat C, Bousquet M, Saint-Pierre M, Lévesque D, Calon F, Rouillard C, Cicchetti F., 2010, Cystamine prevents MPTP-induced toxicity in young adult mice via the up-regulation of the brain-derived neurotrophic factor. *Prog. Neuropsychopharmacol. Biol. Psychiatry.*, 34(1):193-203.

Gill SS, Patel NK, Hotton GR, O'Sullivan K, McCarter R, Bunnage M, Brooks DJ, Svendsen CN, Heywood P., 2003, Direct brain infusion of glial cell line-derived neurotrophic factor in Parkinson disease. *Nat. Med.*, 9(5):589-95.

Gilmore EC, Nowakowski RS, Caviness VS Jr, Herrup K.. 2000. Cell birth, cell death, cell diversity and DNA breaks: how do they all fit together? *Trends Neurosci.*, 23(3):100-5.

Glenner GG, Wong CW., 1984, Alzheimer's disease: initial report of the purification and characterization of a novel cerebrovascular amyloid protein. *Biochem. Biophys. Res. Commun.*, 120(3):885-90.

Glucksmann A, 1951, Cell death in normal vertebrate ontogeny. *Biol. Rev.*, 26:59-86.

Gyárfás T, Knuuttila J, Lindholm P, Rantamäki T, Castrén E., 2010, Regulation of brain-derived neurotrophic factor (BDNF) and cerebral dopamine neurotrophic factor (CDNF) by anti-parkinsonian drug therapy in vivo. *Cell Mol. Neurobiol.*, 30(3):361-8.

Hallböök F, Ibáñez CF, Persson H., 1991, Evolutionary studies of the nerve growth factor family reveal a novel member abundantly expressed in Xenopus ovary. *Neuron.*, 6(5):845-58.

Hamburger V, Levi-Montalcini R., 1949, Proliferation, differentiation and degeneration in the spinal ganglia of the chick embryo under normal and experimental conditions. *J. Exp. Zool.*,111(3):457-501

Hartmann A, Hunot S, Michel PP, Muriel MP, Vyas S, Faucheux BA, Mouatt-Prigent A, Turmel H, Srinivasan A, Ruberg M, Evan GI, Agid Y, Hirsch EC., 2000, Caspase-3: A vulnerability factor and final effector in apoptotic death of dopaminergic neurons in Parkinson's disease. *Proc. Natl. Acad. Sci. USA*, 97(6):2875-80.

Hartmann A, Troadec JD, Hunot S, Kikly K, Faucheux BA, Mouatt-Prigent A, Ruberg M, Agid Y, Hirsch EC., 2001, Caspase-8 is an effector in apoptotic death of dopaminergic neurons in Parkinson's disease, but pathway inhibition results in neuronal necrosis. *J. Neurosci.*, 21(7):2247-55.

Hefti F., 1986, Nerve growth factor promotes survival of septal cholinergic neurons after fimbrial transections., *J. Neurosci.*, 6(8):2155-62.

Hellweg R, Jockers-Scherübl M., 1994, Neurotrophic factors in memory disorders. *Life Sci.*, 55(25-26):2165-9.

Helt CE, Hoernig GR, Albeck DS, Gerhardt GA, Ickes B, Reyland ME, Quissell DO, Strömberg I, Granholm AC., 2001, Neuroprotection of grafted neurons with a GDNF/caspase inhibitor cocktail. *Exp. Neurol*,170(2):258-69.

Higgins GA, Lewis DA, Bahmanyar S, Goldgaber D, Gajdusek DC, Young WG, Morrison JH, Wilson MC., 1988, Differential regulation of amyloid-beta-protein mRNA expression within hippocampal neuronal subpopulations in Alzheimer disease. *Proc. Natl. Acad. Sci. USA*, 85(4):1297-301.

Hohn A, Leibrock J, Bailey K, Barde YA., 1990, Identification and characterization of a novel member of the nerve growth factor/brain-derived neurotrophic factor family. *Nature*, 344(6264):339-41.

Howells DW, Porritt MJ, Wong JY, Batchelor PE, Kalnins R, Hughes AJ, Donnan GA., 2000, Reduced BDNF mRNA expression in the Parkinson's disease substantia nigra. *Exp. Neurol.,* 166(1):127-35.

Hurelbrink CB, Armstrong RJ, Luheshi LM, Dunnett SB, Rosser AE, Barker RA., 2001, Death of dopaminergic neurons in vitro and in nigral grafts: reevaluating the role of caspase activation. *Exp. Neurol*, 171(1):46-58.

Hyman C, Hofer M, Barde YA, Juhasz M, Yancopoulos GD, Squinto SP, Lindsay RM., 1991, BDNF is a neurotrophic factor for dopaminergic neurons of the substantia nigra. *Nature*, 350(6315):230-2.

Ip NY, Ibáñez CF, Nye SH, McClain J, Jones PF, Gies DR, Belluscio L, Le Beau MM, Espinosa R 3rd, Squinto SP, 1992, Mammalian neurotrophin-4: structure, chromosomal localization, tissue distribution, and receptor specificity. *Proc. Natl. Acad. Sci. USA*, 89(7):3060-4.

Ivins KJ, Bui ET, Cotman CW., 1999, Beta-amyloid induces local neurite degeneration in cultured hippocampal neurons: evidence for neuritic apoptosis. *Neurobiol. Dis.*, 5(5):365-78.

Jänicke RU, Sprengart ML, Wati MR, Porter AG., 1998a, Caspase-3 is required for DNA fragmentation and morphological changes associated with apoptosis. *J. Biol. Chem.*, 273(16):9357-60.

Jänicke RU, Ng P, Sprengart ML, Porter AG., 1998b, Caspase-3 is required for alpha-fodrin cleavage but dispensable for cleavage of other death substrates in apoptosis., *J. Biol. Chem.*, 273(25):15540-5.

Jellinger KA., 2000, Cell death mechanisms in Parkinson's disease., *J. Neural. Transm.*, 107(1):1-29.

Kerr JF, Wyllie AH, Currie AR., 1972, Apoptosis: a basic biological phenomenon with wide-ranging implications in tissue kinetics. *Br. J. Cancer*, 26(4):239-57.

Kahns S, Lykkebo S, Jakobsen LD, Nielsen MS, Jensen PH., 2002, Caspase-mediated parkin cleavage in apoptotic cell death. *J. Biol. Chem.*, 277(18):15303-8.

Kirik D, Georgievska B, Rosenblad C, Björklund A., 2001, Delayed infusion of GDNF promotes recovery of motor function in the partial lesion model of Parkinson's disease., *Eur. J. Neurosci.,* 13(8):1589-99.

Kitamura Y, Shimohama S, Kamoshima W, Ota T, Matsuoka Y, Nomura Y, Smith MA, Perry G, Whitehouse PJ, Taniguchi T., 1998, Alteration of proteins regulating apoptosis, Bcl-2, Bcl-x, Bax, Bak, Bad, ICH-1 and CPP32, in Alzheimer's disease. *Brain Res.*,780(2):260-9.

Kordower JH, Palfi S, Chen EY, Ma SY, Sendera T, Cochran EJ, Cochran EJ, Mufson EJ, Penn R, Goetz CG, Comella CD., 1999, Clinicopathological findings following intraventricular glial-derived neurotrophic factor treatment in a patient with Parkinson's disease. *Ann. Neurol.*, 46(3):419-24.

Kruman II, Culmsee C, Chan SL, Kruman Y, Guo Z, Penix L, Mattson MP., 2000, Homocysteine elicits a DNA damage response in neurons that promotes apoptosis and hypersensitivity to excitotoxicity. *J. Neurosci.*, 20(18):6920-6.

Kuwana T, Mackey MR, Perkins G, Ellisman MH, Latterich M, Schneiter R, Green DR, Newmeyer DD., 2002, Bid, Bax, and lipids cooperate to form supramolecular openings in the outer mitochondrial membrane. *Cell.*, 111(3):331-42.

Lang AE, Gill S, Patel NK, Lozano A, Nutt JG, Penn R, Brooks DJ, Hotton G, Moro E, Heywood P, Brodsky MA, Burchiel K, Kelly P, Dalvi A, Scott B, Stacy M, Turner D, Wooten VG, Elias WJ, Laws ER, Dhawan V, Stoessl AJ, Matcham J, Coffey RJ, Traub M., 2006, Randomized controlled trial of intraputamenal glial cell line-derived neurotrophic factor infusion in Parkinson disease., *Ann. Neurol.*, 59(3):459-66.

Laske C, Stellos K, Hoffmann N, Stransky E, Straten G, Eschweiler GW, Leyhe T., 2010, Higher BDNF serum levels predict slower cognitive decline in Alzheimer's disease patients. *Int. J. Neuropsychopharmacol.,* 22:1-6.

Levi-Montalcini R, Hamburger V., 1951, Selective growth stimulating effects of mouse sarcoma on the sensory and sympathetic nervous system of the chick embryo. *J. Exp. Zool.*, 116(2):321-61.

Levivier M, Przedborski S, Bencsics C, Kang UJ., 1995, Intrastriatal implantation of fibroblasts genetically engineered to produce brain-derived neurotrophic factor prevents degeneration of dopaminergic neurons in a rat model of Parkinson's disease. *J. Neurosci.*, 15(12):7810-20.

Li C, Brake WG, Romeo RD, Dunlop JC, Gordon M, Buzescu R, Magarinos AM, Allen PB, Greengard P, Luine V, McEwen BS., 2004, Estrogen alters hippocampal dendritic spine shape and enhances synaptic protein immunoreactivity and spatial memory in female mice. *Proc. Natl. Acad. Sci. USA*, 101(7):2185-90.

Li R, Shen Y, Yang LB, Lue LF, Finch C, Rogers J., 2000, Estrogen enhances uptake of amyloid beta-protein by microglia derived from the human cortex. *J. Neurochem.*, 75(4):1447-54.

Liang XB, Liu XY, Li FQ, Luo Y, Lu J, Zhang WM, Wang XM, Han JS., 2002, Long-term high-frequency electro-acupuncture stimulation prevents neuronal degeneration and upregulates BDNF mRNA in the substantia nigra and ventral tegmental area following medial forebrain bundle axotomy. *Brain Res. Mol. Brain Res.*, 108(1-2):51-9.

Lin LF, Doherty DH, Lile JD, Bektesh S, Collins F., 1993, GDNF: a glial cell line-derived neurotrophic factor for midbrain dopaminergic neurons. *Science*, 260(5111):1130-2.

Liu X, Kim CN, Yang J, Jemmerson R, Wang X., 1996, Induction of apoptotic program in cell-free extracts: requirement for dATP and cytochrome c. *Cell.*, 86(1):147-57.

Luetjens CM, Lankiewicz S, Bui NT, Krohn AJ, Poppe M, Prehn JH., 2001, Up-regulation of Bcl-xL in response to subtoxic beta-amyloid: role in neuronal resistance against apoptotic and oxidative injury. *Neuroscience.*, 102(1):139-50.

Maisonpierre PC, Belluscio L, Squinto S, Ip NY, Furth ME, Lindsay RM, Yancopoulos GD., 1990, Neurotrophin-3: a neurotrophic factor related to NGF and BDNF. *Science*. 1990, 247(4949 Pt 1):1446-51.

Massa V, Savery D, Ybot-Gonzalez P, Ferraro E, Rongvaux A, Cecconi F, Flavell R, Greene ND, Copp AJ., 2009, Apoptosis is not required for mammalian neural tube closure. *Proc. Natl. Acad. Sci. USA*, 106(20):8233-8.

Mayeux R, Marder K, Cote LJ, Denaro J, Hemenegildo N, Mejia H, Tang MX, Lantigua R, Wilder D, Gurland B., 1995, The frequency of idiopathic Parkinson's disease by age, ethnic group, and sex in northern Manhattan, 1988-1993. *Am. J. Epidemiol.*, 142(8):820-7.

Middleton G, Pinon LG, Wyatt S, Davies AM. Bcl-2 accelerates the maturation of early sensory neurons. *Journal of Neuroscience*, 1998, 18(9), 3344-3350.

Middleton G, Wyatt S, Cox A, Korsmeyer SJ, Davies AM. Differences in Bcl-2 and Bax-independent function in regulating apoptosis in sensory neuron populations. *European Journal of Neuroscience*, 2000, 12(3), 819-827.

Middleton G, Wyatt S, Ninkina NN, Davies AM., 2001, Reciprocal developmental changes in the roles of Bcl-w and Bcl-xL in regulating sensory neuron survival. *Development*,128, 447-457.

Middleton GH, 2004, Death of neurons during development. Encyclopedia of Neuroscience 3rd edition: ISBN 0-444-51432-5.

Miller MM, Monjan AA, Buckholtz NS., 2001, Estrogen replacement therapy for the potential treatment or prevention of Alzheimer's disease. *Ann. N. Y. Acad. Sci.,* 949:223-34.

Mizuno Y, Mochizuki H, Sugita Y, Goto K., 1998, Apoptosis in neurodegenerative disorders. *Intern. Med.*, 37(2):192-3.

Mufson EJ, Conner JM, Kordower JH., 1995, Nerve growth factor in Alzheimer's disease: defective retrograde transport to nucleus basalis., *Neuroreport.*, 6(7):1063-6.

Murer MG, Yan Q, Raisman-Vozari R., 2001, Brain-derived neurotrophic factor in the control human brain, and in Alzheimer's disease and Parkinson's disease. *Prog. Neurobiol.*, 63(1):71-124.

Nakagawara A, Nakamura Y, Ikeda H, Hiwasa T, Kuida K, Su MS, Zhao H, Cnaan A, Sakiyama S., 1997, High levels of expression and nuclear localization of interleukin-1 beta converting enzyme (ICE) and CPP32 in favorable human neuroblastomas. *Cancer Res.*, 57(20):4578-84.

Nixon RA., 2006, Autophagy in neurodegenerative disease: friend, foe or turncoat? *Trends Neurosci.*, 29(9):528-35.

Nordeen EJ, Nordeen KW, Sengelaub DR, Arnold AP., 1985, Androgens prevent normally occurring cell death in a sexually dimorphic spinal nucleus. *Science.*, 229(4714):671-3.

Nuñez G, London L, Hockenbery D, Alexander M, McKearn JP, Korsmeyer SJ. ,1990, Deregulated Bcl-2 gene expression selectively prolongs survival of growth factor-deprived hemopoietic cell lines. *J. Immunol.* 1990, 144(9):3602-10.

Nutt JG, Burchiel KJ, Comella CL, Jankovic J, Lang AE, Laws ER Jr, Lozano AM, Penn RD, Simpson RK Jr, Stacy M, Wooten GF; ICV GDNF Study Group., 2003, Randomized, double-blind trial of glial cell line-derived neurotrophic factor (GDNF) in PD. *Neurology.*, 60(1):69-73.

O'Bryant SE, Hobson VL, Hall JR, Barber RC, Zhang S, Johnson L, Diaz-Arrastia R., 2011, Serum Brain-Derived Neurotrophic Factor Levels Are Specifically Associated with Memory Performance among Alzheimer's Disease Cases. *Dement. Geriatr. Cogn.* Disord., 31(1):31-6.

O'Leary DD, Fawcett JW, Cowan WM., 1986, Topographic targeting errors in the retinocollicular projection and their elimination by selective ganglion cell death. *J. Neurosci.*, 6(12):3692-705.

Oldreive CE, Doherty GH., 2010, Effects of tumour necrosis factor-alpha on developing cerebellar granule and Purkinje neurons in vitro. *J. Mol. Neurosci.*, 42(1):44-52.

Olson L, Nordberg A, von Holst H, Bäckman L, Ebendal T, Alafuzoff I, Amberla K, Hartvig P, Herlitz A, Lilja A.,1992, Nerve growth factor affects 11C-nicotine binding, blood flow, EEG, and verbal episodic memory in an Alzheimer patient (case report)., *J. Neural. Transm. Park Dis. Dement. Sect.*, 4(1):79-95.

Oppenheim RW., 1991, Cell death during development of the nervous system. *Annu. Rev. Neurosci.*, 14:453-501.

Oppenheim RW, Flavell RA, Vinsant S, Prevette D, Kuan CY, Rakic P., 2001, Programmed cell death of developing mammalian neurons after genetic deletion of caspases. *J. Neurosci.*, 21(13):4752-60.

Otani H, Erdos M, Leonard WJ., 1993, Tyrosine kinase(s) regulate apoptosis and bcl-2 expression in a growth factor-dependent cell line. *J. Biol. Chem.*, 268(30):22733-6.

Paradis E, Douillard H, Koutroumanis M, Goodyer C, LeBlanc A., 1996, Amyloid beta peptide of Alzheimer's disease downregulates Bcl-2 and upregulates bax expression in human neurons. *J. Neurosci.*, 16(23):7533-9.

Parain K, Murer MG, Yan Q, Faucheux B, Agid Y, Hirsch E, Raisman-Vozari R., 1999, Reduced expression of brain-derived neurotrophic factor protein in Parkinson's disease substantia nigra. *Neuroreport.*, 10(3):557-61.

Park HJ, Lim S, Joo WS, Yin CS, Lee HS, Lee HJ, Seo JC, Leem K, Son YS, Kim YJ, Kim CJ, Kim YS, Chung JH., 2003, Acupuncture prevents 6-hydroxydopamine-induced neuronal death in the nigrostriatal dopaminergic system in the rat Parkinson's disease model. *Exp. Neurol.*, 180(1):93-8.

Parsian A, Sinha R, Racette B, Zhao JH, Perlmutter JS., 2004, Association of a variation in the promoter region of the brain-derived neurotrophic factor gene with familial Parkinson's disease. *Parkinsonism Relat. Disord.*, 10(4):213-9.

Perry EK, Perry RH, BlessedG, Tomlinson BE., 1978, Changes in brain cholinesterases in senile dementia of Alzheimer type., *Neuropathol. Appl. Neurobiol.* 4: 273–277.

Perry EK, Tomlinson BE, Blessed G, Bergmann K, Gibson PH, Perry RH., 1978, Correlation of cholinergic abnormalities with senile plaques and mental test scores in senile dementia. *Br. Med. J.* 2 : 1457–1459.

Pinon LG, Middleton G, Davies AM., 1997, Bcl-2 is required for cranial sensory neuron survival at defined stages of embryonic development. *Development*, 124(20), 4173-4178

Raoul C, Henderson CE, Pettmann B., 1999, Programmed cell death of embryonic motoneurons triggered through the Fas death receptor. *J. Cell Biol.*, 1999, 147(5):1049-62.

Rosenthal A, Goeddel DV, Nguyen T, Lewis M, Shih A, Laramee GR, Nikolics K, Winslow JW., 1990, Primary structure and biological activity of a novel human neurotrophic factor. *Neuron.*, 4(5):767-73.

Satou T, Cummings BJ, Cotman CW., 1995, Immunoreactivity for Bcl-2 protein within neurons in the Alzheimer's disease brain increases with disease severity. *Brain Res.*, 697(1-2):35-43.

Sawada H, Ibi M, Kihara T, Urushitani M, Honda K, Nakanishi M, Akaike A, Shimohama S., 2000, Mechanisms of antiapoptotic effects of estrogens in nigral dopaminergic neurons. *FASEB J.*, 14(9):1202-14.

Schierle GS, Hansson O, Leist M, Nicotera P, Widner H, Brundin P., 1999, Caspase inhibition reduces apoptosis and increases survival of nigral transplants. *Nat. Med.*, 5(1):97-100.

Siegel GJ, Chauhan NB., 2000, Neurotrophic factors in Alzheimer's and Parkinson's disease brain., *Brain Res. Brain Res. Rev.*, 33(2-3):199-227.

Stadelmann C, Brück W, Bancher C, Jellinger K, Lassmann H., 1998, Alzheimer disease: DNA fragmentation indicates increased neuronal vulnerability, but not apoptosis. *J. Neuropathol. Exp. Neurol.*, 57(5):456-64.

Stone DJ, Rozovsky I, Morgan TE, Anderson CP, Finch CE., 1998, Increased synaptic sprouting in response to estrogen via an apolipoprotein E-dependent mechanism: implications for Alzheimer's disease., *J. Neurosci.*, 18(9):3180-5.

Su JH, Satou T, Anderson AJ, Cotman CW., 1996, Up-regulation of Bcl-2 is associated with neuronal DNA damage in Alzheimer's disease. *Neuroreport.*, 7(2):437-40.

Su JH, Deng G, Cotman CW., 1997, Bax protein expression is increased in Alzheimer's brain: correlations with DNA damage, Bcl-2 expression, and brain pathology. *J. Neuropathol. Exp. Neurol.*, 56(1):86-93.

Tang M, Abplanalp W, Ayres S, Subbiah MT., 1996, Superior and distinct antioxidant effects of selected estrogen metabolites on lipid peroxidation. *Metabolism.*,45(4):411-4.

Tomac A, Lindqvist E, Lin LF, Ogren SO, Young D, Hoffer BJ, Olson L., 1995, Protection and repair of the nigrostriatal dopaminergic system by GDNF in vivo. *Nature.*, 373(6512):335-9.

Tompkins MM, Basgall EJ, Zamrini E, Hill WD., 1997, Apoptotic-like changes in Lewy-body-associated disorders and normal aging in substantia nigral neurons. *Am. J. Pathol.*, 150(1):119-31.

Troy CM, Rabacchi SA, Friedman WJ, Frappier TF, Brown K, Shelanski ML., 2000, Caspase-2 mediates neuronal cell death induced by beta-amyloid. *J. Neurosci.*, 20(4): 1386-92.

Tsujimoto Y, Croce CM., 1986, Analysis of the structure, transcripts, and protein products of bcl-2, the gene involved in human follicular lymphoma. *Proc. Natl. Acad. Sci. USA*, 83(14):5214-8.

Tsukahara T, Takeda M, Shimohama S, Ohara O, Hashimoto N., 1995, Effects of brain-derived neurotrophic factor on 1-methyl-4-phenyl-1,2,3,6-tetrahydropyridine-induced parkinsonism in monkeys. *Neurosurgery.*, 37(4):733-9

Van Den Eeden SK, Tanner CM, Bernstein AL, Fross RD, Leimpeter A, Bloch DA, Nelson LM., 2003, Incidence of Parkinson's disease: variation by age, gender, and race/ethnicity., *Am. J. Epidemiol.*, 157(11):1015-22.

Vila M, Jackson-Lewis V, Vukosavic S, Djaldetti R, Liberatore G, Offen D, Korsmeyer SJ, Przedborski S., 2001, Bax ablation prevents dopaminergic neurodegeneration in the 1-methyl- 4-phenyl-1,2,3,6-tetrahydropyridine mouse model of Parkinson's disease. *Proc. Natl. Acad. Sci. USA,* 98(5):2837-42.

Voyvodic JT., 1996, Cell death in cortical development: How much? Why? So what? *Neuron.*, 16(4):693-6.

Weil M, Jacobson MD, Raff MC., 1997, Is programmed cell death required for neural tube closure? *Curr. Biol.*, 7(4):281-4.

Williams LR, Varon S, Peterson GM, Wictorin K, Fischer W, Bjorklund A, Gage FH., 1986, Continuous infusion of nerve growth factor prevents basal forebrain neuronal death after fimbria fornix transection. *Proc. Natl. Acad. Sci. USA*, 83(23):9231-5.

Willis SN, Fletcher JI, Kaufmann T, van Delft MF, Chen L, Czabotar PE, Ierino H, Lee EF, Fairlie WD, Bouillet P, Strasser A, Kluck RM, Adams JM, Huang DC., 2007, Apoptosis initiated when BH3 ligands engage multiple Bcl-2 homologs, not Bax or Bak. *Science.*, 315(5813):856-9.

Xu H, Gouras GK, Greenfield JP, Vincent B, Naslund J, Mazzarelli L, Fried G, Jovanovic JN, Seeger M, Relkin NR, Liao F, Checler F, Buxbaum JD, Chait BT, Thinakaran G, Sisodia SS, Wang R, Greengard P, Gandy S., 1998, Estrogen reduces neuronal generation of Alzheimer beta-amyloid peptides. *Nat. Med.*, 4(4):447-51.

Yang L, Sugama S, Mischak RP, Kiaei M, Bizat N, Brouillet E, Joh TH, Beal MF., 2004, A novel systemically active caspase inhibitor attenuates the toxicities of MPTP, malonate, and 3NP in vivo. *Neurobiol. Dis.*, 17(2):250-9.

Yurek DM, Hipkens SB, Wiegand SJ, Altar CA., 1998, Optimal effectiveness of BDNF for fetal nigral transplants coincides with the ontogenic appearance of BDNF in the striatum., *J. Neurosci.*, 18(15):6040-7.

Zhivotovsky B, Samali A, Gahm A, Orrenius S., 1999, Caspases: their intracellular localization and translocation during apoptosis. *Cell Death Differ.*, 6(7):644-51.

In: Neuronal Cell Apoptosis
Editors: C. J. Schmid and J. L. Wolfe
ISBN: 978-1-61324-633-7

Chapter VII

Neuronal Cell Apoptosis

Fawziah Al Rouq
King Saud University, Department of Physiology,
College of Medicine, Riyadh, Kingdom of Saudi Arabia

Apoptosis

The term apoptosis, from the Greek word for "falling off" of leaves from a tree, is used to describe a process in which a cell actively participate in its own destructive processes. Apoptosis, or programmed cell death, is a physiological process observed in many organs and cells, which may also occur in pathological conditions such as diabetic neuropathy. The apoptotic program is characterized by certain morphological features, that include changes in the plasma membrane such as loss of membrane asymmetry and attachment, condensation of the cytoplasm and nucleus, and internucleosomal cleavage of DNA. In the final stages, the dying cells becomes fragmented into "apoptotic bodies" which are rapidly eliminated by phagocytic cells without eliciting significant inflammatory damage to surrounding cells (Kerr et al., 1972; Kerr et al., 1994; Majno and Joris, 1995).

Apoptosis plays an important role in the regulation of tissue homeostasis, since it can be modulated to support the balance between cell death and cell proliferation (Wyllie, 1985). In the central nervous system, programmed cell death is believed to be responsible for the death of excess neurons during spinal cord development (Oppenheim, 1985).

History and Molecular Mechanism of Apoptosis

Nerve cells, like all the cells of our body, can die in one of two ways, they can either shrink and disappear (apoptosis) or swell and explode (necrosis). Some scientists refer to apoptosis as "cell suicide". It was first described, more than 30 years ago, by scientists who were studying the development of the nervous system in the fetus. They noticed that at an early stage, when the brain starts to be formed, a great many neurons are made. A little later, however, approximately half of these neurons shrink and die (apoptosis). All this occurs before birth, and is a normal phenomenon that allows for accurate and correct nerve cell connections to be made (Parkinson's Diseasee Update Newsletter, 1995).

Following these early observations, scientists discovered that apoptosis occurs not only in the brain, but throughout the body. They also realized that apoptosis continues during our entire life. Each day, a great many of our cells undergo apoptosis. For the most part, for every cell that dies, a cell divides to replace it. Nerve cells, however, are different from all other cells in our body, because they never divide during adult life. Thus, if a nerve cell undergoes apoptosis, it is not replaced. Outside the brain, apoptosis is often benificial. For example, it helps remove damaged or dangerous cells, such as virus-infected cells, or cancer cells (Parkinson's Diseasee Update Newsletter, 1995).

Recent studies suggest that in many, if not all, neurodegenerative disorders including diabetic neuropathy, nerve cells die by apoptosis, and this has triggered renewed interest in the mechanisms that control cell suicide. The mechanism involved in apoptosis have been studied mainly in nerve cells from worms, because this is a relatively easy system to work on (Parkinson's Diseasee Update Newsletter, 1995).

In Caenorhabditis elegans (CE), 131 out of 1090 somatic cells undergo developmental apoptosis. Four death (d) genes named ced-3, ced-4, ced-9 and egglaying defective 1 (egl-1) form the "death machinery" which is responsible for the execution of the cells undergoing programmed cell death (PCD) (Hengartner, 1997). Genetic studies in mutant animals have demonstrated that ced-3 and ced-4 are both required for cell execution. Also they have shown that ced-9 is a survival factor that negatively regulates the activity of ced-3 and ced-4 (Hengartner et al., 1992).

CED-4 acts as an adaptor molecule that regulates the killing activity of CED-3. Ced-4 acts upstream of ced-3 and may be regulated by ced-9 (Shaham and Horvitz, 1996). Loss of egg laying defective-1 (egl-1) function results in survival of all cells that normally undergo apoptosis. EGL-1 might activate apoptosis by binding to CED-9 and inhibiting its activity. Thereby CED-4 is released from the CED-9/CED-4 containing complex (Conradt and Horvitz, 1998).

The mechanisms of apoptosis are regulated by caspases and bcl-2 family (Liang and Fesik, 1997). So, caspases and bcl-2 family explained first, in order to understand the mechanisms of apoptosis.

Caspases

The mammalian homologues of CED-3 are a growing family of related proteases, which are collectively referred to as caspases (cysteine aspartate protease) (Thornberry and Lazebnik, 1998). The first mammalian caspase to be identified was the interleukin-1β-converting enzyme (ICE or caspase 1) (Yuan et al., 1993; Schwartz and Milligan, 1996). Later it was demonstrated that caspase 3 has substrate specificities similar to CED-3 (Chhanabhai et al., 1997). To date the 14 members of the caspase family can be divided into three subfamilies on the basis of the peptide-sequence preferences of their substrates: (i) the ICE-protease subfamily (caspases 1, 4, 5, 13 and 4, and murine caspases 11 and 12); (ii) the CED-3 subfamily (caspases 3, 6, 7, 8, 9 and 10); and (iii) the caspase 2 subfamily.

Each caspase is initially synthesized as a zymogen and requires processing at specific cleavage sites to generate the active enzyme (Stennicke and Salvesen, 1999). The caspases that are the first to be activated, in turn trigger other downstream caspases giving rise to a proteolytic cascade that culminates with the execution of apoptosis.

Bcl-2 Family

Since the first observation that the ced-9 gene encodes a protein that is similar to mammalian B-cell lymphoma-2 (BCL-2) protein (Hengartner and Horvitz, 1994), at least 15 protein transcripts structurally related to CED-9 and BCL-2 have been identified to date, and grouped in the so-called BCL-2 protein family (Adams and Cory, 1998; Chao and Korsmeyer, 1998; Sanchez and Yuan, 2001).

It has been reported that there are 4 domains belonging to bcl-2 family. These domains includes, BH1, BH2, BH3 and BH4. The Bcl-2 family, comprised of both pro-apoptotic and anti-apoptotic members, acts as a checkpoint upstream of caspases and mitochondrial dysfunction (Chao and Korsmeyer, 1998). It has been reported that Bid, a BH3 domain containing protein induces cytochrome c release from mitochondria in response to caspase-8, the apical caspase activated by cell surface death receptors such as Fas and TNF. The cytochrome c releasing activity of Bid was antagonized by Bcl-2 (Luo et al., 1998).

BCL-X_L in mammals interacts with apoptosis activating factor (Apaf1). This interaction occurs via the BH4 domain of bcl-2 and results in inhibition of the association of Apaf1 with procaspase 9, again with blockage of apoptosome formation and caspase 9 activation (Adams and Cory, 2002).

Defective regulation of apoptosis may play a role in the etiology of many diseases, such as cancer, autoimmun disease and degenerative disease of the central nervous system, heart disease (Kerr et al., 1994; Barr and Tomei, 1994) and diabetic neuropathy (Ekshyyan and Aw, 2004; Rai et al., 2005). Therefore, better understanding of the mechanisms of apoptosis may offer new possibilities for the prevention and treatment of diabetic neuropathy.

Apoptotic Death Pathways (Mechanisms of Apoptosis)

There are two major apoptotic death pathways: death receptor and mitochondrial pathways, and these can be differentiated by the relative timing of caspase activation and mitochondrial release of cytochrome C. In the first, which is exemplified by activation of death receptors, an effector caspase is activated prior to mitochondrial alterations. While in the second, cytochrome C is released from the mitochondrial intermembrane space prior to caspase activation (Rai et al., 2005).

Death Receptor Pathway

Death receptors are cell surface receptors that trigger apoptosis. There are several types of death receptors in different tissues, but two members of the tumor necrosis factor receptor

(TNFR) family were recently demonstrated to be involved in neuronal death. Fas (CD95/Apo-1) and the p75 neurotrophin receptor ($p75^{NTR}$) (Raoul et al., 2000).

Fatty Acid Synthase (Fas) System

Historical Background

Fas/APO-1 first appeared in the literature in 1989, when it was described by two independent groups led by Minako Yonehara in Japan (Yonehara et al., 1989), and Peter Krammer in Germany (Trauth et al., 1989). Both teams reported that Fas was a cell surface molecule, expressed on human lymphocytes, which triggered cell death when cross-linked with Fas antibodies or Fas ligand. The relationship between Fas and lymphocyte cell death was reinforced in 1992, when Nagata's group discovered that the lymphoproliferative (lpr) phenotype in mice, characterized by defective lymphocyte apoptosis, was caused by a mutation in Fas (Watanabe-Fukunaga et al., 1992). Then, in 1995, three papers in the same issue of Nature reported that Fas and Fas Ligand expressed on the same cell could interact, resulting in cell "suicide", thus explaining the phenomenon of activation-induced cell death (AICD) in T lymphocytes (Brunner et al., 1995; Ju et al., 1995; Dhein et al., 1995). These findings strengthened the notion that the main physiological role of Fas is the induction of apoptosis in lymphocytes. Throughout the last decade, reviews about Fas with titles such as "The Fas death factor," (Nagata and Golstein, 1995), "A license to kill," (Fraser and Evan, 1996), "Apoptosis by death factor," (Nagata, 1997) and "Death the Fas way" (Sharma et al., 2000) further studied this pathway. However, as early as 1993, reports were documenting non-apoptotic roles for Fas (Owen-Schaub et al., 1993; Alderson et al., 1993).

Fas: The Molecule

Fas is a member of the tumour necrosis factor receptor (TNF-R) superfamily of cytokine receptors and nerve growth factor (NGF) receptor family. Membership in this family is defined by characteristic extracellular cysteine-rich domains (CRDs), which control ligand binding and spontaneous self-assembly into trimers (Chan et al., 2000). A subset of TNF-R superfamily members, including Fas, TNF-R1, and the low affinity neurotrophin receptor ($p75^{NTR}$), are often termed "death receptors" due to their intracellular "death domains" (DD). A DD consists of a series of six α-helices, which can bind other DD-containing proteins through homotypic interactions (Huang et al., 1996). The DDs couple cell surface Fas with intracellular signal transduction cascades via adaptor molecules which also contain DDs. The DDs were originally named to reflect their ability to couple Fas to apoptotic pathways. However, DDs can initiate other pathways, some resulting in proliferation or survival instead of death.

Like other members of the TNFR family, oligomerization of Fas through cross-linking with antibodies or induced by the naturally trimeric FasL is the first event in the Fas-dependent death signaling cascade. The death signaling pathway involves the activation of several cysteine proteases related to the interleukin-1β (IL-1β)-converting enzyme (ICE)

(Enari et al., 1996), recently designated caspases (Alnemri et al., 1996). Transduction of the death signal is accomplished through the "death domain" (DD), an evolutionary conserved, approximately 80-amino acid protein motif at the carboxy terminus of the Fas molecule (Cleveland and Ihle, 1995). One such protein (caspase 8 or FLICE) is of particular interest because it binds to the death effector domain of FADD and upon overexpression induces apoptosis that is blocked by the ICE family inhibitors (Muzio et al., 1996). The FADD/MORT 1 which interacts with Fas through a C-terminal death homology domain and with caspase 8 through an N-terminal FADD domain acts as an adaptor protein (Strasser and Newton, 1999; Chinnaiyan et al., 1995).

Fas Signaling Pathways

Fas monomers spontaneously self-assemble into trimers in the cell membrane (Chan et al., 2000). When bound by Fas Ligand (FasL, CD95L), or by agonistic Fas antibodies, these trimers presumably undergo a conformational change resulting in activation of downstream pathways. Ligand binding may allow the recruitment of new adaptor proteins to the death domain (DD), or may bring previously recruited proteins into proximity, allowing their autocatalytic activation. Four Fas DD-binding molecules have been well characterised: the Fas Associated Death Domain protein (FADD/MORT-1) (Strasser and Newton, 1999), the death associated protein Daxx (Ko et al., 2001), the Receptor Interacting Protein (RIP) (Stanger et al., 1995), and the Fas associated phosphatase (FAP-1) (Sato et al., 1995).

Fas Associated Death Domain Protein (FADD)

Fas-induced death involves clustering of Fas at the cell membrane and recruitment of the Fas-associated death domain (DD)-containing protein (FADD) to the trimerized intracellular DD of the receptor. FADD in turn recruits the proenzymatic form of caspase-8/FADD-like IL-1β-converting enzyme (FLICE), leading to the formation of a multimolecular signaling platform known as the death-inducing signaling complex (DISC). After the formation of the DISC, caspase-8 is activated by autoproteolytic cleavage and initiates the apoptotic cascade (Dutton et al., 2004).

FADD is an adaptor protein that couples Fas to caspase 8 (also known as FADD-like interleukin-l converting enzyme, FLICE). FADD binds to Fas via homotypic DD interactions, and to caspase 8 via a homotypic death effector domain interaction .Activated caspase-8 can activate downstream caspases by direct cleavage or indirectly by cleaving Bid and inducing cytochrome c release from the mitochondria.

In the mitochondrial-initiated pathway, caspase activation is triggered by the formation of a multimeric Apaf-1/cytochrome c complex that is fully functional in recruiting and activating procaspase-9. Activated caspase-9 will then cleave and activate downstream caspases such as caspase-3, -6, and -7 (Budihardjo et al., 1999).

Death Associated XX Protein (Daxx)

Daxx is a transcriptional repressor with no known homologies at present. Daxx is predominantly localised to the nucleus, but it can be translocated to the cytoplasm by the apoptosis signaling kinase 1 (ASK1) (Figure 2) (Ko et al., 2001). ASK1 is required both for the cytoplasmic localization of Daxx, and for the Daxx/Fas interaction (Ko et al., 2001). Daxx can bind to the Fas DD; this interaction is independent of FADD binding (Chang et al., 1999). Daxx can enhance Fas-induced apoptosis by at least two independent mechanisms, namely by activation of Jun kinase (JNK) (Yang et al., 1997), and by transcriptional regulation of genes involved in caspase activation (Torii et al., 1999). Interestingly, like FADD, Daxx may also transduce survival and/or proliferative signals from Fas (Michaelson et al., 1999, Michaelson, 2000).

Receptor Interacting Protein (RIP)

The receptor interacting proteins (RIP, RIP2, RIP3, RIP4) are death domain-containing proteins which, uniquely, possess serine/threonine kinase activity (Stanger et al., 1995; Yu et al., 1999; Meylan et al., 2002). RIP associates with the Fas death domain indirectly (Stanger et al., 1995), but can also interact directly with FADD (Holler et al., 2000; Varfolomeev et al., 1996; Hu and Yang, 2000) and FLIP (Kataoka et al., 2000). In vitro, RIP overexpression results in nuclear factor (NFκB) translocation, JNK activation, and apoptosis (Kelliher et al., 1998; Kasof et al., 2000; Holler et al., 2000).

In vivo RIP appears to function as a survival factor, RIP family proteins may transduce non-apoptotic signal through several pathways, including such as NFκB (Meylan et al., 2002; Kasof et al., 2000; Thome et al., 1998). NFκB activation by RIP is dependent on functional NFκB-inducing kinase (NIK), (Figure 3) (Hu et al., 2000; Malinin et al., 1997; Pazdernik et al., 1999). RIP can also activate caspase-1 directly (Thome et al., 1998). In vivo, activation of caspase-1 allows the processing of pro-interleukin 1B (IL-1B) to active IL-1B, a potent inducer of inflammation (Cerretti et al., 1992).

Fas Associated Phosphatase 1 (FAP1)

Fas is a cell surface receptor that controls a poorly understood signal transduction pathway that leads to cell death by means of apoptosis. A protein tyrosine phosphotase, Fas associated phosphatase (FAP-1), capable of interacting with the cytosolic domain of Fas, was identified. The carboxyl terminal 15 amino acids of Fas are necessary and sufficient for interaction with FAP-1. FAP-1 expression is highest in tissues and cell lines that are relatively resistant to Fas-mediated cytotoxicity. Gene transfer-mediated elevations in FAP-1 partially abolished Fas-induced apoptosis in a T cell line. These findings are consistent with an inhibitory effect of FAP-1 on Fas signal transduction (Sato et al., 1995).

Mitochondrial Pathway

A key event in the mitochondrial pathway is the release of cytochrome c into the cytosol. Cytochrome c is a member of mitochondrial electron-transport chain that is required for generation of ATP. Experiments in cell free systems led to hypothesize that cytochrome c release in mitochondrial apoptosis is either caused by a rupture of the outer mitochondrial membrane and/or by the so-called mitochondrial permeability transition (MPT), that is controlled by a voltage- and Ca^{2+}-sensitive pore, referred to as the permeability transition (PT) pore (Blatt and Glick, 2001).

Subcellular localization studies have shown that the anti-apoptotic members of the B-cell lymphoma (BCL-2) family (BCL-2, BCL-X_L) reside on the mitochondrial outer membrane, while the pro-apoptotic family members (BAD, BAX, BID) may be either cytosolic or present on the cytoplasmic surface of the outer mitochondrial memb(Zimmermann et al., 2001). During internal damage to the cell, i.e. from reactive oxygen species (ROS), pro-apoptotic molecules are activated and translocate to the mitochondria where they induce the release of cytochrome c and apoptosis activating factor-1 (APaf-1) from the intermembrane space. On the other hand, the anti-apoptotic proteins BCL-2 and BCL-X_L work to prevent cytochrome c release from mitochondria, and thereby protect cells from death (Kluck et al., 1997; Yang et al., 1997). Subsequent events involve formation of the apoptosome (cytochorme c, APaf-1, caspase 9 and ATP) that activate caspase 3 leading to apoptosis .

Laboratory Tests for Apoptosis

There are a variety of new reagents and methods for the study of apoptotic cells. The optimal method for identification of apoptotic events will depend on the cell or tissue type, as well as the mode of induction of apoptosis. One of the earliest and most consistently observed features of apoptosis is the induction of series of cyosolic proteases, the caspases. Active caspases cleave numerous intracellular proteins, and contribute to characteristic apoptotic morphology (Stennicke and Salvesen, 1997). Capase-3, in particular, plays a central role in this process. PharMingen features new polyclonal antibodies which preferentially recognize the active form of human and mouse capase-3, providing an important tool for the analysis of caspase-3 activity in apoptotic cells and tissues. Caspase-3 Assay Kit (PharMingen, San Diego, CA, USA) may also be used to identify and quantitate caspase-3 and caspase-3-like activities in apoptotic cell lysates. Alternative earlier markers of apoptosis is the loss of membrane asymmetry, including a redistribution of phosphotidylserine residues to the outer leaflet of the plasma membrane. This may be detected by utilizing the affinity of an antiocoagulant protein, Annexin V, which preferentially binds to negatively charged phospholipids like phosphotidylserine. Annexin V is typically conjugated to fluorescent dyes or viotinylated for easy identification by flow cytometric analysis. Annexin V propidium iodide assay was used in our study. Activation of endonucleases, which cleave DNA at the later stages of apoptosis, often leads to a characteristic "ladder" of DNA fragments which are observed by gel electophoresis. The APO-DIRCT™ and APO-BRDU™ assay allow for incorporation of a fluorescent indicator onto the ends of fragmented DNA, thus providing cells which display this phenotype. The RiboQuant™ Rnase Protection Assay is an exciting new method which may be regulated at the mRNA level as part of the apoptotic process.

These and other methods of analyses are useful for a better understanding of the mechanisms which regulate the apoptotic process (Apoptosis, 1998).

In the current study, DN was diagnosed clinically by Michigan Neuropathy Screening Instrument and electrophysiologically by Nerve Conduction Studies. Serum sFas and sFasL levels were measured to determine the markers of apoptosis. Features of apoptosis in diabetic neuropathy were reproduced in a neuronal tissue culture model using defined conditions by treating the neuronal cell line with the sera of diabetic patients and measure the markers of apoptosis by flow cytometry (Annexin V/propidium iodide assay).

Although there is symptomatic treatment of diabetic neuropathy, the prevention and treatment of diabetic neuropathy as such is still obscure. Dorsal root ganglia, sympathetic ganglin and Schwann cells (SCs), are lost or damaged in patients suffering from diabetic neuropathy. In vitro and in vivo studies have shown features of apoptosis in these cells. It has been observed that high glucose inhibits neurite growth and initiates apoptosis in cultured sympathetic primary neurons, and the addition of insulin-like growth factor – I (IGF-1) a meliorates these changes (Russell and Feldman, 1999).

Hyperglycemia induces caspase 3 activation and monphologic changes in Schwann cells (SCs) consistent with apoptotic death. Over–expression of Bcl-xL or IGF-1, signaling via a phosphaticly linositol 3-Kinase (P1 3Kinase) protects SCs from apoptosis. So, addition of IGF-1 may prevents or a meliorates the apoptosis in neuronal cells line exposed to serum of diabetic patients with neuropathy (Delaney et al., 2001).

Although, there were studies which demonstrate the inhibition of Fas death signals by Fas-associated death-domain like IL-1 beta-converting enzyme-inhibitory proteins (FL1Ps) involvement of FL1Ps in diabetic neuropathy as such is not demenstrated (Tschopp et al., 1998). Only recently there is a study which provide evidence for experssion of FL1P in human pancreatic beta cell and this suggests a novel approach to prevent and treat diabetes by swithching Fas signaling from apoptosis to proliferation (Maedler et al., 2002).

From the above studies we conclude that Fas pathway plays an important role as the initial effector in the development of diabetic neuropathy:

i. High levels of sFas were detected in the sera of diabetic patients with neuropathy.
ii. Serum of diabetic patients with neuropathy induce apoptosis of neuronal cell line.

It seems that apoptosis is involved in the pathogensis of diabetic neuropathy and that activation of the apoptosis cascade is initiated by FAS system. This Fas-pathway could be inhibited in the future studies by using Fas death signal inhibitors.

In: Neuronal Cell Apoptosis
Editors: C. J. Schmid and J. L. Wolfe

ISBN: 978-1-61324-633-7

Chapter VIII

Bcl-2 Homology-3 Domain (BH3)-Only Proteins: Harbingers of Neuronal Demise and Neurodegeneration

Trisha R. Stankiewicz and Daniel A. Linseman*
Department of Biological Sciences and Eleanor Roosevelt Institute,
University of Denver, Denver, Colorado, USA
Research Service, Veterans Affairs Medical Center, Denver, Colorado, USA
Division of Clinical Pharmacology and Toxicology, Department of Medicine and
Neuroscience Program, University of Colorado Denver, Aurora Colorado USA

Abstract

Neuronal apoptosis is an underlying cause of many neurodegenerative diseases such as amyotrophic lateral sclerosis and Parkinson's disease, as well as, in acute neuronal injury such as stroke. The two main pathways that induce apoptosis in neurons are the intrinsic (mitochondrial) and extrinsic death receptor cascades. These pathways are not mutually exclusive and the intrinsic apoptotic cascade can function as an important amplification loop to the extrinsic apoptotic pathway. Evidence also suggests that endoplasmic reticulum (ER) stress is a significant contributor to neuronal apoptosis. Central to each of these pathways is the B-cell lymphoma-2 (Bcl-2) family of proteins which function as crucial regulators of programmed cell death. The Bcl-2 family is comprised of three distinct subgroups: the pro-survival subgroup (e.g., Bcl-2 and Bcl-xL); the pro-apoptotic, multi-domain subgroup (Bax and Bak); and the Bcl-2 homology-3 domain (BH3)-only pro-apoptotic subgroup (e.g., Bim, Bad, and Puma). Regulation of apoptosis is largely mediated by finely tuned interactions between pro-survival and pro-apoptotic Bcl-2 family proteins. An essential role of Bcl-2 family proteins in neuronal survival has been underscored by evidence demonstrating that either loss of pro-survival

* To whom correspondence should be addressed: Department of Biological Sciences and Eleanor Roosevelt Institute, University of Denver, 2199 S. University Blvd., Denver, CO 80208; Tel.: (303) 871-5654; Fax: (303) 871-3471; Email: daniel.linseman@du.edu

Bcl-2 function or gain of Bax/Bak function are causally linked to neuronal apoptosis and neurodegeneration. More recently, a prominent role has been established for BH3-only proteins in neuronal apoptosis associated with chronic neurodegenerative diseases or acute neuronal injury. At least thirteen BH3-only proteins have been described in mammals and evidence suggests that these proteins function as sensors of intracellular and extracellular cell death signals that tip the delicate balance between pro-survival and pro-apoptotic Bcl-2 proteins towards apoptosis. In this review, we discuss the roles of BH3-only proteins in the intrinsic, extrinsic, and ER stress apoptotic cascades, as well as hierarchical BH3-only pathways and regulation of BH3-only proteins. We also examine specific evidence of a role for BH3-only proteins in *in vitro* models of neuronal apoptosis, animal models of neurodegeneration or acute neuronal injury, and in human neuropathology. Given their crucial roles in neuronal apoptosis, BH3-only proteins represent a novel class of therapeutic targets for the treatment of neurodegenerative diseases and acute neuronal traumas such as stroke.

Introduction

Although neuronal cell death can occur through many diverse mechanisms, accumulating evidence suggests that neuronal apoptosis appears to be a major contributor to several devastating neurodegenerative diseases and acute neuronal injury. Such neuronal disorders are characterized by both chronic and acute disease states, and a short list includes a spectrum of disorders such as amyotrophic lateral sclerosis (ALS), Parkinson's Disease, Huntington's Disease, and stroke. In recent years, research has aimed at elucidating the underlying molecular mechanisms of neuronal apoptosis in neuronal cultures, animal models, and human clinical cases.

There are at least three pathways that transmit apoptotic signals in neuronal cells: the intrinsic (mitochondrial) pathway, the extrinsic (death receptor) pathway, and an endoplasmic reticulum (ER) stress-related pathway (Figure 1). The intrinsic pathway occurs through intracellular signaling that arises from cell stressors such as elevations in reactive oxygen species, Ca^{2+} overload, DNA damage, or heat shock. Such signals induce mitochondrial outer membrane permeabilization and activation of executioner caspases. Conversely, the extrinsic pathway is induced by an extracellular signal such as Fas ligand binding to its cognate Fas death receptor, activating a multi-step cascade that is responsible for inducing cell death. Finally, ER stress is an important signal for neuronal apoptosis. The ER is essential for lipid synthesis, protein folding, and protein maturation. Accumulation of misfolded proteins, disruptions in calcium homeostasis, and structural disturbances of the ER can trigger the unfolded protein response (UPR). The UPR disrupts the basic machinery of the ER and if functional activity is not restored, the cell ultimately succumbs to programmed cell death.

Abundant evidence demonstrates that signals of cell survival and cell death are critically regulated by the B-cell lymphoma-2 (Bcl-2) family of proteins. This family is classified into three functionally and structurally distinct categories. The first subfamily includes the pro-survival family members, most notably Bcl-2, Bcl-xL, and Mcl-1. These proteins are critical constituents of the cell, functioning to negatively regulate pro-apoptotic proteins. The second subfamily includes multi-domain pro-apoptotic Bax and Bak which, when activated, act at the mitochondria to induce permeabilization of the outer mitochondrial membrane and release of apoptogenic molecules (e.g. cytochrome C).

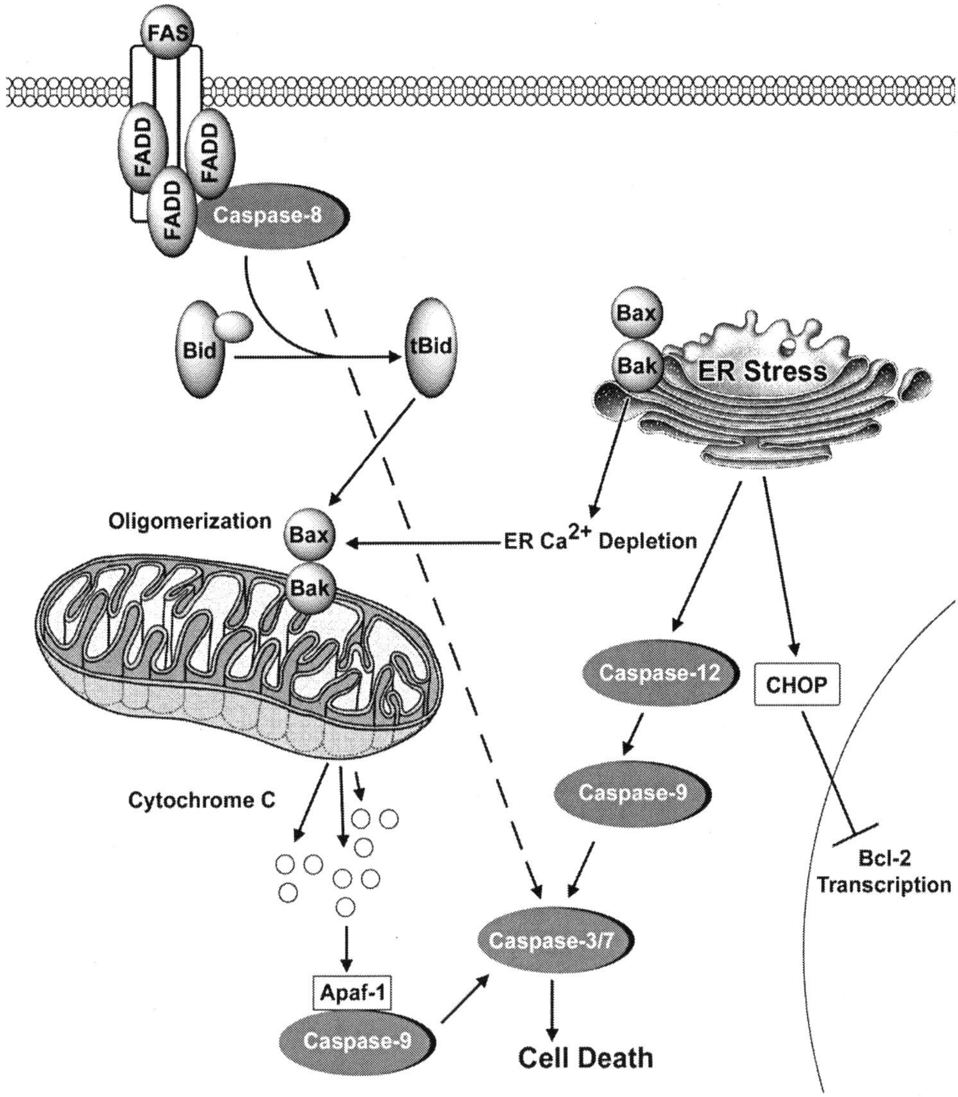

Figure 1. Extrinsic, Intrinsic, and ER Stress-related Apoptotic Pathways.

The final subgroup of the Bcl-2 family comprises the Bcl-2 homology-3 domain (BH3)-only pro-apoptotic proteins. This subfamily includes Bid, Bim, Bad, and Puma, amongst several others. Numerous reports have demonstrated that the balance between pro-survival and pro-apoptotic Bcl-2 family members represents a vital checkpoint in whether or not a cell commits to apoptosis (Danial and Korsmeyer, 2004; Youle and Strasser, 2008).

Although the pro-survival and multi-domain members of the Bcl-2 family play well characterized roles in apoptosis, recent evidence suggests an equally important role for BH3-only proteins in the transmission of death-inducing signals. Indeed, studies demonstrate that BH3-only proteins serve as upstream sensors of intracellular and extracellular death signals that tip the fragile balance between Bcl-2 family proteins towards the initiation of apoptosis (Puthalakath and Strasser, 2002). Following an apoptotic signal, BH3-only proteins are activated transcriptionally or post-translationally and function at the mitochondria to promote the release of apoptogenic factors into the cytosol (Zha *et al.*, 1996; Li *et al.*, 1998; Oda *et al.*, 2000). However, the precise mechanisms by which BH3-only proteins induce apoptosis are unclear and remain under intense investigation.

In this review, we discuss evidence that aberrant expression or activation of BH3-only proteins can trigger apoptosis in neurons, potentially contributing to acute or chronic neuronal disease. A thorough and impressive review on this topic has been published previously; however, more recently published evidence on the subject matter justifies a current analysis of the role of BH3-only proteins in neurodegeneration (Shacka and Roth, 2005). As such, we briefly examine the subdivisions of the Bcl-2 family and data supporting a specific role for BH3-only proteins in the intrinsic, extrinsic, and ER stress-induced apoptotic cascades. We also highlight hierarchical BH3-only pathways, in addition to, regulation of BH3-only proteins. The majority of this review is focused on the involvement of BH3-only proteins in chronic neurodegenerative diseases and acute neuronal traumas. In particular, we discuss specific evidence demonstrating a crucial function of BH3-only proteins in *in vitro* models of neuronal apoptosis, animal models of neurodegeneration or acute neuronal injury, and in human clinical cases. Given their critical roles in neuronal apoptosis, BH3-only proteins may provide new therapeutic targets for the treatment of neurodegenerative diseases and stroke.

Subdivisions of the Bcl-2 Family

Although the overall sequence homology of the Bcl-2 family is relatively low, these proteins contain up to four highly conserved domains, termed Bcl-2 homology (BH) domains 1-4. The subfamilies of Bcl-2 proteins are divided both functionally and structurally based on their BH domains (Fig. 2). The pro-survival proteins contain each of the four BH domains (Fig. 2a). The multi-domain pro-apoptotic proteins contain BH domains 1-3 (Fig. 2b), and as their names suggest, the BH3-only proteins only share a conserved 9-16 residue BH3 domain (Fig. 2c) (Lanave *et al.*, 2004). A variety of techniques such as yeast two-hybrid assays or co-immunoprecipitations have been utilized to demonstrate the importance of the BH3 domain of BH3-only proteins for expressing their pro-apoptotic influence through interactions with other Bcl-2 family members (Zha *et al.*, 1996; Sattler *et al.*, 1997; Cartron *et al.*, 2004).

Pro-Survival Members of the Bcl-2 Family

The major function of the pro-survival members of the Bcl-2 family is to neutralize the pro-apoptotic activity of the multi-domain Bcl-2 family proteins (Borner, 2003). The Bcl-2 protein was the first discovered of the family and is the mammalian homolog of the *C. elegans* pro-survival protein, cell death abnormal-9 (ced-9) (Bakhshi *et al.*, 1985; Hengartner and Horvitz, 1994). The pro-survival role of Bcl-2 is highlighted by studies in Bcl-2-deficient mice, which show a multitude of abnormalities that can be explained by increased cellular death (Veis *et al.*, 1993; Kamada *et al.*, 1995). More recently, Bcl-extra long (Bcl-xL) and Mcl-1 have also been identified as pro-survival proteins of the Bcl-2 family and contain all 4 of the conserved BH domains found in Bcl-2. In addition, Bcl-w and A1 have been described as pro-survival members of the Bcl-2 family; however, they contain less sequence homology within the BH4 domain. Overexpression of any of the five pro-survival Bcl-2 family proteins is sufficient to confer protection against a variety of apoptotic stimuli (Cory *et al.*, 2003). The exact molecular mechanism by which the pro-survival proteins transduce their signals is unclear, but structural studies indicate that the BH1-BH3 domains arrange to expose a hydrophobic surface groove that is required for their pro-survival activity (Yin *et al.*, 1994), via binding to their respective, pro-apoptotic partners (Sattler *et al.*, 1997; Petros *et al.*, 2000; Liu *et al.*, 2003).

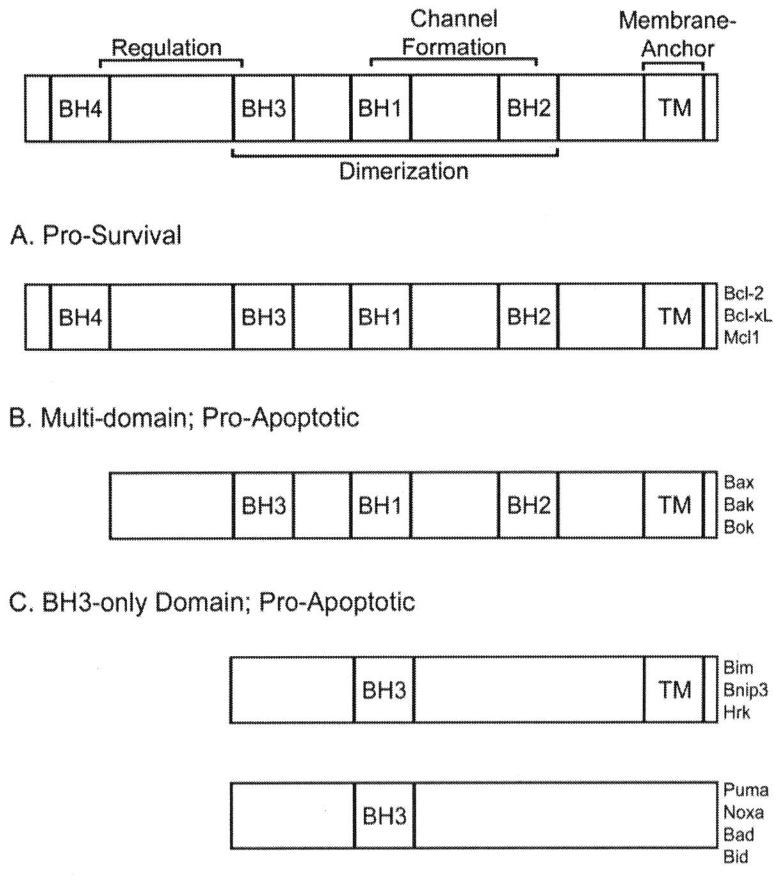

Figure 2. Bcl-2 Family Members.

Multi-Domain Pro-Apoptotic Members of the Bcl-2 Family

The multi-domain pro-apoptotic members of the Bcl-2 family are characterized by their three BH domains (BH1, BH2, and BH3) and they all contain a C-terminal hydrophobic domain (Fesik, 2000). Members of this subfamily include Bax, Bak, and Bok. An additional member, Bcl-X_S, provides an exception to these subfamily characteristics as it differentially comprises the BH3 and BH4 domains in conjunction with the hydrophobic C-terminus (Schultz and Harrington, 2003). Interestingly, the discovery of the 3-dimensional structures of Bax and Bak reveal that they have striking structural similarity to the pro-survival Bcl-2 and

Bcl-xL proteins (Suzuki *et al.*, 2000; Petros *et al.*, 2001; Moldoveanu *et al.*, 2006). The importance of Bax and Bak in apoptotic signaling is emphasized in knock out mouse studies. While deficiency of either Bax or Bak results in only minor abnormalities, doubly deficient mice typically die shortly after birth and have much greater defects in apoptotic signaling in response to apoptotic stimulation, including overexpression of BH3-only proteins (Lindsten *et al.*, 2000). Thus, Bax and Bak are critical downstream mediators of apoptosis with largely overlapping functions (Lindsten *et al.*, 2000; Rathmell *et al.*, 2002). When activated, Bax translocates to the outer mitochondrial membrane where Bak is bound as an integral membrane protein. Here, the proteins oligomerize to form pores that permit the release of apoptogenic factors from the mitochondria, thus inducing apoptosis (Wei *et al.*, 2001).

BH3-Only Members of the Bcl-2 Family

Many studies indicate that the BH3-only subfamily of proteins function as initiators of the mitochondrial apoptotic pathway, linking apoptotic signals to the core apoptotic machinery (Puthalakath and Strasser, 2002). Currently, the BH3-only subfamily consists of numerous pro-apoptotic members: Bik/Blk, Bad, Bid/Bod, Bim, Hrk/Dp5, Noxa, Bnip3, Bnip3L/Nix, Puma, p193, Bmf, Beclin1, and Bcl-G (Opferman and Korsmeyer, 2003; Tsujimoto, 2003; Oberstein *et al.*, 2007). These proteins share sequence homology at their 9-16 amino acid BH3 region (Wong *et al.*, 2008). Overexpression of BH3-only members is sufficient to induce apoptosis (Huang and Strasser, 2000). Many BH3-only proteins exist outside of the mitochondria and require an apoptotic signal to induce their activation and subsequent translocation to the mitochondria (Zha *et al.*, 1996; Li *et al.*, 1998; Puthalakath *et al.* 1999). It is unclear precisely how the BH3-only proteins interact at the mitochondria, yet mutational studies indicate that the BH3 domain is important for initiating apoptosis (Sattler *et al.*, 1997; Petros *et al.*, 2000; Liu *et al.* 2003). The mechanism by which BH3-only proteins initiate apoptosis is controversial and is generally explained by one of two models. The first model asserts that BH3-only proteins induce apoptosis by binding to and neutralizing the ability of pro-survival Bcl-2 proteins to sequester Bax and Bak. The second model suggests that at least some BH3-only proteins (Puma, Bim, tBid) activate Bax and Bak through direct binding (Häcker and Weber, 2006). In fact, the two models are not mutually exclusive and will be discussed in further detail below.

BH3-Only Proteins as Regulators of Apoptosis

Extrinsic Pathway

The extrinsic apoptotic pathway is aptly named as it requires an extracellular signal to induce apoptosis. This cascade is activated when a death ligand (e.g. Fas ligand) binds to a transmembrane death receptor (e.g. Fas) and leads to receptor oligomerization and recruitment of the Fas-associated death domain (FADD) adaptor protein. The death domain of the adaptor protein recruits pro-caspase-8 and forms the death inducing signaling complex (DISC). DISC formation triggers proteolysis of pro-caspase-8 to its active form, caspase-8, which then cleaves and activates two critical executioners of apoptosis, pro-caspase-3 and pro-caspase-7. An alternate function of caspase-8 is cleavage of BH3-only protein Bid to an active form termed truncated Bid (tBid). Truncated bid translocates to the mitochondria and initiates the intrinsic apoptotic pathway by inducing the oligomerization of Bax and Bak, discussed below. Thus, tBid is an important player that links the extrinsic and intrinsic apoptotic cascades (Li *et al.*, 1998; Yin *et al.*, 2006).

Intrinsic Pathway

Central to the intrinsic apoptotic pathway is the mitochondria whose outer membrane's permeabilization is a critical determinant in executing the death program. The permeability of the mitochondria is tightly regulated by the balance of pro-survival and pro-apoptotic Bcl-2 family members. Under normal conditions, pro-survival proteins such as Bcl-2 and Bcl-xL function to neutralize the inherent pro-death activity of multi-domain pro-apoptotic family members. Intrinsic apoptosis is provoked by apoptotic signals that originate within the cell, such as DNA damage or oxidative stress. Following apoptotic stimulation, BH3-only proteins respond by either inactivating pro-survival Bcl-2 members or by directly activating Bax and Bak to form homo- or heteroligomeric pores in the outer mitochondrial membrane. This allows subsequent release of apoptogenic factors such as cytochrome C (cytC) and Smac/Diablo. CytC binds monomeric apoptotic protease-activating factor-1 (APAF-1), inducing a conformational change that permits heptadimerization and recruitment of pro-caspase-9 into a large complex known as the apoptosome (Zou *et al.*, 1999). Once activated, pro-caspase-9 activates the death executioners caspase-3 and caspase-7 (Hakem *et al.*, 1998).

ER Stress-Induced Pathway

The apoptotic pathway induced by ER stress is the least characterized of the three apoptotic cascades in mammals. Some major functions of the ER include protein folding and processing and calcium storage and signaling. When these functions are impaired, unfolded proteins accumulate in the ER and contribute to ER stress. Following ER stress, the unfolded protein response (UPR) is activated to restore functionality to the organelle. When ER stress is too severe to restore function, the cell will undergo apoptosis. Transcription of CHOP/GADD153 is induced in cells experiencing ER stress and evidence suggests it contributes to this apoptotic pathway at least in part by repressing the promoter region of Bcl-2 and thereby sensitizing cells to the pro-death effects of BH3-only proteins (McCullough *et al.*, 2001). An addititional role of CHOP/GADD153 may be induction of Bim, Puma, and Noxa (Li *et al.*, 2006; Puthalakath *et al.*, 2007; Galehdar *et al.*, 2010). Another mechanism by which ER stress may induce apoptosis is through a unique pathway of caspase activation. Pro-caspase-12 is present within the ER and is cleaved in response to ER stress; however, the mechanism of proteolysis is unclear. Following activation, caspase-12 directly activates caspase-9 to induce apoptosis independently of the mitochondria in response to ER stress (Morishima *et al.*, 2002). However, it has been reported that caspase-12 does not play a role in the ER stress-induced apoptotic pathways in human white blood cells (Saleh *et al.*, 2004).

Thus, it is unclear whether caspase-12 plays a substantial role in human neurodegenerative disease.

In addition to the above mitochondria-independent mechanism of apoptosis, the proximity of mitochondria to the ER within a cell suggests that crosstalk between the two organelles may play a significant role in apoptosis. Bcl-2 and Bax, in cooperation with inositol 1,4,5-triphosphate receptors (IP3Rs), regulate the release of ER calcium into the cytosol (Chen *et al.*, 2004). As the sensitivity of the mitochondria to trigger apoptosis is responsive to changes in intracellular calcium levels, interplay between the two organelles during apoptotic stimulation has been suggested (Danial and Korsmeyer, 2004; Verkhratsky *et al.*, 2005). Some studies suggest that ER stress may induce the oligomerization of Bax and Bak on the ER membrane, leading to Ca^{2+} store depletion and apoptosis. Consistent with this theory, it has been demonstrated that ER stress induces loss of mitochondrial membrane potential and release of cytC (Rao *et al.*, 2004). Furthermore, overexpression of pro-survival Bcl-2 and Bcl-xL are sufficient to protect the cell from ER stress-induced apoptosis and Bax/Bak doubly deficient cells are resistant to many ER stressors, suggesting that ER initiated death signals are amplified and integrated at the level of the mitochondria (Distelhorst *et al.*, 1996; Wei *et al.*, 2001).

Involvement of BH3-Only Proteins in Apoptosis

It has become evident that tBid is an important BH3-only protein linking the extrinsic apoptotic cascade to the intrinsic apoptotic amplification loop. Indeed, in some cell types, Bid deficiency results in an inability to induce Fas-receptor ligand apoptosis, highlighting its importance in mediating extrinsic apoptosis (Yin *et al.*, 2006). Additional studies underline the importance of other BH3-only proteins in regulating the apoptotic cascades discussed above. While the exact mechanism is unknown, Bad functions to lower the threshold required for tBid to induce mitochondrial membrane permeabilization (Howells *et al.*, 2010). Further, DNA damage-induced apoptosis relies on the upregulation of Puma in a p53-dependent manner (Jeffers *et al.*, 2003; Villunger *et al.*, 2003). Both Puma and Noxa have been highlighted in studies of gamma and UV radiation-induced apoptosis. However, while each insult induces p53, only Puma is essential for gamma radiation-induced apoptosis, whereas Noxa is required for propagating an apoptotic signal in response to UV radiation (Naik *et al.*, 2007). These findings suggest that different cellular insults may trigger the intrinsic apoptotic cascade through differential activation of BH3-only proteins. Consistent with a role for Noxa in intrinsic apoptosis, overexpressed Noxa localizes to the mitochondria and is sufficient to induce apoptosis in some cell types (Oda *et al.*, 2000; Schuler *et al.*, 2003). A role for Hrk/DP5 in intrinsic apoptosis has also been established as neurons from mice deficient for Hrk/DP5 display increased resistance to nerve growth factor withdrawal, although to a lesser degree than mice deficient for Bax, thus indicating that other BH3-only proteins are more likely involved (Imaizumi *et al.*, 2004). Furthermore, studies indicate that Bim is required for intrinsic apoptosis in response to growth factor withdrawal in a variety of cell types, including neurons (Bouillet *et al.*, 2002; Putcha *et al.*, 2003). Although a weaker apoptotic stimulus, Bnip3 has also been linked to intrinsic apoptosis through activation of Bax and Bak and induced mitochondrial permeability (Kubli *et al.*, 2007). It is clear that overwhelming evidence is accumulating to support a critical role for BH3-only proteins in both extrinsic and intrinsic apoptosis.

Increasing evidence suggests that ER stress may be sufficient to trigger the mitochondrial apoptotic cascade (Annis *et al.*, 2001; Wei *et al.*, 2001). Following treatment with the ER stressor, tunicamycin, there is an increase in UPR stress responders, such as BIP and GRP94. In addition, there is an induction of Puma, activation of the mitochondrial apoptotic pathway, and subsequent release of CytC in neuroblastoma cells, suggesting a functional role for Puma as a mechanistic link between ER stress and mitochondrial apoptosis. Consistent with these results, neuronal apoptosis occurs following overexpression of Puma while cells deficient in Puma are resistant to ER stress-induced apoptosis (Reimertz *et al.*, 2003). One mechanism of inducing ER stress is through activation of CHOP/GADD153 and recent evidence suggests that CHOP/GADD153 may be responsible for the upregulation of Puma, Bim, and Noxa, and subsequent activation of apoptosis (Li *et al.*, 2006; Puthalakath *et al.*, 2007; Galehdar *et al.*, 2010). While both Puma and Bim are implicated in ER stress, Bim appears to play a more important role in thymocytes, whereas Puma is more critical in neurons (Reimertz *et al.*, 2003; Puthalakath *et al.*, 2007). Collectively, these data suggest a key role for BH3-only proteins in mediating ER stress-induced apoptosis.

Hierarchical Interactions between Bcl-2 Family Members

The Bcl-2 family of proteins forms a complex network of interactions which maintain the delicate balance between cell survival and apoptosis; yet, interactions between family members remain incompletely understood. Various studies have demonstrated that Bax and Bak are downstream effectors of BH3-only-initiated apoptosis as cells doubly deficient in these pro-death proteins are resistant to a variety of apoptotic stimuli, including overexpression of BH3-only proteins (Wei *et al.*, 2001; Zong *et al.*, 2001). Furthermore, overexpression of Bcl-2 or Bcl-xL is sufficient to block the apoptotic signal transmitted by expression of BH3-only proteins or by Bax and/or Bak (Gross *et al.*, 1999). This evidence suggests that BH3-only proteins function to initiate and transmit apoptotic signals to the mitochondria and/or ER through direct interactions with either Bcl-2/Bcl-xL and/or direct interactions Bax/Bak.

There are currently two opposing models to describe the mechanism by which BH3-only proteins activate Bax and Bak: the displacement model and the direct activation model (Fig. 3). According to the displacement model (Fig. 3a), Bax and Bak are sequestered by pro-survival Bcl-2 members in a reversible manner. Upon receiving an apoptotic signal, BH3-only proteins bind pro-survival proteins to induce the release of Bax and Bak to stimulate apoptosis (Willis *et al.*, 2005; 2007). The BH3-only proteins Bim, tBid, and Puma can bind all pro-survival Bcl-2 family members while Bad and Noxa only bind to specific classes of Bcl-2 family proteins (Chen *et al.*, 2005). Indeed, studies suggest that Bax and Bak induce apoptosis without engaging BH3-only proteins (Willis *et al.*, 2007). Further, Bik was shown to neutralize Bcl-xL to displace Bax and induce apoptosis (Gillisen *et al.*, 2007). This model suggests a simple hierarchy of Bcl-2 family proteins; BH3-only proteins lie upstream of Bcl-2-like proteins, which function upstream of multi-domain pro-apoptotic Bax and Bak.

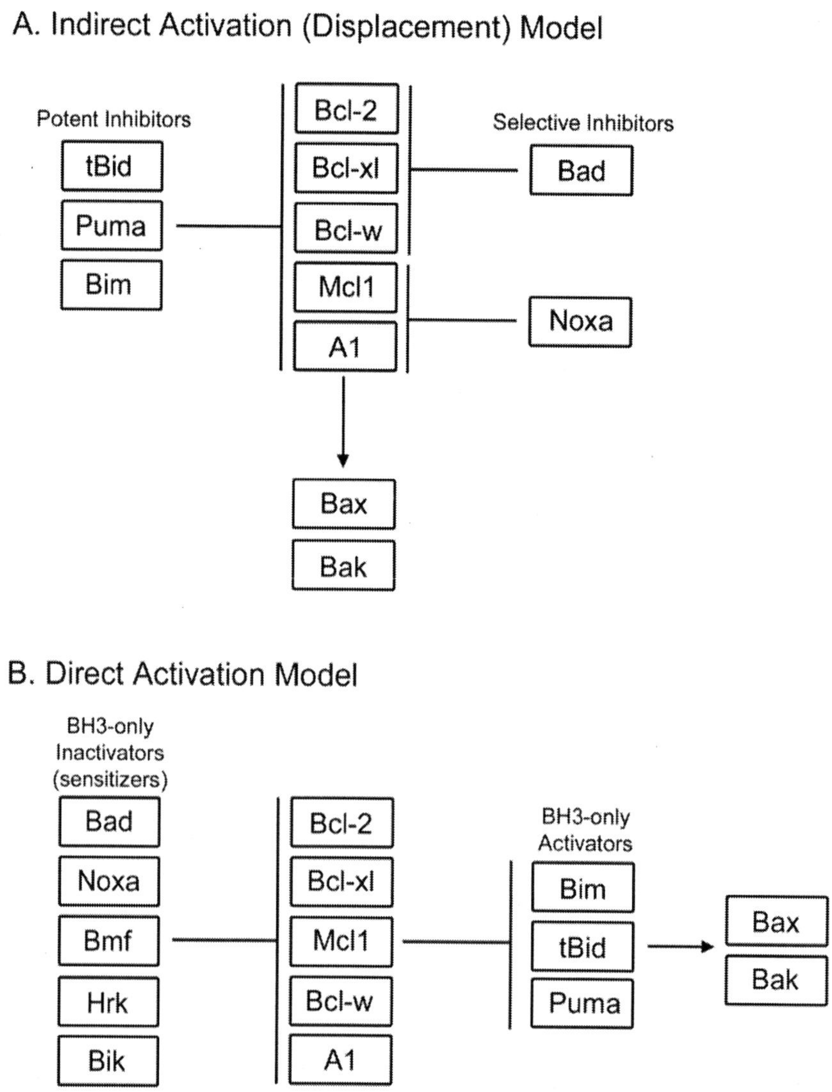

Figure 3. BH3-only Activation of Bax/Bak.

Conversely, the direct activation model suggests that select BH3-only proteins have the inherent ability to directly activate Bax and Bak (Fig. 3b). Under this model, those BH3-only proteins that do not have the ability to directly bind Bax and Bak are termed "sensitizers" or "inactivators" and they function to neutralize the pro-survival Bcl-2 proteins (Letai *et al.*, 2002; Kuwana *et al.*, 2005). Yet, overwhelming evidence suggests that BH3-only proteins have the ability to bind to Bak and Bax directly. Several reports suggest that peptides representing the BH3 domain of Bid or Bim oligomerize with Bax and Bak to induce cytC release (Letai *et al.*, 2002). Further, Kim *et al.* (2006) found similar results when examining full length BH3-only proteins. These data have been substantiated by additional accounts and collectively these reports have been instrumental in establishing a hierarchy of Bcl-2 family interactions (Kuwana *et al.*, 2005). As proposed by the direct activation model, evidence suggests that two classes of BH3-only proteins exist with respect to apoptosis: those that act as inactivators of Bcl-2-like proteins and those that directly activate Bax and Bak. Recent studies indicate that inactivators (Bad, Bmf, Bik/Blk, Hrk/Dp5, and Noxa) function upstream to inhibit Bcl-2, Bcl-xL, and Mcl-1. Conversely, BH3-only activators (tBid, Bim, Puma) are sequestered and held in an inactive state by Bcl-2, Bcl-xL, and Mcl1. Upon apoptotic stimulation, the BH3-only inactivators bind to Bcl-2-like proteins to induce the release of BH3-only activators, resulting in activation of Bax and Bak and ensuing cytC release (Kim *et al.*, 2006). To further support this model, mutagenesis of Bax and Bak demonstrated that they are not directly regulated by Bcl-2-like proteins. The model proposed by Kim *et al.* (2006) also suggests that pro-survival Bcl-2 members are regulated differentially by BH3-only inactivators. Indeed, Bad displaces tBid, Bim, and Puma from Bcl-2 and Bcl-xL. On the other hand, Noxa displaces Puma from Mcl-1. Collectively, these data support a model where BH3-only activators function downstream of both BH3-only inactivators and pro-survival Bcl-2 family members. Regardless of which model is correct, it is clear that BH3-only proteins are essential regulators of Bax/Bak activation and consequent apoptosis.

Regulation of BH3-Only Proteins

Transcriptional Regulation

The decision to live or succumb to apoptosis relies on the ratio of pro-survival Bcl-2 family members to pro-apoptotic family members, and in particular, BH3-only proteins. As such, regulation of BH3-only proteins represents an important mechanism by which cells maintain the ability to finely tune the interactions between pro-death and pro-survival cellular signals. The cell-type specific expression of BH3-only proteins suggests complex regulation at the transcriptional and post-translational levels (Table 1). Numerous BH3-only proteins are clearly regulated at the transcriptional level including Bim, Puma, Noxa, Bmf, Bnip3, Bik, and Hrk/DP5. The transcription factor p53 is implicated in the transcriptional induction of several BH3-only proteins. Specifically, p53 induces the expression of Puma, Noxa, and Bim (Han *et al.* 2001, Yu *et al.* 2001; Cregan *et al.*, 2004; Erlacher *et al.*, 2005). Although all three of these BH3-only proteins are regulated by p53, the transcription factor appears to influence the expression of Puma more extensively (Chipuk *et al.*, 2005). Interestingly, histone deactylases suppress p53-dependent Puma upregulation and prevent apoptosis in cortical mouse neurons (Uo *et al.*, 2009). Hypoxia-inducible factor-1 (HIF-1) has also been demonstrated to upregulate Bnip3 and Noxa in some cell lines, providing a role for BH3-only proteins in hypoxia-induced death (Sowter *et al.*, 2001; Kim *et al.*, 2004). Under conditions of ER stress, Bim appears to be transcriptionally upregulated via a CHOP/GADD153-dependent mechanism (Puthalakath *et al.*, 2007). Although following nerve growth factor withdrawal in superior cervical ganglia neurons, Bim is induced in a JNK-dependent manner (Putcha *et al.*, 2003). Forkhead transcription factor like 1 (FKHR1) has also been described as a positive transcriptional regulator of both Bim and Noxa in neuronal cells (Linseman *et al.*, 2002; Obexer *et al.*, 2007). Indeed, inactivation of FKHR1 via insulin like growth factor-1 (IGF-1) treatment reduced the expression of Bim and protected cerebellar granule neurons from apoptosis (Linseman *et al.*, 2002). Interestingly, a novel mechanism of regulation at the RNA processing level has been suggested for Bim in neurons because although each of the 3 known Bim isoforms are from the same transcript, only Bim_{EL} is induced in response to growth factor withdrawal in neurons (Gilley *et al.*, 2003). The diverse mechanisms that exist to regulate of Bim expression suggest complex cell type- and/or stimulus-specific regulation of BH3-only proteins.

Table 1. Regulation of BH3-Only Proteins

Protein	Mode of Regulation	Stimulus	Known Regulatory Factors	References
Bid	Posttranslational	Death receptor activation	Caspase 8, GranzymeB, Calpain, Lysozyme	20, 56, 83, 139
Bim	Transcriptional	ER Stress, nerve growth factor withdrawal, Aβ toxicity	CHOP, FOXO, p53	87, 118, 119
	Posttranslational	UV light, Nerve growth factor withdrawal, Growth factor stimulation	Dynein light chain, Ubiquitin ERK1/2, JNK	117, 119
Bik	Translational	Alternative reading frame (ARF) tumor suppressor overexpression	CTBP2	75
	Posttranslational	Unknown	Unknown kinase	150
Bnip3	Transcriptional	Hypoxia	HIF-1	137
Puma	Transcriptional	DNA Damage	p53, CTBP1/2	45, 52, 168
Noxa	Transcriptional	Etoposide treatment	p53, HIF-1, CTBP	27, 45, 71
Hrk	Transcriptional	Interleukin-3 (IL-3), Aβ toxicity	DREAM	64, 128
Bmf	Translational	Alternative reading frame (ARF) tumor suppressor overexpression	CTBP2	75
	Posttranslational	Anoikis	Myosin V	119, 120
Bad	Posttranslational	Cytokine or growth factor withdrawal	ERK1/2, AKT, 14-3-3 chaperone proteins	134, 169

Hrk/DP5 is upregulated following treatment of cortical neurons with Aβ and its expression peaks when the cells are committed to die (Imaizumi *et al.*, 1999). Furthermore, Hrk/DP5 is repressed transcriptionally and is downregulated through binding of DREAM to a sequence in its 3' untranslated region (Sanz *et al.*, 2001). In addition to DREAM, C-terminal-binding proteins (CTBP1 and CTBP2) play a key role in the transcriptional repression of BH3-only proteins. Knock out of both CTBP1 and CTBP2 in epithelial cells results in upregulation of Noxa, in addition to, hypersensitivity to apoptotic stimuli. These results have been confirmed by microarray analysis, which showed an increase in the expression of both Puma and Noxa in CTBP-deficient cells (Grooteclaes *et al.*, 2003). Further, CTBP2 has also been implicated in the transcriptional repression of Bik and Bmf (Kovi *et al.*, 2010). The sensitivity of apoptosis to the ratio of pro-survival to pro-apoptotic proteins suggests that transcriptional regulation of BH3-only proteins is an important mechanism mediating programmed cell death.

Post-translational Regulation

Many different mechanisms of post-translational regulation have been suggested for BH3-only proteins, such as cleavage, phosphorylation, localization, and degradation. As previously discussed, cleavage of Bid represents a crucial step in linking the extrinsic apoptotic pathway to the intrinsic apoptotic cascade. When the extrinsic apoptotic pathway is stimulated, ligation of death receptors induces activation of caspase-8, which, in turn, cleaves and activates Bid (Li *et al.*, 1998). In addition to caspase-8, evidence also suggests that granzymeB, lysozyme, and calpain each have the ability to cleave Bid (Heiben *et al.*, 2000; Chen *et al.*, 2001; Stoka *et al.*, 2001). Cleavage of Bid is sufficient to activate its pro-death activity since the myristoylated fragment readily localizes to membranes such as the mitochondria where it interacts with Bax/Bak to induce cell death (Zha *et al.*, 2000).

In addition to proteolytic cleavage, phosphorylation of BH3-only proteins has been shown to regulate their pro-apoptotic activity. While the mechanism is incompletely understood, Bad, Bim, Bmf, and Bik have been shown to be regulated by phosphorylation. In response to growth factors like IL-3 or EGF, Bad can be phosphorylated at Ser 112 by ERK/Map kinases or at Ser 136 by Akt. Phosphorylated Bad binds with a higher affinity to cytosolic 14-3-3 chaperone proteins than to pro-survival Bcl-2 family members, thus abrogating its pro-death activity. Dephosphorylation of both residues is required to restore the apoptotic potential of Bad (Zha *et al.*, 1996; She *et al.*, 2005). In addition to Bad, the pro-apoptotic activity of Bim is also modulated by phosphorylation at several serine residues (including Ser55, 65, 69, and 100). Pro-survival signals such as growth factor stimulation provoke ERK1/2 to phosphorylate Bim which targets it for proteosomal degradation (Ley *et al.*, 2003). Conversely, the pro-apoptotic activity of Bim can be stimulated by phosphorylation, as growth factor withdrawal in neurons induces phosphorylation at Ser65 in a JNK-dependent manner (Putcha *et al.*, 2003). Similarly to JNK-dependent phosphorylation of Bim, studies suggest that phosphorylation of the BH3-only protein Bik increases its ability to initiate apoptosis (Verma *et al.*, 2001). While the phosphorylation of BH3-only proteins is complex, these data suggest that it is an important mechanism for both activating and inactivating the pro-death activity of BH3-only proteins, most likely in a context-specific manner.

Select BH3-only proteins are held in an inactive state through interactions with cytoskeletal structures. Of the three known isoforms of Bim (Bim_L, Bim_{EL}, and Bim_S), Bim_{EL} and Bim_L associate with the dynein light chain 2 and are thereby sequestered to the microtubule-associated dynein motor complex. Both isoforms are freed upon dephosphoylation following serum withdrawal and translocate to the mitochondria to interact with Bcl-2 and Bcl-xL to induce apoptosis (Puthalakath *et al.*, 1999). Similarly, Bmf associates with dynein light chain 2 and is sequestered by myosin V motors. Certain signals, such as anoikis, free Bmf to translocate to the mitochondria to interact with Bcl-2 (Puthalakath *et al.*, 2001). This mechanism of posttranslational regulation demonstrates the importance of the subcellular localization of BH3-only proteins.

Finally, degradation of BH3-only proteins is an important mechanism to maintain the balance between pro-survival and pro-apoptotic proteins. As mentioned above, Bim is targeted for ubiquitination and proteosomal degradation following phosphorylation by ERK1/2 (Luciano *et al.* 2003; Harada *et al.* 2004). Further, following growth factor withdrawal in osteoclasts, it has been demonstrated that Bim ubiquitination is reduced, increasing the amount of Bim expression and inducing apoptosis (Akiyama *et al.* 2003). While this list of regulatory mechanisms is not exhaustive, it serves to highlight the multitude of diverse methods for regulation of BH3-only proteins (Table 1).

Implications of BH3-Only Proteins as Key Players in Neurodegenerative Disease

Amyotrophic Lateral Sclerosis

The complexity of BH3-only protein regulation allows for delicate control over the apoptotic process, but dysregulation has been implicated in several neuronal diseases (Table 2). Amyotrophic lateral sclerosis (ALS) is a fatal neuromuscular disorder characterized by the progressive and selective loss of upper and lower motor neurons accompanied by skeletal muscle atrophy, ultimately leading to paralysis and death due to upper respiratory failure. In addition, defects in protein quality control have been suggested as an underlying cause of ALS, thus linking ER stress-induced apoptosis to neuronal death in models of ALS. Indeed, it has been reported that the BH3-only protein Puma is upregulated and sufficient to execute ER stress-induced apoptosis in cultured motoneurons. Consistent with this result, Puma$^{-/-}$ cultured motoneurons are protected against ER stress (Kieran *et al.*, 2007).

Table 2. BH3-only proteins and neurodegeneration: highlights in *in vivo* animal models and human disease

BH3-only Protein	System	Disease	References
Bid	G93A mSOD1 mice	Amytrophic Lateral Sclerosis	48
	MPTP-treated mice	Parkinson's Disease	153
	R1/6, R2/6, Tet/HD94 mice	Huntington's Disease	40, 171
	Ischemic injury (mice)	Cerebral Ischemia	115
Bim	G93A mSOD1 mice	Amytrophic Lateral Sclerosis	58
	MPTP-treated mice	Parkinson's Disease	111
	R1/6, R2/6, Tet/HD94 mice	Huntington's Disease	40, 171
	Ischemic injury (mice)	Cerebral Ischemia	144
Bnip3	Transient focal ischemia (rats)	Cerebral Ischemia	3
Bad	G93A mSOD1 mice	Amytrophic Lateral Sclerosis	154
Puma	G93A mSOD1 mice	Amyotrophic Lateral Sclerosis	69
	6-OHDA-treated mice	Parkinson's Disease	8
	Transient global ischemia (rats)	Cerebral Ischemia	144
Noxa	Ischemic injury (rats)	Cerebral Ischemia	71
	Neuronal axotomy (mice)	Nerve/brain injury	72
Hrk	Post-mortem human anterior horn spine	Amytrophic Lateral Sclerosis	135
	Optic nerve transection (rats)	Nerve/brain injury	162
	PolyQ-AR neurotoxicity (mice)	Spinal Muscular Atrophy	167

The role of BH3-only proteins in the G93A mutant hSOD1 (msod1) transgenic mouse model of familial ALS has been examined recently. Kieran *et al.* (2007) investigated the G93A mouse model for BH3-only protein expression and found that prior to symptom onset,

Puma is upregulated. More impressively, this group revealed that Puma$^{-/-}$ mice show improved motoneuron survival and delayed disease onset. Yet, these mice did not demonstrate improved lifespan, indicating that additional factors may be responsible for later stages of disease progression. Similarly, Bim is upregulated in this same mouse model and ablation of Bim increases lifespan and decreases neuronal cell death in the ventral horn of G93A mSOD1 mice (Hetz *et al.*, 2007). An additional BH3-only protein, Bad, is reportedly increased in G93A mSOD1 symptomatic mice as well (Vukosavic *et al.*, 1999). Another study showed that tBid is increased in the spinal cord of G93A mSOD1 mice, suggesting a role for extrinsic apoptosis in ALS (Guegan *et al.*, 2002). These data emphasize the potential role of BH3-only proteins in the progression of ALS in a mouse model of familial ALS.

While the involvement of BH3-only proteins in human ALS cases has not been studied comprehensively, Shinoe *et al.* (2001) report that Hrk/DP5 expression is upregulated in spinal motor neurons from post-mortem ALS patients. These authors found an increase in the levels of Hrk/DP5 mRNA and protein in the anterior horn of ALS patients. Further, Hrk/DP5 was shown to co-precipitate with Bcl-2 from ALS patient spinal cord. These data suggest that Hrk/DP5 may function to inactivate the pro-survival activity of Bcl-2. Interestingly, a decrease in Bcl-2 has also been reported in post-mortem ALS patients (Mu *et al.*, 1996; Martin *et al.*, 1999). In addition, motoneurons from the anterior horn of post-mortem sporadic ALS patients show increased ER stress and defects in protein degradation, supporting a role for ER stress and the associated BH3-only proteins, Puma and Bim. In further support of a role for BH3-only proteins in ALS pathology, p53 is elevated in the spinal cord of ALS patients; therefore, Puma, Noxa, and Bim may also be elevated (Erlacher *et al.*, 2005; Eve *et al.*, 2007). These data suggest that BH3-only proteins may play significant roles in the etiology of human ALS and may serve as molecular targets for future ALS therapies.

Parkinson's Disease

Parkinson's disease (PD) is a devastating neurological disorder characterized by degeneration of dopaminergic (DA) neurons in the substantia nigra pars compacta. In the 1-methyl-4-phenyl-1,2,3,6-tetrahydropyridine (MPTP) mouse model of PD, DA neuro-degeneration can be largely accounted for by complex I deficiency leading to Bax activation and mitochondrial apoptosis (Vila and Przedhorski, 2001; Perier *et al.* 2005). Several reports have demonstrated BH3-only protein activation and subsequent apoptosis in various *in vitro* models of PD which use either MPTP or 6-hydroxydopamine (6-OHDA) to induce mitochondrial dysfunction, oxidative stress, and cell death. Under these conditions, a dominant role for Puma has been implicated in DA neuron death. Exposure to 6-OHDA induces Puma expression in primary ventral midbrain cultures and genetic deletion of Puma in mice prevents DA neuron death in these models (Bernstein *et al.*, 2011). Further, treatment with 6-OHDA induces the expression of Puma mRNA and protein in a p53-dependent manner, consistent with the notion that Puma is a key transcriptional target of p53 (Biswas *et al.*, 2005b; Kieran *et al.*, 2007). However, evidence suggests that the co-activation of p53 and Puma is not always observed in PD models. For example, in a mouse model of complex I blockade, p53 can induce death without the activation of Puma (Perier *et al.*, 2007). Further, Puma is induced in SH-SY5Y neuroblastoma cells exposed to 6-OHDA by a mechanism that is independent of p53, but is dependent on p38 MAPK (Gomez-Lazaro *et al.*, 2008). As Puma is a known activator of Bax, this BH3-only protein may induce apoptosis by direct activation of Bax-dependent cytC release, regardless of the mechanism by which it is induced. Further, the herbicide paraquat has also been used to induce DA cell death and recent evidence implicates the BH3-only proteins Bid, Bnip3, and Noxa in paraquat-induced toxicity (Fei *et al.*, 2008). Taken together, these data suggest that BH3-only proteins play a significant role in the DA neuron death that underlies neurodegeneration associated with PD.

While the precise mechanism leading to DA neuron death in PD remains largely unknown, several animal models have been developed that mimic the disease progression and symptoms. Many of these models simulate the disease using agents or neurotoxins that selectively kill catecholaminergic neurons or induce oxidative stress (Betarbet *et al.*, 2002). For instance, MPTP treatment of mice results in cytC release, caspase-3, -8, and -9 activation, and truncation of BH3-only Bid. In the substantia nigra, caspase-9 is activated downstream of cytC release and its inhibition prevents the activation of caspase-3 and caspase-8. These data indicate that MPTP induces caspase-8 activation and cleavage of Bid downstream of caspase-9 in treated mice (Viswanath, 2001). In addition to Bid, a role for Bim has been elucidated in animal models of PD. Studies by Perier *et al.* (2007) using MPTP as an insult in wild type and $p53^{-/-}$ mice suggest that two molecular pathways are responsible for inducing mitochondrial-dependent death in DA neurons. One pathway induces the expression of p53 and subsequent transcriptional induction of Bax. A second pathway is involved in Bax translocation to the mitochondria in a manner dependent on JNK/c-Jun-mediated induction of Bim (Perier *et al.*, 2007). However, the involvement of Bim in PD models remains controversial, as some *in vitro* studies indicate that Bim interference is not protective (Biswas *et al.*, 2005b, Steckley *et al.*, 2007).

Although a specific role for BH3-only proteins has not been examined directly in post-mortem human PD brains, several elements of their associated apoptotic cascades have been observed in PD tissue. Consistent with animal models of PD, complex I deficiency, JNK activation, Bax activation, and activation of caspase-3 have been reported in post-mortem PD brain (Tatton, 2000; Hartmann *et al.*, 2000; Ferrer *et al.*, 2001; Keeney *et al.*, 2006). Furthermore, p53 is elevated in the striatum of PD patients and could function to upregulate its BH3-only targets Puma, Noxa, or Bim (Erlacher *et al.*, 2005; Mogi *et al.*, 2007). Although data accumulated in diverse animal models of PD suggest a significant role of BH3-only proteins in PD, the specific involvement of these pro-death proteins in human PD cases is largely unexplored and warrants further investigation.

Alzheimer's Disease

Alzheimer's disease (AD) is a neurodegenerative disorder which manifests itself as progressive neuronal loss, gradual decline in cognitive function, and eventual dementia. The extensive loss of neurons in the AD brain has led many to speculate that apoptosis plays a role in the etiology of this disease; however, reports have produced conflicting results (Lassman *et al.*, 1995; Su *et al.*, 1997; Roth *et al.*, 2001). Nonetheless, many studies have focused on the potential role of Bcl-2 family members in this disease. AD is characterized by accumulation of β-amyloid plaques and neurofibrillary tangles, which are thought to play a critical role in the etiology of the disease. Recent studies indicate that the precursor to β-amyloid plaques, Aβ peptide, may utilize BH3-only proteins to induce apoptosis in neuronal cultures. Indeed, Biswas *et al.* (2007) have shown that Bim is induced by Aβ in hippocampal and cortical neuronal cultures in a cdk4-dependent manner. Further, $Bim^{-/-}$ cortical neurons are protected from Aβ-induced apoptosis. These data are consistent with evidence showing that Bim is induced following cdk4 activation under conditions of growth factor withdrawal (Puthalakath and Strasser, 2002; Biswas *et al.*, 2005a). It was also reported that Aβ induces Hrk/DP5 expression in cortical neuron cultures (Imaizumi *et al.*, 1999). An additional feature of patients with AD is increased ER stress. Thus, an additional possibility is that Bim induction may be a result of an ER stress-related pathway which ultimately induces apoptosis via the stress sensitive kinase JNK (Urano *et al.*, 2000). Surprisingly, in contrast to the above *in vitro* studies, BH3-only proteins have not been investigated in mouse models of AD. Further, of the known BH3-only proteins, only Bim has been evaluated directly in post-mortem AD tissue. Biswas *et al.* (2007) reported Bim upregulation in entorhinal cortical neurons of AD brains. Moreover, neurons overexpressing Bim also have altered expression of cdk4. In addition, indirect evidence suggests a possible role of BH3-only proteins in AD. For example, p53 is elevated in the temporal and frontal lobes of AD brains, suggesting that it may function to transcriptionally upregulate Puma, Noxa, and/or Bim. Finally, altered expression of JNK is observed in post-mortem AD brains, consistent with a possible role for Bim in AD disease progression (Pei *et al.*, 2001). Further investigation into the mechanism of BH3-only induction in AD models and human clinical cases is clearly required and may provide important therapeutic insights into chronic neurodegenerative disease.

Huntington's Disease

Huntington's disease (HD) is characterized by a decrease in muscle coordination, cognitive decline, and dementia. It is a type of polyglutamine (polyQ) disease that results from a neurotoxic mutation within the huntingtin gene resulting in an expanded number of CAG repeats. The underlying pathology of HD involves the death of medium spiny GABAergic neurons of the striatum. A role for Bad has been established within primary striatal neurons subjected to 3-nitroprionic acid, a neurotoxin model of HD, as an increase in membrane-bound Bad was observed following treatment (Galas *et al.*, 2004). The R6/1 mouse produces a mutant huntingtin protein and has been examined for expression of BH3-only proteins. Bid is elevated in the striatum of the R6/1 mouse and both Bid and Bim_{EL} are elevated in the striatum of the R6/1/$BDNF^{-/-}$ mouse. Bid appears to be upregulated early in disease progression, while increased Bim_{EL} appears later. In this model, the levels of Bax, Bak, Bad, Bcl-2, and Bcl-xL remain unmodified, suggesting a possible mechanism of inducing death through specific upregulation of BH3-only proteins. The Tet/HD94 and R6/2 mouse models were also examined and revealed elevations in Bid and Bim_{EL} expression (Zhang *et al.*, 2003; García-Martínez *et al.*, 2007).

Additional studies in mice expressing mutant huntingtin demonstrated that Bax translocates to the mitochondria within striatal neurons following upregulation and activation of Bim_{EL} (Leon *et al.*, 2010). Although only minimally investigated at this point, these data suggest a central role for BH3-only proteins in animal models of HD. Future studies should focus on examining the expression of BH3-only proteins in human HD to validate the findings observed *in vitro* and in animal models.

Ischemia/Reperfusion

Cerebral ischemia and reperfusion (I/R) have devastating effects on the brain and increasing evidence suggests that neuronal death may be induced by activation of the mitochondrial apoptotic pathway. Animal models of cerebral I/R recapitulate many of the pathological processes of acute neuronal injury such as stroke. Similar to various other diseases discussed, the JNK pathway appears to be involved in the upregulation of BH3-only proteins. Specifically, Hrk/DP5 levels are shown to be increased in a rat model of I/R (Guan *et al.*, 2006). Consistent with this report, a recent investigation of the localization of JNK family members following transient middle cerebral artery occlusion revealed that JNK2 and JNK3 translocate to the mitochondria following ischemic stimulation (Zhao and Herdgen, 2009). As an alternate mechanism to JNK activation, the E2F1 transcription factor has also been implicated in neuronal death following ischemia (MacManus *et al.*, 1999; 2003). E2F1 has been shown to induce cell death through induction of p53 (Hsieh *et al.*, 2002) and overexpression of E2F1 induces apoptosis of cultured cortical neurons (Hou *et al.*, 2000). Finally, Bim induction in trophic factor-deprived neurons can occur via an E2F1-dependent mechanism (Biswas *et al.*, 2005a). These findings suggest that ischemic neuronal death may be due, in part, to activation of transcription effectors such as E2F1 or p53 and induction of their downstream BH3-only targets Puma and Bim.

Additional studies also support a key role for BH3-only proteins in neuronal death due to ischemic or I/R injury. In a hippocampal cell line, Puma was upregulated following the overexpression of hypoxia inducible factor-1α (HIF-1α) while siRNA targeting of Puma was protective. Bnip3 was similarly upregulated in this model, although its overexpression alone was not sufficient to induce apoptosis (Aminova *et al.*, 2008).

Induction of transient global ischemia in rats reportedly upregulates Puma in hippocampal neurons where it localizes to mitochondria and is bound to Bcl-xL and Bax. Consistent with the known regulation of Puma by p53, pfithrin-α blocked the induction of Puma (Niizuma *et al.*, 2009). More recent *in vivo* studies have also generated evidence that Puma, and less substantially, Bim, are induced in the hippocampus of mice following ischemic injury. Indeed, genetic knock down of either BH3-only protein conferred at least partial protection (Ness *et al.*, 2006; Tsuchiya *et al.*, 2010). In addition to Puma and Bim, a strong correlation exists between Bid knock down and infarct size following ischemic injury and neurons from Bid-deficient mice are resistant to ischemia *in vivo* (Plesnila *et al.*, 2001). There is also evidence to suggest that Bnip3 accumulates in striatal and cortical neurons following transient focal ischemia and associated induction of HIF-1 in rats (Althaus et al, 2006). HIF-1 also induces Noxa under ischemic conditions and oligonucleotides against Noxa decrease infarct size and prevent hypoxic-induced death in rat brain tissue (Kim *et al.*, 2004) Thus, BH3-only proteins are critical mediators of pro-apoptotic signals in response to neuronal ischemia injury in animal models. As of yet, the role of BH3-only proteins in ischemic or I/R brain injury in human (e.g. stoke) has not been evaluated.

Nerve/Brain Injury

Peripheral nerve injury and traumatic brain injury can often lead to apoptotic neuronal death and alterations in the expression of BH3-only proteins have been observed. Following acute spinal cord injury (SCI), Hrk/DP5 is induced and localizes to neurons and oligodendrocytes undergoing apoptosis in the murine spinal cord. Moreover, administration of a JNK inhibitor prevents both the induction of Hrk/DP5 and caspase-3 activation (Yin *et al.*, 2005). Further, following neuronal axotomy in the adult mouse, Noxa is upregulated in a partially p53-dependent manner in motor neurons and Noxa deficiency rescues motor neurons from axotomy-induced death (Kiryu-Seo *et al.*, 2005).

Both Bim and Hrk/DP5 have been shown to be upregulated in retinal ganglion cells following optic nerve transection (Wakabayashi *et al.*, 2002; Näpänkangas *et al.*, 2003). It is evident that BH3-only proteins have the potential to induce cell death following nerve/brain injury, yet further studies in animal models and humans will be critical in determining the exact role these pro-death proteins play in traumatic nerve injury.

Spinal Bulbar Muscular Atrophy

Spinal bulbar muscular atrophy is an inherited disorder characterized by progressive neuromuscular degeneration and weakness that eventually leads to death. It is caused by a polyglutamine repeat expansion in the androgen receptor. Although the role of BH3-only proteins has not been extensively studied in models of this disease, evidence from mouse models suggests that JNK-dependent activation of Hrk/DP5 may activate apoptosis of neurons in a Bax-dependent manner. Indeed, neurons null for Hrk/DP5, but not Bim, were significantly protected against apoptosis induced by expression of a polyQ androgen receptor constituent (Young *et al.*, 2009).

The role of BH3-only proteins has only recently been examined in spinal bulbar muscular atrophy and further examination of these pro-death proteins is necessary in animal models and human clinical cases to clarify their potential involvement in this disease.

Fetal Alcohol Syndrome

Fetal alcohol syndrome results in craniofacial malformations, a reduction in brain mass, and neurobehavioral abnormalities (Streissguth *et al.*, 2000). Although the exact mechanism by which brain cell death occurs is unclear, ethanol-induced neuronal apoptosis in the developing mouse brain is marked by both caspase-3 activation and classical apoptotic morphology (Olney *et al.*, 2002). Bcl-2 proteins have recently been implicated in ethanol-induced neuronal apoptosis as Bax-deficient mice are protected from ethanol exposure (Young *et al.*, 2003). Further, Puma has been identified as a mediator of cell death, independent of the transcription factor p53, following ethanol exposure in neonatal mice (Ghosh *et al.*, 2009).

These data are consistent with other reports showing that Puma can be upregulated in a p53-independent manner. Moreover, the above studies suggest a central role for both Puma and Bax in neuronal loss associated with fetal alcohol syndrome.

Cerebellar Ataxia

Some studies have focused on the Lurcher mouse model of cerebellar ataxia, a spontaneous mutant characterized by primary and selective early degeneration of cerebellar Purkinje neurons and later secondary loss of cerebellar granule neurons and inferior olivary neurons. Although the modes of cell death in each neuronal type are not certain, several sources of data indicate that Bax-dependent apoptosis is a key mechanism, at least for cerebellar granule neurons (Selimi *et al.*, 2000). Although few studies have been done on BH3-only proteins in this mouse model, the roles of Hrk/DP5 and Bim have been examined. Cerebellar granule neurons show enhanced expression of Hrk/DP5 after exposure to apoptotic stimuli *in vitro*; however, these cells still underwent apoptosis in the absence of Hrk/DP5. Furthermore, ablation of Hrk/DP5 is not sufficient to inhibit neuronal degeneration in the Lurcher mouse (Coultas *et al.*, 2007). Bim is expressed in cerebellar Purkinje neurons, yet reduced Bim expression also fails to alter the Lurcher mouse phenotype (Bouillet *et al.*, 2003). Although the relatively high levels of Hrk/DP5 and Bim expression do not appear to play a role in this model of neurodegeneration, further studies may identify an important role for other BH3-only proteins in the neurodegenerative phenotype of the Lurcher mouse.

Miscellaneous Studies

Oxidative Stress

Many studies examined more general features of multiple neurological disorders to elucidate in a broader sense how BH3-only proteins mediate neuronal death. A principal feature of many neurodegenerative disorders (e.g., ALS, PD, AD, stoke) is oxidative stress (Margaill *et al.*, 2005; Andersen, 2004). Induction of oxidative stress in primary cortical neurons triggers increases in the expression of Bim, Noxa, and Puma (Steckley *et al.*, 2007). Furthermore, Puma-deficient neurons are resistant to many forms of oxidative stress, while neurons deficient for either Bim or Noxa remain sensitive to oxidative stress-induced apoptosis (Steckley *et al.*, 2007). Additional data supporting a primary role for Puma in oxidative stress induced-apoptosis include zinc-treated neuroblastoma cells which upregulate Puma via activation of p53 and ERK pathways (Hara *et al.*, 2009). Thus, BH3-only proteins, and in particular, Puma, appear to play a key role in cell death induced by oxidative stress. The clinical merits of inhibiting BH3-only (e.g. Puma) activation are supported by the fact that oxidative stress is a major contributing factor to many neurodegenerative diseases, as well as stroke.

Excitotoxicity

BH3-only proteins have also been studied in excitotoxic neuronal death, another key feature of multiple neurodegenerative diseases. For instance, kainic acid induces excitotoxicity and nerve cell degeneration in the hippocampus and amygdala of rat brain. Although Bax is upregulated following kainic acid treatment, the BH3-only proteins Hrk/DP5 and Bim are uncharacteristically downregulated (Korhonen *et al.*, 2003). In contrast, a more recent study highlighted an important role for Bim in neuronal excitotoxicity. Following adenosine monophosphate protein kinase (AMPK) activation in neocortical neurons, Bim expression increases and induces apoptosis in a JNK-dependent manner (Concannon *et al.*, 2010). To further support a role for BH3-only proteins in excitotoxicity, this same group demonstrated that Puma regulates ER stress-induced apoptosis in response to N-methyl-D-aspartic acid (NMDA) receptor-mediated excitotoxicity (Concannon *et al.*, 2008). These results suggest that BH3-only protein regulation in response to excitotoxic injury is complex and may occur in a stimulus and cell-specific manner. Further, many neurodegenerative diseases are characterized by oxidative stress and excitotoxicity, both of which are reported to induce DNA damage and potentially lead to the activation of p53 transcriptional targets Puma, Noxa, and Bim (Culmsee and Matson, 2005; Erlacher *et al.*, 2005). Thus, p53 may be a common critical element in the activation of BH3-only proteins in several neurodegenerative diseases.

BH3-Only Proteins as Novel Therapeutic Targets for Neurodegeneration

Emerging data support a significant role for BH3-only proteins in causing the neuronal cell death that underlies acute and chronic neurodegenerative disorders. A major goal of current research is focused on the development of novel "antagonists" of BH3-only activity to inhibit the neuronal apoptosis associated with neurodegeneration. As the decision between cell survival and succumbing to apoptosis depends on the balance of Bcl-2 family proteins, ameliorating cell death may be accomplished by either increasing the activity of pro-survival Bcl-2 proteins or decreasing the activity of multi-domain or BH3-only pro-death proteins. To this end, targeted delivery of pro-survival Bcl-2 family proteins as a therapeutic approach has been reviewed elsewhere and will not be discussed here (Shacka and Roth, 2005). In the following section, we review recent studies on BH3-only "antagonists" that interact with and limit the pro-apoptotic activity of BH3-only proteins. We also discuss growth factor pathways which could be stimulated to suppress BH3-only activation, and finally, we touch on RNA interference as a potential strategy to downregulate BH3-only proteins in disease. These approaches represent new avenues of therapeutic design for neurodegeneration.

Strategies for Inhibition: BH3-Only "Antagonists"

In recent years, several diverse approaches designed to inhibit the pro-death function of BH3-only proteins have been examined as potential therapeutics for neurodegenerative disease (Table 3). A promising bidentate derivative of Bid has been developed which suppresses the translocation of tBid to the mitochondria and prevents the release of apoptogenic factors into the cytosol (Becattini *et al.*, 2006). This compound is particularly appealing for treatment of diseases that demonstrate truncation of Bid, such as familial mSOD1 ALS (Guegan *et al.*, 2003). *In vivo* evidence suggests that molecular inhibitors of JNK may also be beneficial in inhibiting neurodegeneration following cerebral ischemia. For example, a cell permeable peptide inhibitor of JNK, D-JNKI1, rescues neurons from cerebral ischemia (Repici and Borsello, 2006). Another selective inhibitor of JNK, SP600125, prevents the enhanced expression of Bim and neuronal apoptosis induced during cerebral ischemia (Okuno *et al.*, 2004). An additional compound, tauroursodeoxycholic acid (TUDCA), has been shown to be neuroprotective in animal models of diseases such as HD and PD (Keene *et al.*, 2001; Duan *et al.*, 2002). TUDCA was also shown to prevent apoptosis in neuronal cells subjected to Aβ toxicity and in human endothelial cells (Solá *et al.*, 2003).

Furthermore, evidence suggests that TUDCA may protect neurons through inhibition of the BH3-only protein Bad, as was shown in excitotoxic cortical neuron death (Castro *et al.*, 2004). These data suggest that compounds designed to either directly or indirectly inhibit the pro-death activity of BH3-only proteins (i.e., BH3-only "antagonists") may represent a novel therapeutic approach to inhibit neuronal cell death in an array of neurodegenerative diseases.

Table 3. BH3-only Protein "Antagonists"

BH3-only Protein "Antagonist"	Relevant Disease Models	Cellular Effects	References
Bidentate Derivative of Bid	ALS	Prevents tBid translocation to mitochondria	7
D-JNKI1	Cerebral Ischemia	JNK inhibition	125
SP600125	Cerebral Ischemia	Attenuated induction of Bim expression	107
Tudca	Cerebral Ischemia, Neuronal excitoxicity	Increased phosphorylation of Bad	32, 67
Estrogens	PD, AD, stroke, traumatic brain injury	Inactivate Bad through PI3K/AKT pathway	98, 159, 161
TGF-β	Cerebral Ischemia	Inhibited Bad, decreased caspase-3	173
NGF	Unexamined	Increased Bim phosphorylation	10
IGF-1	AD, cerebellar ataxia, ALS	Decreased Bim expression	87
VEGF-β	Ocular neurodegeneration, stroke	Decreased Bmf, Hrk, Bad, Bim	85

Estrogens as Indirect BH3-Only "Antagonists"

Estrogens have been studied at length as therapeutics in models of PD, AD, stroke, and traumatic brain injury. Although widely shown to induce the expression of Bcl-2 and Bcl-xL, more recent evidence suggests that estrogens may also confer neuroprotection through alteration of BH3-only activity. In a rat model of middle cerebral artery occlusion, estrogen inactivated Bad through induction of the PI3K/AKT pathway and effectively reduced infarct volume and neuronal apoptosis (Won *et al.*, 2005). This is consistent with a more recent study that found that striatal dopaminergic neurons from estrogen receptor knockout mice succumb to MPTP-induced apoptosis more readily than wild type mice via a Bad-dependent mechanism (Morissette *et al.*, 2008). Furthermore, in primary cerebrocortical cultures subjected to Aβ peptide, estrogen conferred neuroprotection by preventing an increase in Bim expression, suggesting a possible role for estrogen treatment of AD (Yao *et al.*, 2007). Thus, estrogens indirectly suppress the activity of BH3-only proteins and protect neurons from apoptosis.

Growth Factors

As trophic factor withdrawal is known to induce neuronal apoptosis through alterations in the expression of many BH3-only proteins, these factors have also been evaluated as neuroprotective compounds. Adenoviral delivery of transforming growth factor β (TGF-β) reduced infarct size in mice subjected to ischemic conditions; these effects were attributed to the inactivation of BH3-only Bad and decreased levels of active caspase-3 (Zhu *et al.*, 2002). In a similar manner, Nerve growth factor has been shown to prevent apoptosis through a reduction in the activity of Bim (Biswas *et al.*, 2002). Insulin-like growth factor-1 (IGF-1), has also been reported to protect neurons via activation of the PI3K/AKT pathway which functions to induce phosphorylation of a forkhead transcription factor (FKHRL1), resulting in decreased levels of Bim expression (Linseman *et al.*, 2002). The PI3K/AKT pathway also suppresses Bad activity in neurons and may function to rescue cells through inactivation of this BH3-only protein (Datta *et al.*, 2000). Interestingly, vascular endothelial growth factor β (VEGF-β) has been identified as a pro-survival factor in mouse models of ocular neurodegeneration and stroke, and appears to rescue neurons through suppression of p53 and several BH3-only proteins, including Bmf, Hrk/DP5, Bad, and Bim (Li *et al.*, 2008). Results from these studies provide a framework for the development of trophic factor treatment options for neurodegenerative diseases and acute neuronal injury.

RNA Interference

Lastly, a new therapeutic avenue is suggested by recent evidence highlighting a role for RNA interference in the treatment of neurodegeneration. Kole *et al.* (2011) demonstrated that a specific miRNA, miR-29b, inhibits neuronal cell death in response to diverse apoptotic stimuli, including nerve growth factor deprivation and tunicamycin treatment. These authors further report that miR-29b inhibits apoptosis upstream of CytC by suppressing the translation of Bim, Bmf, Puma, and Hrk/DP5. Although identified in cardiomyocytes rather than in neurons, another miRNA, miR-24, also represses BH3-only proteins (i.e., Bim) and inhibits apoptosis (Qian *et al.*, 2011). These studies highlight a novel strategy for implementing RNA interference of BH3-only proteins as a new therapeutic approach for the treatment of neurodegenerative diseases and acute neuronal injury.

Conclusion

In recent years, significant advances have been made in elucidating the role of BH3-only proteins in *in vitro* paradigms of neuronal injury and animals models of neurodegenerative diseases, stroke, and neuronal trauma. Moreover, studies are beginning to focus on evaluation of BH3-only expression and/or activation in human disease, stroke, and trauma, providing a solid foundation for the design of future treatment strategies. While our understanding of how BH3-only proteins function to induce apoptosis has grown tremendously, identifying the precise molecular mechanisms of their activity remains challenging. Furthermore, only a few studies have examined the role of BH3-only proteins in human disease, and this area of investigation has relied entirely on evaluation of post-mortem tissue samples. This area of study necessitates further investigation, as well as, development of reliable assays to assess the role of BH3-only proteins *during* disease progression (e.g., evaluation of tissue biopsies or CSF for BH3-only proteins). In addition, as the role of BH3-only proteins in neurodegenerative disease becomes further resolved, more focus should be given to developing novel strategies to abrogate the pro-apoptotic activity of these proteins. In conclusion, the BH3-only family of proteins represents a novel therapeutic target for the treatment of neurodegenerative diseases, stroke, and other types of neuronal injury.

Acknowledgment

The authors sincerely thank Dr. Kevin Roth and Arindam Ghosh (Department of Pathology, University of Alabama at Birmingham, Birmingham, AL, USA), and Dr. Sean Cregan (Department of Physiology and Pharmacology, University of Western Ontario, London, Ontario, Canada) for critical reviews of the manuscript.

The authors are funded by grants from the Department of Veterans Affairs (Merit Review award to DAL) and the National Institute of Neurological Disorders and Stroke (R01NS062766 to DAL).

References

Akiyama, T., et al., Regulation of osteoclast apoptosis by ubiquitylation of proapoptotic BH3-only Bcl-2 family member Bim. *Embo J.,* 2003. 22(24): p. 6653-64.

Althaus, J., et al., Expression of the gene encoding the pro-apoptotic BNIP3 protein and stimulation of hypoxia-inducible factor-1alpha (HIF-1alpha) protein following focal cerebral ischemia in rats. *Neurochem. Int,* 2006. 48(8): p. 687-95.

Aminova, L.R., A. Siddiq, and R.R. Ratan, Antioxidants, HIF prolyl hydroxylase inhibitors or short interfering RNAs to BNIP3 or PUMA, can prevent prodeath effects of the transcriptional activator, HIF-1alpha, in a mouse hippocampal neuronal line. *Antioxid Redox Signal,* 2008. 10(12): p. 1989-98.

Andersen, J.K., Oxidative stress in neurodegeneration: cause or consequence? *Nat. Med.,* 2004. 10 Suppl: p. S18-25.

Annis, M.G., et al., Endoplasmic reticulum localized Bcl-2 prevents apoptosis when redistribution of cytochrome c is a late event. *Oncogene,* 2001. 20(16): p. 1939-52.

Bakhshi, A., et al., Cloning the chromosomal breakpoint of t(14;18) human lymphomas: clustering around JH on chromosome 14 and near a transcriptional unit on 18. *Cell,* 1985. 41(3): p. 899-906.

Becattini, B., et al., Structure-activity relationships by interligand NOE-based design and synthesis of antiapoptotic compounds targeting Bid. *Proc. Natl. Acad. Sci. USA,* 2006. 103(33): p. 12602-6.

Bernstein, A.I., et al., 6-OHDA generated ROS induces DNA damage and p53- and PUMA-dependent cell death. *Mol Neurodegener.* 6(1): p. 2.

Betarbet, R., et al., Mechanistic approaches to Parkinson's disease pathogenesis. *Brain Pathol.,* 2002. 12(4): p. 499-510.

Biswas, S.C. and L.A. Greene, Nerve growth factor (NGF) down-regulates the Bcl-2 homology 3 (BH3) domain-only protein Bim and suppresses its proapoptotic activity by phosphorylation. *J. Biol. Chem.,* 2002. 277(51): p. 49511-6.

Biswas, S.C., D.X. Liu, and L.A. Greene, Bim is a direct target of a neuronal E2F-dependent apoptotic pathway. *J. Neurosci.,* 2005. 25(37): p. 8349-58.

Biswas, S.C., et al., Puma and p53 play required roles in death evoked in a cellular model of Parkinson disease. *Neurochem Res*, 2005. 30(6-7): p. 839-45.

Biswas, S.C., et al., Bim is elevated in Alzheimer's disease neurons and is required for beta-amyloid-induced neuronal apoptosis. *J. Neurosci*, 2007. 27(4): p. 893-900.

Borner, C., The Bcl-2 protein family: sensors and checkpoints for life-or-death decisions. *Mol. Immunol.*, 2003. 39(11): p. 615-47.

Bouillet, P., et al., BH3-only Bcl-2 family member Bim is required for apoptosis of autoreactive thymocytes. *Nature*, 2002. 415(6874): p. 922-6.

Bouillet, P., et al., Loss of pro-apoptotic BH3-only Bcl-2 family member Bim does not protect mutant Lurcher mice from neurodegeneration. *J. Neurosci Res.*, 2003. 74(5): p. 777-81.

Cartron, P.F., et al., The first alpha helix of Bax plays a necessary role in its ligand-induced activation by the BH3-only proteins Bid and PUMA. *Mol. Cell*, 2004. 16(5): p. 807-18.

Castro, R.E., et al., The bile acid tauroursodeoxycholic acid modulates phosphorylation and translocation of bad via phosphatidylinositol 3-kinase in glutamate-induced apoptosis of rat cortical neurons. *J. Pharmacol. Exp Ther*, 2004. 311(2): p. 845-52.

Chen, L., et al., Differential targeting of prosurvival Bcl-2 proteins by their BH3-only ligands allows complementary apoptotic function. *Mol Cell*, 2005. 17(3): p. 393-403.

Chen, M., et al., Bid is cleaved by calpain to an active fragment in vitro and during myocardial ischemia/reperfusion. *J. Biol. Chem.*, 2001. 276(33): p. 30724-8.

Chen, R., et al., Bcl-2 functionally interacts with inositol 1,4,5-trisphosphate receptors to regulate calcium release from the ER in response to inositol 1,4,5-trisphosphate. *J Cell Biol*, 2004. 166(2): p. 193-203.

Chipuk, J.E., et al., PUMA couples the nuclear and cytoplasmic proapoptotic function of p53. *Science*, 2005. 309(5741): p. 1732-5.

Concannon, C.G., et al., AMP kinase-mediated activation of the BH3-only protein Bim couples energy depletion to stress-induced apoptosis. *J. Cell Biol.*, 2010. 189(1): p. 83-94.

Concannon, C.G., et al., NMDA receptor-mediated excitotoxic neuronal apoptosis in vitro and in vivo occurs in an ER stress and PUMA independent manner. *J. Neurochem.*, 2008. 105(3): p. 891-903.

Cory, S., D.C. Huang, and J.M. Adams, The Bcl-2 family: roles in cell survival and oncogenesis. *Oncogene*, 2003. 22(53): p. 8590-607.

Coultas, L., et al., Hrk/DP5 contributes to the apoptosis of select neuronal populations but is dispensable for haematopoietic cell apoptosis. *J Cell Sci*, 2007. 120(Pt 12): p. 2044-52.

Cregan, S.P., et al., p53 activation domain 1 is essential for PUMA upregulation and p53-mediated neuronal cell death. *J. Neurosci.*, 2004. 24(44): p. 10003-12.

Culmsee, C. and M.P. Mattson, p53 in neuronal apoptosis. *Biochem. Biophys. Res. Commun,* 2005. 331(3): p. 761-77.

Danial, N.N. and S.J. Korsmeyer, Cell death: critical control points. *Cell*, 2004. 116(2): p. 205-19.

Datta, S.R., et al., 14-3-3 proteins and survival kinases cooperate to inactivate BAD by BH3 domain phosphorylation. *Mol. Cell*, 2000. 6(1): p. 41-51.

Distelhorst, C.W. and T.S. McCormick, Bcl-2 acts subsequent to and independent of Ca2+ fluxes to inhibit apoptosis in thapsigargin- and glucocorticoid-treated mouse lymphoma cells. *Cell Calcium*, 1996. 19(6): p. 473-83.

Duan, W.M., et al., Tauroursodeoxycholic acid improves the survival and function of nigral transplants in a rat model of Parkinson's disease. *Cell Transplant*, 2002. 11(3): p. 195-205.

Erlacher, M., et al., BH3-only proteins Puma and Bim are rate-limiting for gamma-radiation- and glucocorticoid-induced apoptosis of lymphoid cells in vivo. *Blood*, 2005. 106(13): p. 4131-8.

Eve, D.J., J.S. Dennis, and B.A. Citron, Transcription factor p53 in degenerating spinal cords. *Brain Res.*, 2007. 1150: p. 174-81.

Fei, Q., et al., Paraquat neurotoxicity is mediated by a Bak-dependent mechanism. *J. Biol. Chem.*, 2008. 283(6): p. 3357-64.

Ferrer, I., et al., Active, phosphorylation-dependent MAP kinases, MAPK/ERK, SAPK/JNK and p38, and specific transcription factor substrates are differentially expressed following systemic administration of kainic acid to the adult rat. *Acta Neuropathol*, 2002. 103(4): p. 391-407.

Fesik, S.W., Insights into programmed cell death through structural biology. *Cell*, 2000. 103(2): p. 273-82.

Galas, M.C., et al., Death of cortical and striatal neurons induced by mitochondrial defect involves differential molecular mechanisms. *Neurobiol Dis*, 2004. 15(1): p. 152-9.

Galehdar, Z., et al., Neuronal apoptosis induced by endoplasmic reticulum stress is regulated by ATF4-CHOP-mediated induction of the Bcl-2 homology 3-only member PUMA. *J Neurosci*, 2010. 30(50): p. 16938-48.

García-Martínez, J.M., et al., BH3-only proteins Bid and Bim(EL) are differentially involved in neuronal dysfunction in mouse models of Huntington's disease. *J Neurosci Res*, 2007. 85(12): p. 2756-69.

Ghosh, A.P., et al., The proapoptotic BH3-only, Bcl-2 family member, Puma is critical for acute ethanol-induced neuronal apoptosis. *J. Neuropathol. Exp. Neurol.*, 2009. 68(7): p. 747-56.

Gilley, J., P.J. Coffer, and J. Ham, FOXO transcription factors directly activate bim gene expression and promote apoptosis in sympathetic neurons. *J. Cell Biol.*, 2003. 162(4): p. 613-22.

Gillissen, B., et al., Mcl-1 determines the Bax dependency of Nbk/Bik-induced apoptosis. *J Cell Biol*, 2007. 179(4): p. 701-15.

Gomez-Lazaro, M., et al., 6-Hydroxydopamine activates the mitochondrial apoptosis pathway through p38 MAPK-mediated, p53-independent activation of Bax and PUMA. *J. Neurochem.*, 2008. 104(6): p. 1599-612.

Grooteclaes, M., et al., C-terminal-binding protein corepresses epithelial and proapoptotic gene expression programs. *Proc. Natl. Acad. Sci. USA*, 2003. 100(8): p. 4568-73.

Gross, A., J.M. McDonnell, and S.J. Korsmeyer, BCL-2 family members and the mitochondria in apoptosis. *Genes. Dev.*, 1999. 13(15): p. 1899-911.

Guan, Q.H., et al., Brain ischemia/reperfusion-induced expression of DP5 and its interaction with Bcl-2, thus freeing Bax from Bcl-2/Bax dimmers are mediated by c-Jun N-terminal kinase (JNK) pathway. *Neurosci. Lett, 2006.* 393(2-3): p. 153-61.

Guegan, C. and S. Przedborski, Programmed cell death in amyotrophic lateral sclerosis. *J. Clin. Invest*, 2003. 111(2): p. 153-61.

Guegan, C., et al., Instrumental activation of bid by caspase-1 in a transgenic mouse model of ALS. *Mol Cell Neurosci*, 2002. 20(4): p. 553-62.

Häcker, G. and A. Weber, BH3-only proteins trigger cytochrome c release, but how? Arch Biochem Biophys, 2007. 462(2): p. 150-5.

Hakem, R., et al., Differential requirement for caspase 9 in apoptotic pathways in vivo. *Cell*, 1998. 94(3): p. 339-52.

Han, J., et al., Expression of bbc3, a pro-apoptotic BH3-only gene, is regulated by diverse cell death and survival signals. *Proc. Natl. Acad. Sci. USA*, 2001. 98(20): p. 11318-23.

Hara, H., T. Kamiya, and T. Adachi, Zinc induces expression of the BH3-only protein PUMA through p53 and ERK pathways in SH-SY5Y neuroblastoma cells. *Neurochem. Res,* 2009. 34(8): p. 1498-506.

Harada, H., et al., Survival factor-induced extracellular signal-regulated kinase phosphorylates BIM, inhibiting its association with BAX and proapoptotic activity. *Proc. Natl. Acad. Sci. USA*, 2004. 101(43): p. 15313-7.

Hartmann, A., et al., Caspase-3: A vulnerability factor and final effector in apoptotic death of dopaminergic neurons in Parkinson's disease. *Proc. Natl. Acad. Sci. USA,* 2000. 97(6): p. 2875-80.

Heibein, J.A., et al., Granzyme B-mediated cytochrome c release is regulated by the Bcl-2 family members bid and Bax. *J. Exp. Med*, 2000. 192(10): p. 1391-402.

Hengartner, M.O. and H.R. Horvitz, C. elegans cell survival gene ced-9 encodes a functional homolog of the mammalian proto-oncogene bcl-2. *Cell*, 1994. 76(4): p. 665-76.

Hetz, C., et al., The proapoptotic BCL-2 family member BIM mediates motoneuron loss in a model of amyotrophic lateral sclerosis. *Cell Death Differ*, 2007. 14(7): p. 1386-9.

Hou, S.T., et al., The transcription factor E2F1 modulates apoptosis of neurons. *J. Neurochem.,* 2000. 75(1): p. 91-100.

Howells, C.C., et al., The Bcl-2-associated death promoter (BAD) lowers the threshold at which the Bcl-2-interacting domain death agonist (BID) triggers mitochondria disintegration. *J. Theor. Biol., 2010.*

Hsieh, J.K., et al., Novel function of the cyclin A binding site of E2F in regulating p53-induced apoptosis in response to DNA damage. *Mol Cell Biol*, 2002. 22(1): p. 78-93.

Huang, D.C. and A. Strasser, BH3-Only proteins-essential initiators of apoptotic cell death. *Cell, 2000.* 103(6): p. 839-42.

Imaizumi, K., et al., Critical role for DP5/Harakiri, a Bcl-2 homology domain 3-only Bcl-2 family member, in axotomy-induced neuronal cell death. *J. Neurosci.*, 2004. 24(15): p. 3721-5.

Imaizumi, K., et al., The cell death-promoting gene DP5, which interacts with the BCL2 family, is induced during neuronal apoptosis following exposure to amyloid beta protein. *J. Biol. Chem.*, 1999. 274(12): p. 7975-81.

Jeffers, J.R., et al., Puma is an essential mediator of p53-dependent and -independent apoptotic pathways. *Cancer Cell*, 2003. 4(4): p. 321-8.

Kamada, S., et al., bcl-2 deficiency in mice leads to pleiotropic abnormalities: accelerated lymphoid cell death in thymus and spleen, polycystic kidney, hair hypopigmentation, and distorted small intestine. *Cancer Res*, 1995. 55(2): p. 354-9.

Keene, C.D., et al., A bile acid protects against motor and cognitive deficits and reduces striatal degeneration in the 3-nitropropionic acid model of Huntington's disease. *Exp. Neurol.*, 2001. 171(2): p. 351-60.

Keeney, P.M., et al., Parkinson's disease brain mitochondrial complex I has oxidatively damaged subunits and is functionally impaired and misassembled. *J. Neurosci,* 2006. 26(19): p. 5256-64.

Kieran, D., et al., Deletion of the BH3-only protein puma protects motoneurons from ER stress-induced apoptosis and delays motoneuron loss in ALS mice. *Proc. Natl. Acad. Sci. USA,* 2007. 104(51): p. 20606-11.

Kim, H., et al., Hierarchical regulation of mitochondrion-dependent apoptosis by BCL-2 subfamilies. *Nat Cell Biol,* 2006. 8(12): p. 1348-58.

Kim, J.Y., et al., BH3-only protein Noxa is a mediator of hypoxic cell death induced by hypoxia-inducible factor 1alpha. *J. Exp. Med.*, 2004. 199(1): p. 113-24.

Kiryu-Seo, S., et al., Noxa is a critical mediator of p53-dependent motor neuron death after nerve injury in adult mouse. *J. Neurosci.*, 2005. 25(6): p. 1442-7.

Kole, A.J., et al., miR-29b is activated during neuronal maturation and targets BH3-only genes to restrict apoptosis. *Genes Dev.* 25(2): p. 125-30.

Korhonen, L., et al., Increase in Bcl-2 phosphorylation and reduced levels of BH3-only Bcl-2 family proteins in kainic acid-mediated neuronal death in the rat brain. *Eur J Neurosci,* 2003. 18(5): p. 1121-34.

Kovi, R.C., et al., An ARF/CtBP2 complex regulates BH3-only gene expression and p53-independent apoptosis. *Cell Death Differ.* 17(3): p. 513-21.

Kubli, D.A., J.E. Ycaza, and A.B. Gustafsson, Bnip3 mediates mitochondrial dysfunction and cell death through Bax and Bak. *Biochem. J.*, 2007. 405(3): p. 407-15.

Kuwana, T., et al., BH3 domains of BH3-only proteins differentially regulate Bax-mediated mitochondrial membrane permeabilization both directly and indirectly. *Mol Cell,* 2005. 17(4): p. 525-35.

Lanave, C., M. Santamaria, and C. Saccone, Comparative genomics: the evolutionary history of the Bcl-2 family. *Gene*, 2004. 333: p. 71-9.

Lassmann, H., et al., Cell death in Alzheimer's disease evaluated by DNA fragmentation in situ. *Acta Neuropathol.*, 1995. 89(1): p. 35-41.

Leon, R., et al., BimEL as a possible molecular link between proteasome dysfunction and cell death induced by mutant huntingtin. *Eur. J. Neurosci.* 31(11): p. 1915-25.

Letai, A., et al., Distinct BH3 domains either sensitize or activate mitochondrial apoptosis, serving as prototype cancer therapeutics. *Cancer Cell*, 2002. 2(3): p. 183-92.

Ley, R., et al., Activation of the ERK1/2 signaling pathway promotes phosphorylation and proteasome-dependent degradation of the BH3-only protein, Bim. *J. Biol. Chem.,* 2003. 278(21): p. 18811-6.

Li, H., et al., Cleavage of BID by caspase 8 mediates the mitochondrial damage in the Fas pathway of apoptosis. *Cell*, 1998. 94(4): p. 491-501.

Li, J., B. Lee, and A.S. Lee, Endoplasmic reticulum stress-induced apoptosis: multiple pathways and activation of p53-up-regulated modulator of apoptosis (PUMA) and NOXA by p53. *J. Biol. Chem.*, 2006. 281(11): p. 7260-70.

Li, Y., et al., VEGF-B inhibits apoptosis via VEGFR-1-mediated suppression of the expression of BH3-only protein genes in mice and rats. *J. Clin. Invest*, 2008. 118(3): p. 913-23.

Lindsten, T., et al., The combined functions of proapoptotic Bcl-2 family members bak and bax are essential for normal development of multiple tissues. *Mol. Cell*, 2000. 6(6): p. 1389-99.

Linseman, D.A., et al., Insulin-like growth factor-I blocks Bcl-2 interacting mediator of cell death (Bim) induction and intrinsic death signaling in cerebellar granule neurons. *J. Neurosci,* 2002. 22(21): p. 9287-97.

Liu, X., et al., The structure of a Bcl-xL/Bim fragment complex: implications for Bim function. *Immunity*, 2003. 19(3): p. 341-52.

Luciano, F., et al., Phosphorylation of Bim-EL by Erk1/2 on serine 69 promotes its degradation via the proteasome pathway and regulates its proapoptotic function. *Oncogene*, 2003. 22(43): p. 6785-93.

MacManus, J.P., et al., Absence of the transcription factor E2F1 attenuates brain injury and improves behavior after focal ischemia in mice. *J. Cereb Blood Flow Metab*, 2003. 23(9): p. 1020-8.

MacManus, J.P., et al., Decreased brain infarct following focal ischemia in mice lacking the transcription factor E2F1. *Neuroreport*, 1999. 10(13): p. 2711-4.

Margaill, I., M. Plotkine, and D. Lerouet, Antioxidant strategies in the treatment of stroke. *Free Radic Biol. Med.*, 2005. 39(4): p. 429-43.

Martin, L.J., Neuronal death in amyotrophic lateral sclerosis is apoptosis: possible contribution of a programmed cell death mechanism. *J Neuropathol Exp Neurol*, 1999. 58(5): p. 459-71.

McCullough, K.D., et al., Gadd153 sensitizes cells to endoplasmic reticulum stress by down-regulating Bcl2 and perturbing the cellular redox state. *Mol. Cell Biol.,* 2001. 21(4): p. 1249-59.

Mogi, M., et al., p53 protein, interferon-gamma, and NF-kappaB levels are elevated in the parkinsonian brain. *Neurosci. Lett.,* 2007. 414(1): p. 94-7.

Moldoveanu, T., et al., The X-ray structure of a BAK homodimer reveals an inhibitory zinc binding site. *Mol. Cell,* 2006. 24(5): p. 677-88.

Morishima, N., et al., An endoplasmic reticulum stress-specific caspase cascade in apoptosis. Cytochrome c-independent activation of caspase-9 by caspase-12. *J. Biol. Chem.,* 2002. 277(37): p. 34287-94.

Morissette, M., et al., Contribution of estrogen receptors alpha and beta to the effects of estradiol in the brain. *J. Steroid. Biochem. Mol. Biol.,* 2008. 108(3-5): p. 327-38.

Mu, X., et al., Altered expression of bcl-2 and bax mRNA in amyotrophic lateral sclerosis spinal cord motor neurons. *Ann. Neurol,* 1996. 40(3): p. 379-86.

Naik, E., et al., Ultraviolet radiation triggers apoptosis of fibroblasts and skin keratinocytes mainly via the BH3-only protein Noxa. *J. Cell Biol.*, 2007. 176(4): p. 415-24.

Näpänkangas, U., et al., Rat retinal ganglion cells upregulate the pro-apoptotic BH3-only protein Bim after optic nerve transection. *Brain Res. Mol. Brain Res*, 2003. 120(1): p. 30-7.

Ness, J.M., et al., Selective involvement of BH3-only Bcl-2 family members Bim and Bad in neonatal hypoxia-ischemia. *Brain Res, 2006.* 1099(1): p. 150-9.

Niizuma, K., et al., Potential role of PUMA in delayed death of hippocampal CA1 neurons after transient global cerebral ischemia. *Stroke*, 2009. 40(2): p. 618-25.

Oberstein, A., P.D. Jeffrey, and Y. Shi, Crystal structure of the Bcl-XL-Beclin 1 peptide complex: Beclin 1 is a novel BH3-only protein. *J. Biol. Chem.*, 2007. 282(17): p. 13123-32.

Obexer, P., et al., FKHRL1-mediated expression of Noxa and Bim induces apoptosis via the mitochondria in neuroblastoma cells. *Cell Death Differ*, 2007. 14(3): p. 534-47.

Oda, E., et al., Noxa, a BH3-only member of the Bcl-2 family and candidate mediator of p53-induced apoptosis. *Science*, 2000. 288(5468): p. 1053-8.

Okuno, S., et al., The c-Jun N-terminal protein kinase signaling pathway mediates Bax activation and subsequent neuronal apoptosis through interaction with Bim after transient focal cerebral ischemia. *J. Neurosci.*, 2004. 24(36): p. 7879-87.

Olney, J.W., et al., Ethanol-induced apoptotic neurodegeneration in the developing C57BL/6 mouse brain. *Brain Res. Dev. Brain Res*, 2002. 133(2): p. 115-26.

Opferman, J.T. and S.J. Korsmeyer, Apoptosis in the development and maintenance of the immune system. *Nat. Immunol.*, 2003. 4(5): p. 410-5.

Pei, J.J., et al., Localization of active forms of C-jun kinase (JNK) and p38 kinase in Alzheimer's disease brains at different stages of neurofibrillary degeneration. *J Alzheimers Dis*, 2001. 3(1): p. 41-48.

Perier, C., et al., Two molecular pathways initiate mitochondria-dependent dopaminergic neurodegeneration in experimental Parkinson's disease. *Proc. Natl. Acad. Sci. USA,* 2007. 104(19): p. 8161-6.

Perier, C., et al., Complex I deficiency primes Bax-dependent neuronal apoptosis through mitochondrial oxidative damage. *Proc. Natl. Acad. Sci. USA*, 2005. 102(52): p. 19126-31.

Petros, A.M., et al., Solution structure of the antiapoptotic protein bcl-2. *Proc. Natl. Acad. Sci. USA*, 2001. 98(6): p. 3012-7.

Petros, A.M., et al., Rationale for Bcl-xL/Bad peptide complex formation from structure, mutagenesis, and biophysical studies. *Protein. Sci*, 2000. 9(12): p. 2528-34.

Plesnila, N., et al., BID mediates neuronal cell death after oxygen/ glucose deprivation and focal cerebral ischemia. *Proc. Natl. Acad. Sci. USA,* 2001. 98(26): p. 15318-23.

Putcha, G.V., et al., JNK-mediated BIM phosphorylation potentiates BAX-dependent apoptosis. *Neuron*, 2003. 38(6): p. 899-914.

Puthalakath, H., et al., The proapoptotic activity of the Bcl-2 family member Bim is regulated by interaction with the dynein motor complex. *Mol. Cell,* 1999. 3(3): p. 287-96.

Puthalakath, H., et al., ER stress triggers apoptosis by activating BH3-only protein Bim. *Cell*, 2007. 129(7): p. 1337-49.

Puthalakath, H. and A. Strasser, Keeping killers on a tight leash: transcriptional and post-translational control of the pro-apoptotic activity of BH3-only proteins. *Cell Death Differ*, 2002. 9(5): p. 505-12.

Puthalakath, H., et al., Bmf: a proapoptotic BH3-only protein regulated by interaction with the myosin V actin motor complex, activated by anoikis. *Science*, 2001. 293(5536): p. 1829-32.

Qian, L., et al., miR-24 inhibits apoptosis and represses Bim in mouse cardiomyocytes. *J Exp Med.*

Rao, R.V., H.M. Ellerby, and D.E. Bredesen, Coupling endoplasmic reticulum stress to the cell death program. *Cell Death Differ*, 2004. 11(4): p. 372-80.

Rathmell, J.C., et al., Deficiency in Bak and Bax perturbs thymic selection and lymphoid homeostasis. *Nat. Immunol*, 2002. 3(10): p. 932-9.

Reimertz, C., et al., Gene expression during ER stress-induced apoptosis in neurons: induction of the BH3-only protein Bbc3/PUMA and activation of the mitochondrial apoptosis pathway. *J. Cell Biol,* 2003. 162(4): p. 587-97.

Repici, M. and T. Borsello, JNK pathway as therapeutic target to prevent degeneration in the central nervous system. *Adv. Exp. Med. Biol.*, 2006. 588: p. 145-55.

Roth, K.A., Caspases, apoptosis, and Alzheimer disease: causation, correlation, and confusion. *J. Neuropathol Exp. Neurol.*, 2001. 60(9): p. 829-38.

Saleh, M., et al., Differential modulation of endotoxin responsiveness by human caspase-12 polymorphisms. *Nature, 2004.* 429(6987): p. 75-9.

Sanz, C., et al., Interleukin 3-dependent activation of DREAM is involved in transcriptional silencing of the apoptotic Hrk gene in hematopoietic progenitor cells. *Embo J*, 2001. 20(9): p. 2286-92.

Sattler, M., et al., Structure of Bcl-xL-Bak peptide complex: recognition between regulators of apoptosis. *Science*, 1997. 275(5302): p. 983-6.

Schuler, M., et al., p53 triggers apoptosis in oncogene-expressing fibroblasts by the induction of Noxa and mitochondrial Bax translocation. *Cell Death Differ*, 2003. 10(4): p. 451-60.

Schultz, D.R. and W.J. Harrington, Jr., Apoptosis: programmed cell death at a molecular level. *Semin. Arthritis Rheum*, 2003. 32(6): p. 345-69.

Selimi, F., M.W. Vogel, and J. Mariani, Bax inactivation in lurcher mutants rescues cerebellar granule cells but not purkinje cells or inferior olivary neurons. *J. Neurosci*, 2000. 20(14): p. 5339-45.

Shacka, J.J. and K.A. Roth, Regulation of neuronal cell death and neurodegeneration by members of the Bcl-2 family: therapeutic implications. *Curr. Drug Targets CNS Neurol Disord*, 2005. 4(1): p. 25-39.

She, Q.B., et al., The BAD protein integrates survival signaling by EGFR/MAPK and PI3K/Akt kinase pathways in PTEN-deficient tumor cells. *Cancer Cell*, 2005. 8(4): p. 287-97.

Shinoe, T., et al., Upregulation of the pro-apoptotic BH3-only peptide harakiri in spinal neurons of amyotrophic lateral sclerosis patients. *Neurosci. Lett*, 2001. 313(3): p. 153-7.

Solá, S., et al., Tauroursodeoxycholic acid prevents amyloid-beta peptide-induced neuronal death via a phosphatidylinositol 3-kinase-dependent signaling pathway. *Mol. Med.*, 2003. 9(9-12): p. 226-34.

Sowter, H.M., et al., HIF-1-dependent regulation of hypoxic induction of the cell death factors BNIP3 and NIX in human tumors. *Cancer Res*, 2001. 61(18): p. 6669-73.

Steckley, D., et al., Puma is a dominant regulator of oxidative stress induced Bax activation and neuronal apoptosis. *J. Neurosci.*, 2007. 27(47): p. 12989-99.

Stoka, V., et al., Lysosomal protease pathways to apoptosis. Cleavage of bid, not pro-caspases, is the most likely route. *J. Biol. Chem*, 2001. 276(5): p. 3149-57.

Streissguth, A.P. and K. O'Malley, Neuropsychiatric implications and long-term consequences of fetal alcohol spectrum disorders. *Semin. Clin. Neuropsychiatry*, 2000. 5(3): p. 177-90.

Su, J.H., G. Deng, and C.W. Cotman, Bax protein expression is increased in Alzheimer's brain: correlations with DNA damage, Bcl-2 expression, and brain pathology. *J. Neuropathol. Exp. Neurol.*, 1997. 56(1): p. 86-93.

Suzuki, M., R.J. Youle, and N. Tjandra, Structure of Bax: coregulation of dimer formation and intracellular localization. *Cell*, 2000. 103(4): p. 645-54.

Tatton, N.A., Increased caspase 3 and Bax immunoreactivity accompany nuclear GAPDH translocation and neuronal apoptosis in Parkinson's disease. *Exp. Neurol*, 2000. 166(1): p. 29-43.

Tsuchiya, T., et al., Bcl-2 homology domain 3-only proteins Puma and Bim mediate the vulnerability of CA1 hippocampal neurons to proteasome inhibition in vivo. *Eur. J. Neurosci.* 33(3): p. 401-8.

Tsujimoto, Y., Cell death regulation by the Bcl-2 protein family in the mitochondria. *J. Cell Physiol.*, 2003. 195(2): p. 158-67.

Uo, T., T.D. Veenstra, and R.S. Morrison, Histone deacetylase inhibitors prevent p53-dependent and p53-independent Bax-mediated neuronal apoptosis through two distinct mechanisms. *J. Neurosci., 2009.* 29(9): p. 2824-32.

Urano, F., et al., Coupling of stress in the ER to activation of JNK protein kinases by transmembrane protein kinase IRE1. *Science*, 2000. 287(5453): p. 664-6.

Veis, D.J., et al., Bcl-2-deficient mice demonstrate fulminant lymphoid apoptosis, polycystic kidneys, and hypopigmented hair. *Cell*, 1993. 75(2): p. 229-40.

Verkhratsky, A., Physiology and pathophysiology of the calcium store in the endoplasmic reticulum of neurons. *Physiol. Rev*, 2005. 85(1): p. 201-79.

Verma, S., L.J. Zhao, and G. Chinnadurai, Phosphorylation of the pro-apoptotic protein BIK: mapping of phosphorylation sites and effect on apoptosis. *J. Biol. Chem.*, 2001. 276(7): p. 4671-6.

Vila, M., et al., Bax ablation prevents dopaminergic neurodegeneration in the 1-methyl- 4-phenyl-1,2,3,6-tetrahydropyridine mouse model of Parkinson's disease. *Proc. Natl. Acad. Sci. USA*, 2001. 98(5): p. 2837-42.

Villunger, A., et al., p53- and drug-induced apoptotic responses mediated by BH3-only proteins puma and noxa. *Science*, 2003. 302(5647): p. 1036-8.

Viswanath, V., et al., Caspase-9 activation results in downstream caspase-8 activation and bid cleavage in 1-methyl-4-phenyl-1,2,3,6-tetrahydropyridine-induced Parkinson's disease. *J. Neurosci*, 2001. 21(24): p. 9519-28.

Vukosavic, S., et al., Bax and Bcl-2 interaction in a transgenic mouse model of familial amyotrophic lateral sclerosis. *J. Neurochem*, 1999. 73(6): p. 2460-8.

Wakabayashi, T., J. Kosaka, and S. Hommura, Up-regulation of Hrk, a regulator of cell death, in retinal ganglion cells of axotomized rat retina. *Neurosci. Lett.*, 2002. 318(2): p. 77-80.

Wei, M.C., et al., Proapoptotic BAX and BAK: a requisite gateway to mitochondrial dysfunction and death. *Science*, 2001. 292(5517): p. 727-30.

Willis, S.N., et al., Proapoptotic Bak is sequestered by Mcl-1 and Bcl-xL, but not Bcl-2, until displaced by BH3-only proteins. *Genes Dev*, 2005. 19(11): p. 1294-305.

Willis, S.N., et al., Apoptosis initiated when BH3 ligands engage multiple Bcl-2 homologs, not Bax or Bak. *Science*, 2007. 315(5813): p. 856-9.

Won, C.K., et al., Estradiol prevents the injury-induced decrease of Akt activation and Bad phosphorylation. *Neurosci. Lett*, 2005. 387(2): p. 115-9.

Wong, W.W. and H. Puthalakath, Bcl-2 family proteins: the sentinels of the mitochondrial apoptosis pathway. *IUBMB Life*, 2008. 60(6): p. 390-7.

Yao, M., T.V. Nguyen, and C.J. Pike, Estrogen regulates Bcl-w and Bim expression: role in protection against beta-amyloid peptide-induced neuronal death. *J. Neurosci*, 2007. 27(6): p. 1422-33.

Yin, K.J., et al., JNK activation contributes to DP5 induction and apoptosis following traumatic spinal cord injury. *Neurobiol. Dis*, 2005. 20(3): p. 881-9.

Yin, X.M., Bid, a BH3-only multi-functional molecule, is at the cross road of life and death. *Gene*, 2006. 369: p. 7-19.

Yin, X.M., Z.N. Oltvai, and S.J. Korsmeyer, BH1 and BH2 domains of Bcl-2 are required for inhibition of apoptosis and heterodimerization with Bax. *Nature*, 1994. 369(6478): p. 321-3.

Youle, R.J. and A. Strasser, The BCL-2 protein family: opposing activities that mediate cell death. *Nat. Rev. Mol. Cell Biol*, 2008. 9(1): p. 47-59.

Young, C., et al., Ethanol-induced neuronal apoptosis in vivo requires BAX in the developing mouse brain. *Cell Death Differ, 2003.* 10(10): p.1148-55.

Young, J.E., et al., Polyglutamine-expanded androgen receptor truncation fragments activate a Bax-dependent apoptotic cascade mediated by DP5/Hrk. *J. Neurosci*, 2009. 29(7): p. 1987-97.

Yu, J., et al., PUMA induces the rapid apoptosis of colorectal cancer cells. *Mol. Cell,* 2001. 7(3): p. 673-82.

Zha, J., et al., Serine phosphorylation of death agonist BAD in response to survival factor results in binding to 14-3-3 not BCL-X(L). *Cell*, 1996. 87(4): p. 619-28.

Zha, J., et al., Posttranslational N-myristoylation of BID as a molecular switch for targeting mitochondria and apoptosis. *Science*, 2000. 290(5497): p. 1761-5.

Zhang, Y., et al., Sequential activation of individual caspases, and of alterations in Bcl-2 proapoptotic signals in a mouse model of Huntington's disease. *J. Neurochem*, 2003. 87(5): p. 1184-92.

Zhao, Y. and T. Herdegen, Cerebral ischemia provokes a profound exchange of activated JNK isoforms in brain mitochondria. *Mol. Cell Neurosci*, 2009. 41(2): p. 186-95.

Zhu, Y., et al., Transforming growth factor-beta 1 increases bad phosphorylation and protects neurons against damage. *J. Neurosci*, 2002. 22(10): p. 3898-909.

Zong, W.X., et al., BH3-only proteins that bind pro-survival Bcl-2 family members fail to induce apoptosis in the absence of Bax and Bak. *Genes Dev.,* 2001. 15(12): p. 1481-6.

Zou, H., et al., An APAF-1.cytochrome c multimeric complex is a functional apoptosome that activates procaspase-9. *J. Biol. Chem.*, 1999. 274(17): p. 11549-56.

Index

A

B

C

D

E

F

G

H

I

J

K

L

M

N

O

P

Q

R

S

V

W

X

Y

Z